Sustainable Civil Engineering

This reference text establishes linkages between the user industries and the providers of clean technologies and sustainable materials for a rapid transformation of the small and medium-sized enterprises (SMEs). The text covers several aspects of sustainable applications, including clean technologies, climate change and its effects, sustainable buildings (smart cities), sustainability in road construction, sustainable use of geosynthetics, innovative materials, and sustainable construction practices. The text will be useful for senior undergraduate students, graduate students, and researchers in the fields of civil engineering and other infrastructure-related professionals and planners.

This book:

- Discusses clean technologies and sustainable materials in depth
- Covers concepts of sustainability in road construction and water-retaining structures
- Examines environmental policies and practices
- Discusses climate change and its effects in a comprehensive manner
- Covers sustainable buildings including smart cities

As this book discusses concepts related to sustainable civil engineering practices in a single volume, it will be an ideal reference text for everyone aiming at developments of sustainable infrastructures.

Sustainable Civil Engineering

Principles and Applications

Edited by
Varinder Singh Kanwar
Sanjay Kumar Shukla
Siby John
Harpreet Singh Kandra

CRC Press
Taylor & Francis Group
Boca Raton London New York

CRC Press is an imprint of the
Taylor & Francis Group, an **informa** business

Front cover image: D-Krab/Shutterstock

First edition published 2023
by CRC Press
6000 Broken Sound Parkway NW, Suite 300, Boca Raton, FL 33487-2742

and by CRC Press
4 Park Square, Milton Park, Abingdon, Oxon, OX14 4RN

CRC Press is an imprint of Taylor & Francis Group, LLC

Library of Congress Cataloging-in-Publication Data
Names: S. Kanwar, Varinder, editor. | Shukla, Sanjay Kumar, editor. |
John, Siby, editor. | Kandra, Harpreet, editor.
Title: Sustainable civil engineering : principles and applications / edited by Varinder S Kanwar, Sanjay K Shukla, Siby John and Harpreet Kandra.
Description: First edition. | Boca Raton : CRC Press, 2023. |
Includes bibliographical references and index.
Identifiers: LCCN 2022057774 | ISBN 9780367751579 (hardback) |
ISBN 9781032436630 (paperback) | ISBN 9781003368335 (ebook)
Subjects: LCSH: Civil engineering–Environmental aspects. | Sustainable engineering. |
Sustainable construction.
Classification: LCC TD195.C54 S98 2023 | DDC 624.028/6–dc23/eng/20230111
LC record available at https://lccn.loc.gov/2022057774

ISBN: 978-0-367-75157-9 (hbk)
ISBN: 978-1-032-43663-0 (pbk)
ISBN: 978-1-003-36833-5 (ebk)

DOI: 10.1201/9781003368335

Typeset in Sabon
by codeMantra

Contents

Preface

The world is currently facing several problems related to infrastructure and environment as a result of several factors, with the changing climate, increasing population, and urbanization being the most significant causal factors. Current population growth and increasing quality of life are exerting pressures on our natural resources, while also increasing the divide between the haves and the have-nots. There is a great, and often unknown, unmeasured and unreported impacts of often inequitable development on the flora and fauna.

There is widespread discussion on how to continue our life without affecting the overall environment around us, especially after the recent pandemic and natural disasters observed worldwide. The concept and practice of sustainable development has been one such approach.

This book, titled *Sustainable Civil Engineering – Principles and Application,* presents several applications of sustainability in 17 chapters. Diverse subjects covered in the book range from utilization of wastes and natural products in construction; innovative construction practice in foundations; carbon-neutral construction; energy-efficient and smart buildings; landslide monitoring; rainwater harvesting systems and stormwater treatment; asset management; to Internet of Things. These chapters have come from different parts of the world, representing both developing and developed economies.

We do hope that this book will be beneficial to students, researchers, and professionals working in to enhance use of sustainable practices in civil engineering.

We thank all the authors for their contributions to this book. We also thank the staff of CRC Press for their full support and cooperation at all the stages of the publication of this book.

Comments and suggestions from the readers and users of this book are most welcome.

Dr Varinder Singh Kanwar, *Chitkara University, Himachal Pradesh, India*
Dr Sanjay Kumar Shukla, *Edith Cowan University, Perth, Australia*
Dr Harpreet Singh Kandra, *Federation University Australia*
Dr Siby John, *Punjab Engineering College, Chandigarh, India*

About the Editors

Prof. (Dr.) Varinder Singh Kanwar is the Vice Chancellor, Chitkara University, Himachal Pradesh, India. With a Ph.D. in Civil Engineering, M.E. in Structural Engineering from Thapar University Patiala, and Post Graduate Diploma in Rural Development from IGNOU, Dr. Varinder S. Kanwar carries more than 28 years of research, teaching, and administrative experience. He is a fellow of Institution of Engineers (India), a Senior Member of IEEE, and a Life Member of ISTE, ICI, IGS, and IRC. He has written 8 books, 70 research papers in national and international journals, and 17 research papers in conferences. He has guided 3 M.E. and 8 Ph.D. students, filed 10 patents in India, and completed 4 with, 7 ongoing government-funded research projects worth more than 3 crores INR. His major research areas are Health Monitoring of Structures, Sustainable Civil Engineering Materials and Practices, Environmental Engineering, and Alternate Construction Materials. Dr. Kanwar has received several awards from state and national bodies.

Dr. Sanjay Kumar Shukla is a world-renowned expert in Civil (Geotechnical) Engineering. He is Founding Editor-in-Chief of *International Journal of Geosynthetics and Ground Engineering, Springer Nature*, Switzerland, and Founding Research Group Leader (Geotechnical and Geoenvironmental Engineering) at Edith Cowan University (ECU), Perth, Australia. He holds the Distinguished Professorship of Civil Engineering at nine universities, including Indian Institute of Technology Madras, Chennai and Delhi Technological University, Delhi, India. He has over 25 years of experience in teaching, research, consultancy, administration/management, and professional engagement. He is a registered Chartered Professional Engineer in Civil and Geotechnical Engineering, Engineers Australia, Asia Pacific Economic Cooperation (APEC) Engineer in Civil Engineering, and International Professional Engineer in Civil Engineering, International Engineering Association. He collaborates with several world-class universities, research institutions, industries, and individuals on academic and field projects. As a consulting geotechnical engineer, he has successfully provided solutions to the challenging field problems faced by many engineering organizations. He has authored more than 300 research papers and technical articles, including over 190 refereed journal publications. He is also an author/editor of 24 books, including 7 textbooks and 23 book chapters. During 2020–2022, his ICE textbooks, namely *Core Principles of Soil Mechanics* and *Core Concepts of Geotechnical Engineering*, have been ranked #1 by Amazon. His research and academic works have been cited well. Shukla's generalized expressions for seismic active thrust (2015) and seismic passive resistance (2013) are routinely used by practising engineers worldwide for designing the retaining structures. Shukla's wraparound reinforcement technique, developed during 2007–2008, is a well-established sustainable ground improvement technique. He is among the top 2% researchers in his field for single-year 2020 impact at the global level, as per the list published by Elsevier in 2022. He has been honoured with several awards, including 2021 ECU Aspire Award by the Business Events Perth, Australia, and the most prestigious IGS Award 2018 by the International Geosynthetics Society (IGS), USA, in recognition of his outstanding contribution to the development and use of geosynthetics. He serves on the editorial boards of several international journals. He is a fellow of American Society of Civil Engineers and Engineers Australia, a life fellow of Institution of Engineers (India) and Indian Geotechnical Society, and a member of several other professional bodies.

Dr. Harpreet Singh Kandra is an academic and researcher with experience in water management, environmental management, environmental reporting, data management, and policy analysis. He also has over 20 years of teaching, training, and project experience in Australia and India. Harpreet has demonstrated experience of working in many projects involving multidisciplinary teams. He has published over 60 papers, reports, and conference publications. He has been awarded scholarships to pursue research degree at Monash University on stormwater treatment and management. Harpreet has a Chartered Engineer status in Civil Engineering and Environmental Engineering from Engineers Australia. He also received the Vice Chancellor's Award for Contributions to Student Learning at Federation University Australia, 2020. Harpreet also works with several community groups in Melbourne on aspects of environment, waste management, and social cohesion and was the Australian Citizen of the Year in Cardinia Shire Council in 2019. He has been inducted in the 1st Victorian Multicultural Honour Roll, 2022.

Dr. Siby John is presently Professor of Civil Engineering and Deputy Director at Punjab Engineering College, Chandigarh. Dr. John has more than 35 years of academic and research experience. From 2004 to 2006, Dr. John worked as Dean of Punjab Technical University, Jalandhar.

Prof. John has authored more than 100 peer-reviewed articles including original research papers and book chapters. He has more than 150 publications in conferences and seminars. He has also authored four books in the field of 'Environmental Pollution/Engineering'. Dr. John has received several awards and scholarships. He is the recipient of the prestigious *Rekha Nandi and Bhupesh Nandi Best Paper Award* for his publication in the *Environmental Engineering Journal* of the Institution of Engineers India in 2007–2008. He is also recipient of the Eminent Environmental Engineer Award for his eminence and contribution to the profession of Environmental Engineering in the year 2015 and distinguished faculty award in 2016. He has edited a special issue of the *International Journal of Environmental Science and Engineering* on 'Environment and Infrastructure'. He is currently the Chairperson of the State Environmental Assessment Committee of Chandigarh.

List of Contributors

Iqbal Asif
University of South Australia
Mawson Lakes, Australia

Cara D. Beal
School of Engineering and Built
    Environment
Griffith University
Queensland, Australia

Simon Beecham
University of South Australia
Mawson Lakes, Australia

Danny Byrne
College of Engineering and Science
Victoria University
Melbourne, Victoria, Australia

Ayon Chakraborty
Federation University
Ballarat, Australia

G. Chattopadhyay
Institute of Innovation, Sciences
    and Sustainability (IISS)
Federation University
Churchill, Australia

T. A. Choudhury
Federation University Australia
Gippsland, Victoria, Australia

Susanga Costa
Deakin University
Waurn Ponds, Australia

Ana Deletic
Faculty of Engineering
Queensland University of
    Technology
Queensland, Australia

Khang Dinh
Federation University Australia
Gippsland, Victoria, Australia

Fernanda Bessa Ferreira
CONSTRUCT, Faculty of
    Engineering
University of Porto
Porto, Portugal

Thulo Ram Gurung
KBR
Queensland, Australia

Richards H.S.
Institute of Innovation, Sciences
    and Sustainability (IISS)
Federation University
Churchill, Australia

Vijayalaxmi J.
School of Planning and
 Architecture
Vijayawada, India

Siby John
Punjab Engineering College
 (Deemed to be University)
Chandigarh, India

Roohollah Kalatehjari
Built Environment Engineering
 Department, School of Future
 Environments
Auckland University of Technology
Auckland, New Zealand

Harpreet Singh Kandra
School of Engineering, Information
 Technology and Physical
 Sciences
Federation University
Churchill, Australia

Abhishek Kanoungo
Chitkara University
Himachal Pradesh, India

Shristi Kanoungo
Punjab Engineering College
 (Deemed to be University)
Chandigarh, India

Manvi Kanwar
Department of Civil Engineering
Chitkara University
Himachal Pradesh, India

Varinder Singh Kanwar
Chitkara University
Himachal Pradesh, India

Jayantha Kodikara
Monash University
Clayton, Australia

Pramod Kumar
Punjab Engineering College
 (Deemed to be University)
Chandigarh, India

Suryani Lim
Federation University Australia
Gippsland, Victoria, Australia

Maria de Lurdes Lopes
CONSTRUCT, Faculty of
 Engineering
University of Porto
Porto, Portugal

David McCarthy
Monash University
Clayton, Australia

Shobha Muthukumaran
Institute for Sustainable Industries
 and Liveable Cities
Victoria University
Melbourne, Victoria, Australia

Islam N.
Swan Hill Rural City Council
Swan Hill, Australia

Dimuth Navarata
College of Engineering and
 Science, Institute for Sustainable
 Industries and Liveable Cities
Victoria University
Melbourne, Victoria, Australia

Hua-Fu Pei
Department of Geotechnical
 Engineering
Dalian University of Technology
Dalian, China

Paulo Pereira
CONSTRUCT, Faculty of
 Engineering
University of Porto
Porto, Portugal

Sateesh Pisini
Department of Civil Engineering
Fiji National University
Suva, Fiji

C. Prakasam
Department of Geography, School
   of Earth Sciences
Assam University
Karbi Anglong, India

Aravinth R.
Bharathi Vidyapeeth University
Pune, India

Saravanan R.
EcoFirst
Tata Consulting Engineering
   Limited
Bangalore, India

Md. Mizanur Rahman
University of South Australia
Mawson Lakes, Australia

Muhammad Nouman Amjad Raja
Department of Civil Engineering,
   University of Management and
   Technology
Lahore, Pakistan

Ahmad Safuan A. Rashid
Centre of Tropical Engineering,
   Faculty of Engineering
Universiti Teknologi Malaysia
Johor, Malaysia

Dilan Robert
RMIT University
Melbourne, Australia

Ashok K. Sharma
Institute for Sustainable Industries
   and Liveable Cities, College of
   Engineering and Science
Victoria University
Melbourne, Victoria, Australia

M.K. Sharma
Environmental Hydrology Division
National Institute of Hydrology
Roorkee, India

Bin Shi
School of Earth Sciences and
   Engineering
Nanjing University
Nanjing, China

Mohammad Gharehzadeh Shirazi
Centre of Tropical Engineering,
   Faculty of Engineering
Universiti Teknologi Malaysia
Johor, Malaysia

Sanjay Kumar Shukla
Geotechnical and
   Geoenvironmental Engineering
   Research Group
Edith Cowan University
Perth, Australia

Rodney A. Stewart
Griffith School of Engineering and
   Built Environment
Griffith University
Gold Coast, Queensland, Australia

Dao-Yuan Tan
Department of Civil and
   Environmental Engineering
The Hong Kong Polytechnic
   University
Hung Hom, Kowloon, Hong Kong,
   China

Ankush Tanta
Department of Civil Engineering
Chitkara University
Himachal Pradesh, India

Swetha Thammadi
Department of Civil Engineering
Fiji National University
Suva, Fiji

Aishwarya Sudheer Vadukkumchery
School of Planning and
   Architecture
Vijayawada, India

Castorina Silva Vieira
CONSTRUCT, Faculty of
   Engineering
University of Porto
Porto, Portugal

Suzanne Wilkinson
School of Built Environment
Massey University
Auckland, New Zealand

Xiao Ye
School of Earth Sciences and
   Engineering
Nanjing University
Nanjing, China

Hong-Hu Zhu
School of Earth Sciences and
   Engineering
Nanjing University
Nanjing, China

Chapter 1

Recycled construction and demolition waste as backfill material for geosynthetic-reinforced structures

Fernanda Bessa Ferreira, Castorina Silva Vieira,
Paulo Pereira, and Maria de Lurdes Lopes
University of Porto

CONTENTS

1.1 INTRODUCTION

Environment preservation and the sustainable usage of natural resources are key challenges faced by the international community nowadays. In particular, the construction sector is among the main contributors to the global consumption of natural resources, accounting for about 50% of all extracted materials (European Commission 2020). In addition, construction and demolition (C&D) waste, which consists of the debris produced during excavation, site preparation, construction, maintenance, rehabilitation and demolition of civil infrastructures, is one of the heaviest and most voluminous waste streams in developed and developing countries alike, including the European Union where it represents over 35% of all waste produced (European Commission 2020). The adoption of responsible approaches for

DOI: 10.1201/9781003368335-1

1

design and construction, including efficient C&D waste reuse and recycling, is imperative and can make a decisive contribution to achieving sustainable development and climate neutrality. Reuse and recycling of C&D waste not only reduces the environmental impacts caused by natural resource extraction and processing, but also facilitates more effective waste management in the construction industry, reducing the amounts of wastes ending in landfills (Vieira and Pereira 2015b; Arulrajah et al. 2017, 2020).

In this context, various studies have recently been carried out to evaluate the suitability of recycled C&D wastes as substitute soils and aggregates in different civil engineering projects, such as transportation infrastructure, structural fills, earth retaining walls, ground improvement works and concrete production (Leite et al. 2011; Arulrajah et al. 2013, 2020; Santos et al. 2014; Silva et al. 2014; Vieira et al. 2016; Henzinger and Heyer 2018; Pereira et al. 2019; Pourkhorshidi et al. 2020; Lu et al. 2021; Naeini et al. 2021; Yaghoubi et al. 2021). One application that has raised particular attention among researchers and practitioners is the use of recycled C&D waste as an alternative fill material in the construction of geosynthetic-reinforced systems, including retaining walls, slopes and embankments (Santos et al. 2013; Arulrajah et al. 2014; Santos et al. 2014; Vieira et al. 2016, 2020a; Ferreira et al. 2023).

Geosynthetics are polymeric materials that have been successfully employed in civil and environmental engineering applications over the past decades. Geogrids and geotextiles (including high-strength geotextiles, also termed as "geocomposite reinforcements") have been typically used for soil reinforcement and stabilisation (Greenway et al. 1999; Allen and Bathurst 2002; Allen et al. 2002; Benjamim et al. 2007; Palmeira 2009; Ferreira et al. 2013, 2015, 2016a, 2020a; Rahman et al. 2014; Maghool et al. 2020; Karnam Prabhakara et al. 2021a, 2021b). However, polymeric materials tend to exhibit time-dependent response (creep and stress relaxation) when embedded in soil and exposed to tensile loads. The viscoelastic characteristics of geosynthetic products are susceptible to influence the behaviour of the reinforced systems over their design lifetimes, and hence, the long-term mechanical properties of geosynthetics as well as soil-geosynthetic interfaces are of critical significance for permanent reinforcement applications (Lopes et al. 1994; Min et al. 1995; Leshchinsky et al. 1997; Li and Rowe 2001; Kongkitkul et al. 2007; Bathurst et al. 2012; Miyata et al. 2014; Cardile et al. 2021; Ferreira et al. 2021b).

This chapter reviews recent research work conducted at the University of Porto to investigate the feasibility of using fine-grained mixed recycled C&D wastes in the construction of geosynthetic-reinforced structures, with particular emphasis on the long-term behaviour. The recycled C&D materials used in this research were produced from wastes resulting primarily from the demolition of masonry fences, maintenance and/or rehabilitation of buildings and recovering of C&D wastes from illegal deposits. The recycling process involved various phases, including the elimination of metals and light contaminants (such as rubber, plastic, cork, wood and foam) in a trommel and the subsequent grain size separation through conveyor belts and a set of sieves. The materials used in the current research consisted of the finer fraction (0–10 mm) obtained during the aforementioned recycling process, and were mainly composed of soil, with some unbound and hydraulically bound aggregates, concrete and mortar products.

1.2 TENSILE AND CREEP BEHAVIOUR OF GEOSYNTHETICS

1.2.1 Exposure of geosynthetics to recycled C&D waste

To evaluate the potential chemical and environmental degradation of geosynthetics resulting from the exposure of these materials to fine-grained recycled C&D wastes and a clayey sand (reference backfill) under actual environmental conditions, small trial embankments (3.0 m by 2.0 m in plan and 0.45 m in height) were constructed with embedded geosynthetic samples.

Figure 1.1 presents the gradation curves of the selected backfill materials. Two different geosynthetics were employed, specifically a uniaxial geocomposite reinforcement (GCR1), also termed as a "high-strength geotextile", and a uniaxial high-density polyethylene (HDPE) geogrid (GGR). The geocomposite reinforcement consisted of high-modulus polyester (PET) fibres attached to a nonwoven polypropylene (PP) geotextile backing, with a mass per unit area of 340 g/m². The nominal tensile strength and corresponding

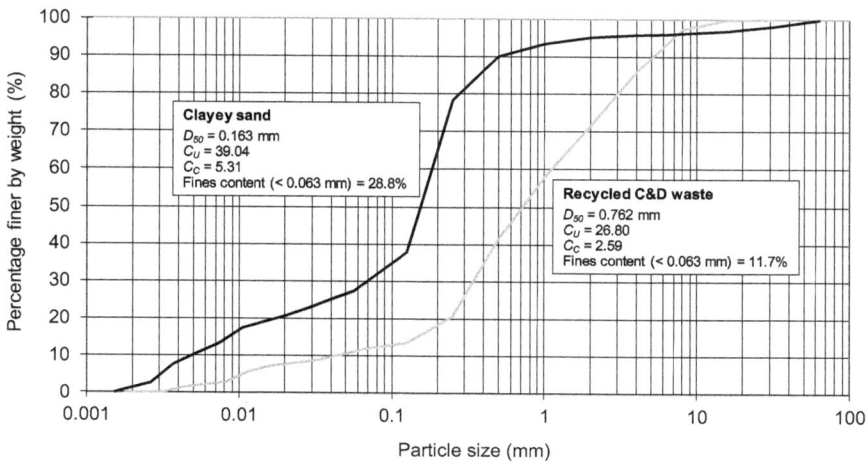

Figure 1.1 Particle size distribution curves of the different backfills.

Source: Modified after Ferreira et al. (2021a).

Table 1.1 Physical and mechanical properties of the geosynthetics

	GCR1	GGR	GCR2
Raw material	PP & PET	HDPE	PP & PET
Mass per unit area (g/m²)	340	450	260
Aperture dimensions (mm)	-	16×219	-
Width of longitudinal members (mm)	-	6	-
Width of transverse members (mm)	-	16	-
Thickness of longitudinal members (mm)	-	1.1	-
Thickness of transverse members (mm)	-	2.5–2.7	-
Mean value of the tensile strength[a] (kN/m)	75	68	40
Elongation at maximum load[a] (%)	10	11±3	10

[a] Based on wide-width tensile tests (machine direction) according to EN ISO 10319 (values provided by the manufacturer).

elongation based on wide-width tensile tests (EN ISO 10319) reported by the manufacturer were 75 kN/m and 10%, respectively (machine direction). The HDPE geogrid was an extruded geogrid with a mass per unit area of 450 g/m². The nominal tensile strength and elongation of this geogrid were 68 kN/m and 11±3%, respectively, according to the manufacturer's specifications (Table 1.1).

The geosynthetic samples were installed without overlapping at different levels within the embankments, at a vertical spacing of 0.20 m (Figure 1.2a). On top of each geosynthetic layer, a first backfill layer was placed manually to prevent any damage. A lightweight compaction procedure was employed to avoid wind and rain erosion, while minimising mechanical damage of the geosynthetics. The slopes of the trial embankments were compacted and then covered with a coarser recycled C&D material to mitigate erosion due to rainwater (Figure 1.2b).

The exhumation of geosynthetic samples was performed manually after different exposure periods, specifically: (i) right after installation (i.e. after embankment construction), (ii) after 6 months, (iii) after 12 months and (iv) after 24 months. During exhumation, special care was taken to prevent additional damage to the samples (Figure 1.2c). It was found that irrespective of the backfill type (i.e. natural soil or recycled C&D waste), some geosynthetic samples had been crossed by vegetation roots during the exposure period (Figure 1.2d). The exhumed samples were then stored at constant temperature (about 20°C) up to tensile and creep rupture testing. In fact, changes in temperature may affect the measured tensile properties of geosynthetics. Therefore, conditioning and testing in a standard atmosphere is recommended by relevant geosynthetic standards (e.g. EN ISO 10319:2015 and EN ISO 13431:1999). Further details about the construction procedures of these experimental embankments and exhumation of the geosynthetics samples can be found in previous publications (Vieira and Pereira 2015a, 2021).

Figure 1.2 Exposure of geosynthetics to recycled C&D wastes. (a) Installation of geosynthetic samples. (b) Completed trial embankment. (c) Exhumation of geosynthetic samples. (d) Vegetation roots crossing the geosynthetic.

1.2.2 Short-term tensile behaviour

To characterise the short-term tensile response of virgin and exhumed specimens of the geocomposite reinforcement and the geogrid, a series of wide-width tensile tests were carried out in accordance with the Standard EN ISO 10319:2015. Figure 1.3 compares the average values (i.e. from five specimens tested under repeatability conditions) of the ultimate tensile strength (Figure 1.3a), secant stiffness at 2% strain (Figure 1.3b) and secant stiffness at maximum tensile load (Figure 1.3c) for virgin specimens, specimens that were exhumed right after C&D waste embankment construction and those exhumed after a 24-month exposure to the clayey sand and the recycled C&D waste.

The exposure of the geocomposite reinforcement to the fill materials induced a significant degradation of the ultimate tensile strength (by up to 26.8%) and stiffness at maximum load (by up to 24.6%). However, the reduction of the secant stiffness at 2% strain was less significant (by up to 7.6%). The degradation of the ultimate tensile strength of the exhumed geocomposite specimens is associated with the construction process, including material handling and installation, and the chemical degradation caused by the long-term exposure to

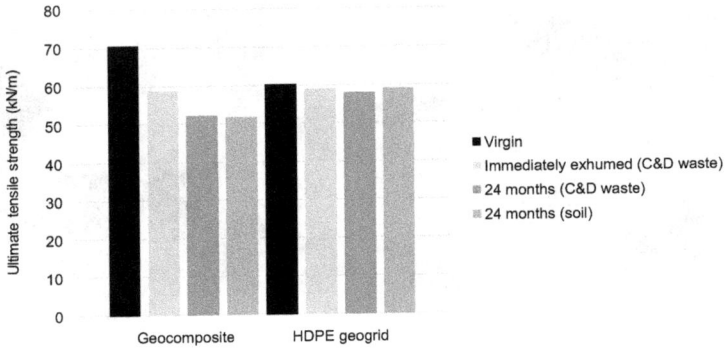

(a)

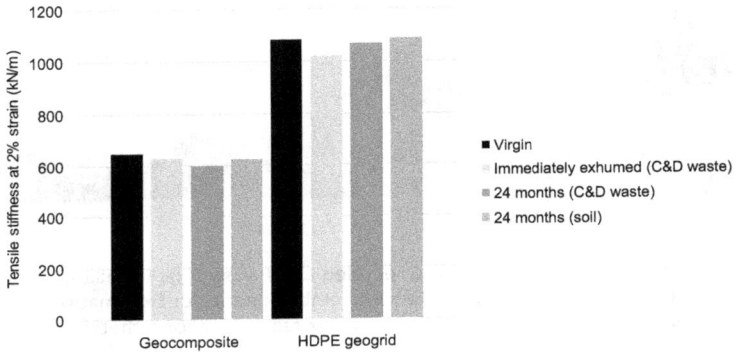

(b)

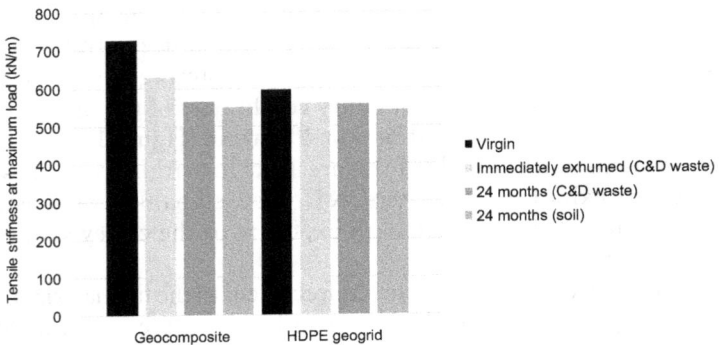

(c)

Figure 1.3 Comparison of tensile strength properties for virgin and exhumed specimens (average values). (a) Ultimate tensile strength. (b) Tensile stiffness at 2% strain. (c) Tensile stiffness at maximum load.

the fill materials. In fact, the specimens exhumed right after installation experienced a tensile strength loss of 16.7%, which may be related to less effective connection of the polyester fibres to the geotextile backing as a result of material handling and installation. The tensile strength loss for specimens exhumed 24 months after embankment construction was about 9%–10% greater. Accordingly, only 9%–10% of the tensile strength loss can be attributed to the long-term contact with the backfill materials, whereas the remaining reduction is associated with the construction and handling processes.

Figure 1.3 also shows that the tensile strength properties of the geogrid, particularly the ultimate tensile strength and the stiffness at 2% strain, were not significantly affected by the different exposure conditions (maximum reductions of 4.0% and 5.7%, respectively). The tensile stiffness at maximum load decreased by up to 9.0%.

Comparing the results obtained for specimens that were previously exposed to the soil and the recycled C&D waste for 24 months, it can be concluded that the different backfill materials induced similar effects on the tensile properties of the geosynthetics. This implies that the recycled material did not cause any additional degradation when compared with the reference soil, thus supporting the suitability of recycled C&D wastes as a sustainable replacement for traditional fill materials of geosynthetic-reinforced systems.

1.2.3 Creep behaviour

The long-term tensile response of geosynthetics is generally evaluated by in-isolation tensile creep and creep rupture tests (Allen and Bathurst 1996; Zornberg et al. 2004; Bueno et al. 2005; Miyata et al. 2014; Pinho-Lopes et al. 2018; Deng and Huangfu 2021; Ferreira et al. 2021a).

The long-term tensile response of the geocomposite reinforcement (GCR1) was evaluated in this study by a series of creep rupture tests performed according to the Standard EN ISO 13431:1999. Virgin specimens as well as those previously exposed to the recycled C&D waste and the reference soil (clayey sand) for 24 months were used in these tests. The specimens (300 mm long×100 mm wide) were subjected to a constant load imposed by weights acting through a lever system under a constant temperature (20°C). The applied tensile force remained constant up to the specimen rupture, and the time to creep rupture was automatically recorded. A video-extensometer was employed to monitor the evolution of geosynthetic strains throughout the tests, by using two reference points 200 mm apart. The virgin specimens were tested under load levels ranging from 68% to 90%, whereas the exhumed specimens from the recycled C&D waste and the reference soil were subjected to load levels ranging from 63% to 73% and 65% to 73%, respectively. It is noteworthy that the abovementioned load levels were expressed as the applied creep load normalised with respect to the ultimate tensile strength of virgin specimens estimated by short-term tensile tests (see Section 1.2.2).

Figure 1.4a illustrates the relationships between the creep loads (expressed as a proportion of the ultimate tensile strength of virgin specimens) and

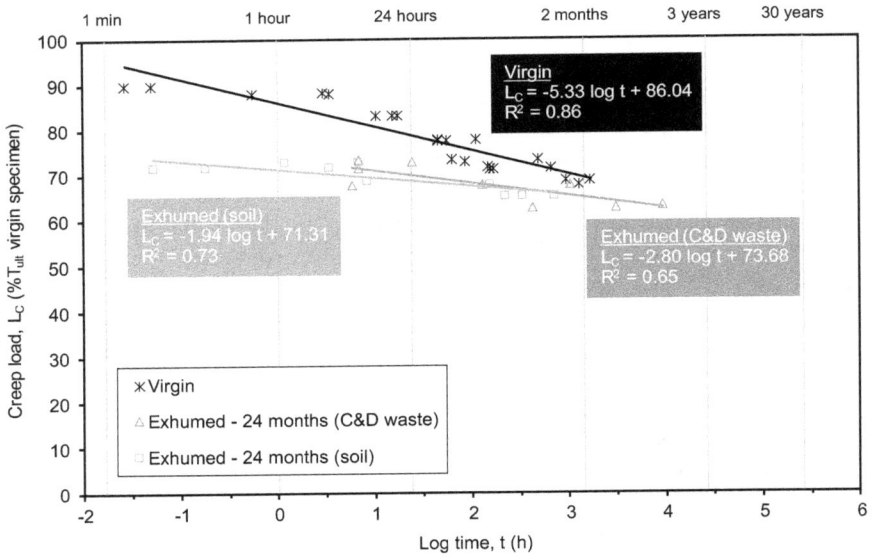

(a)

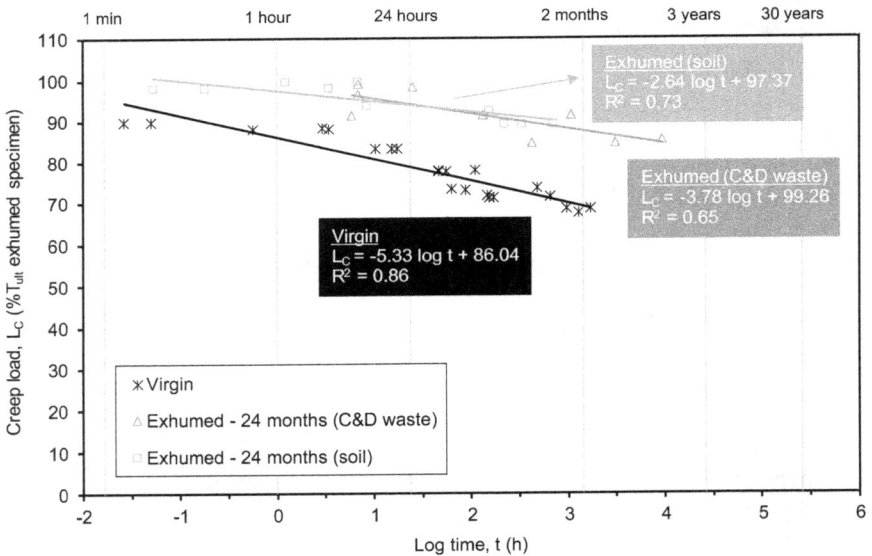

(b)

Figure 1.4 Applied creep loads plotted against time to rupture and associated rupture curves. (a) Creep loads with respect to T_{ult} of virgin specimens. (b) Creep loads with respect to T_{ult} of exhumed specimens.

Source: (a) Modified after Ferreira et al. (2021a).

the time to rupture of the different specimens in a semi-logarithmic scale. In Figure 1.4b, the applied creep loads are normalised with respect to the ultimate tensile strength of each specimen type (i.e. virgin, exhumed from the recycled C&D waste or the reference soil). The creep rupture envelopes (log-linear fits to the test data) that are generally used to estimate the long-term strength (i.e. the retained strength for a particular design lifetime), as well as the respective equations and coefficients of determination (R^2), are also included in these graphs.

Figure 1.4a shows that when subjected to the same tensile force (i.e. absolute value), the virgin specimens exhibited greater tensile strength properties, in comparison with the exhumed specimens, which is consistent with the tensile test results presented in Section 1.2.2. However, considering the extrapolation of the creep rupture curves to estimate the long-term available strength, the specimens previously subjected to the fill materials would lead to more optimistic estimates of the retained strength, which is likely attributed to the intrusion of fine grains into the geosynthetic pores. This finding suggests that the use of virgin specimens (i.e. the conventional procedure) to estimate the long-term tensile strength of geosynthetic products through creep rupture tests may be considered as a conservative approach.

As shown in Figure 1.4b, when the creep loads are expressed as a proportion of the ultimate tensile strength of each specimen type, the exhumed specimens can be considered to exhibit higher performance than the virgin specimens, irrespective of the applied creep load.

Furthermore, the variability of results was more pronounced for the exhumed specimens, as reflected by the lower coefficients of determination (Figure 1.4). As expected, the effects of the exposure of the geosynthetics to the fill materials were not completely homogeneous. Given that the results were fairly similar irrespective of the fill material, it can also be concluded that the exposure to the recycled waste did not induce any further degradation in the long-term strength properties of this geocomposite reinforcement.

As mentioned earlier, the axial strains of the specimens were continuously monitored during the tests. Figure 1.5 presents the variation of the axial strain with log time obtained in selected tests on virgin and exhumed specimens up to creep rupture. Note that all of the creep loads are expressed in terms of the ultimate tensile strength of virgin specimens. For load levels ranging from 68% to 73%, the virgin specimens exhibited ductile response with rupture strains higher than 10%. At the first stage of the test, the strains increased almost linearly with log time (i.e. primary creep), but the rupture was preceded by a significant increment in the axial strain rate (i.e. tertiary creep). In contrast, the exhumed specimens exhibited brittle response, with the creep curves generally showing approximately linear strains (in a semi-log scale) up to rupture. However, the virgin specimens subjected to higher load levels (>73%) also exhibited brittle behaviour. As mentioned in Section 1.2.2, the maximum tensile strength of the virgin

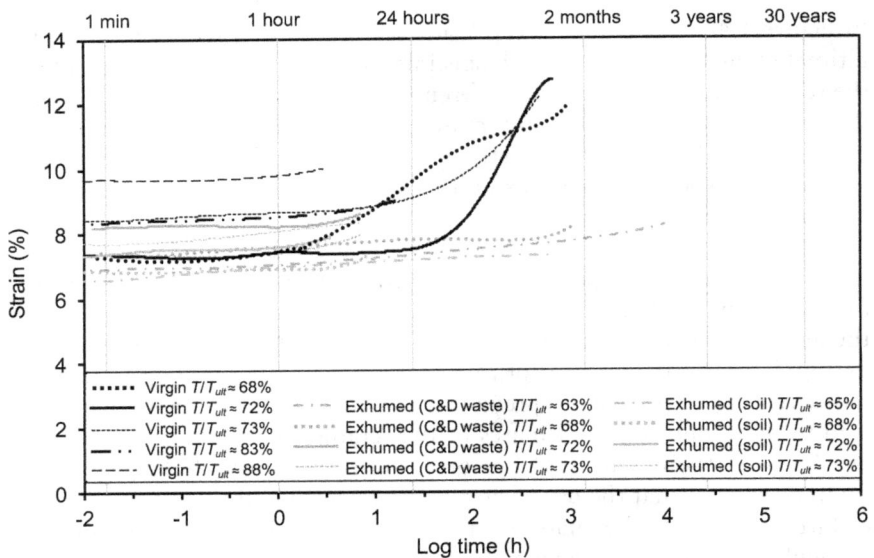

Figure 1.5 Evolution of axial strain as a function of log time from creep rupture tests on virgin and exhumed specimens.

Source: Modified after Ferreira et al. (2021a).

specimens was higher than that of the exhumed specimens, and thus for the latter, any specific load level (as represented in Figure 1.5) was closer to the maximum tensile strength. In fact, all the exhumed specimens were tested under creep loads ≥85% of their ultimate tensile strength, which exceeds the value beyond which the virgin specimens exhibited brittle response. The brittle response of the exhumed specimens can thus be justified by the fact that the applied creep loads exceeded the threshold level beyond which the geosynthetic starts to show brittle behaviour.

1.3 RECYCLED C&D MATERIAL-GEOSYNTHETIC INTERFACE BEHAVIOUR

1.3.1 Pullout behaviour

1.3.1.1 Monotonic tests

Understanding soil-geosynthetic interface response is of critical significance for the design and stability analysis of geosynthetic-reinforced soil systems (Palmeira 2009; Lopes 2012; Moraci et al. 2014; Prashanth et al. 2016; Hegde and Roy 2018; Ferreira et al. 2020b). Pullout tests are commonly used to characterise the interaction between soils and geosynthetics in the anchorage zone of geosynthetic-reinforced soil structures. The test yields the geosynthetic pullout resistance, which is an essential parameter required by design codes of geosynthetic-reinforced systems.

To evaluate the pullout response of a uniaxial extruded geogrid in a recycled C&D material and to assess the effects of the rate of displacement (i.e. pullout rate or clamp velocity) and geosynthetic specimen size on the pullout test results, a series of laboratory pullout tests was conducted using a large-scale prototype pullout test apparatus. The geogrid used (GGR) was a HDPE uniaxial geogrid with the properties listed in Table 1.1. The recycled C&D waste was compacted to a dry density of 16.1 kN/m^3 (80% of the maximum dry density) at the optimum water content (w_{opt}=9%), as estimated by the Modified Proctor test (EN 13286-2:2010). The pullout tests were conducted under a low vertical pressure of 10 kPa at the geosynthetic level, so as to resemble low depths, where pullout failure is most susceptible to take place in geosynthetic-reinforced soil systems.

The effect of the displacement rate on the evolution of pullout force with front displacement and on the displacements measured over the reinforcement length at maximum pullout load is illustrated in Figure 1.6a and b, respectively. The tests were conducted on 200 mm×600 mm specimens at varying pullout rates (1, 2 and 4 mm/min). Figure 1.6a shows that the pullout resistance measured in the tests is dependent upon the displacement rate, such that the faster the displacement rate, the higher the peak pullout force. With the increase in the rate of displacement (from 1 to 4 mm/min), the peak pullout force increased by up to 16%. In addition, the front displacement at peak was significantly higher for the displacement rates of 2 and 4 mm/min, when compared with that for the lowest rate. Irrespective of the pullout rate, the failure resulted from the reinforcement rupture (i.e. internal failure in tension).

The displacements measured over the geogrid length when the peak pullout load was mobilised (Figure 1.6b) show that the strains were particularly relevant along the front section of the reinforcement and tended to reduce with the distance to the front end. Furthermore, the displacements along the geogrid were greater when higher displacement rates (2 and 4 mm/min) were imposed. In short, the variation in the pullout rate from 1 to 2 mm/min affected the pullout test results to a considerable extent, but when the pullout rate was further increased, no significant influence was observed. These results are in good agreement with previous studies, which have demonstrated that due to the intrinsic viscous characteristics of geosynthetic products, the peak strength is affected by the loading rate and typically increases with the loading rate at failure (Lopes and Ladeira 1996; Hirakawa et al. 2003; Ferreira et al. 2020b). It is noteworthy that according to the American Standard (ASTM D6706-01:2013), if excess pore water pressures are not expected, the pullout load shall be applied at a rate of 1±0.1 mm/min. However, the European Standard EN 13738:2004

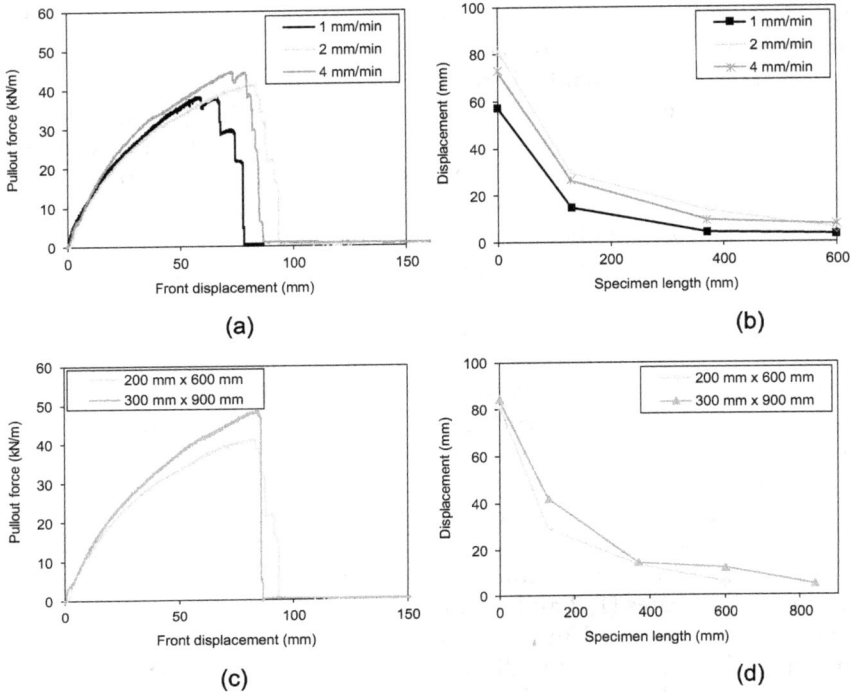

Figure 1.6 Monotonic pullout test results for a HDPE uniaxial geogrid embedded in a recycled C&D waste ($\sigma = 10$ kPa). (a) and (b) Effect of displacement rate (specimen size of 200 mm×600 mm). (c) and (d) Effect of specimen size (displacement rate of 2 mm/min).

Source: Modified after Vieira et al. (2020b).

recommends the use of a higher displacement rate (2 ± 0.2 mm/min) in geosynthetic pullout tests involving free draining soils. Given that the pullout rate is susceptible to influence the results, such parameter should be taken into account when comparing the results from pullout tests carried out by different researchers worldwide.

Figure 1.6c and d illustrates the influence of the specimen dimensions on the pullout response of the geogrid. The tests were performed at a rate of displacement of 2 mm/min on 200 mm×600 mm and 300 mm×900 mm specimens, in accordance with the EN 13738:2004. Figure 1.6c indicates that the measured peak load capacity increased with the specimen dimensions, which may be attributed to the number of transverse bars of the geogrid specimen (3 or 4 bars for specimen lengths of 600 or 900 mm, respectively). Given the relevant contribution of the passive resistance mechanism to the global pullout capacity of the geogrid, considerably higher peak pullout resistance (17%) was attained for the longer specimen.

As shown in Figure 1.6d, regardless of the specimen size, the displacements mobilised at maximum pullout load at both the front and rear ends of the geogrid were identical and the full length of the specimens was mobilised. It can also be observed that for the longer specimen, the deformations were significantly greater throughout the first two sections, when compared to those at the back half of the reinforcement length.

The pullout interaction coefficient is a key parameter used in the design of geosynthetic-reinforced systems. It can be determined as the ratio of the maximum shear stress reached at the backfill-geosynthetic interface during a pullout test to the maximum shear stress mobilised in a direct shear test performed on the backfill material under identical normal stress. The pullout interaction coefficients for the studied interface ranged from 1.13 to 1.57. These values are identical to or greater than those generally presented in the literature for interfaces between soils and geogrids (Goodhue et al. 2001; Mohiuddin 2003; Tang et al. 2008; Hsieh et al. 2011; Ferreira et al. 2016d). In addition, the obtained values exceed the pullout interaction coefficients typically adopted in the design of geosynthetic-reinforced systems in the absence of experimental data, which is an important indicator of the suitability of this recycled C&D waste as a substitute fill material in geosynthetic-reinforced soil structures (Vieira et al. 2016; Vieira 2020).

1.3.1.2 Effect of cyclic loading

In addition to static loading, geosynthetic-reinforced soil systems may be subjected to repeated loading and unloading cycles over their design lifetime, in which case the knowledge about the soil-geosynthetic interface behaviour under cyclic loading is of great importance (Ferreira et al. 2016b, 2020b; Liu et al. 2016; Cardile et al. 2019; Mahigir et al. 2021). The same applies to geosynthetic-reinforced systems (e.g. walls and slopes) built with alternative fill materials such as recycled C&D waste.

To examine the monotonic, cyclic and post-cyclic pullout response of two geosynthetics (a geocomposite reinforcement and a geogrid) in a recycled C&D waste, monotonic and multistage tests were conducted using a large-scale pullout test device. The properties of the geocomposite reinforcement (GCR1) and the HDPE geogrid (GGR) used in this study have been described earlier (Table 1.1). The C&D material was compacted to the target dry density of 16.1 kN/m³ (i.e. 80% of the maximum Modified Proctor dry density) and at the optimum water content (w_{opt}=9%) (EN 13286-2:2010).

The multistage tests involved three sequential procedures: (i) a constant loading rate of 0.2 kN/min was applied until the pullout load reached a predefined value (pre-cyclic pullout load, P_C) expressed as a percentage of the peak pullout resistance (P_R) mobilised in the monotonic (benchmark) test; (ii) a sinusoidal cyclic tensile load with constant frequency (f_C=0.1 Hz) and amplitude (A_C) was applied for n=100 cycles; and (iii) a constant loading rate (0.2 kN/min) was imposed until the end of the test. The influence of the amplitude level (A_C/P_R) and pre-cyclic pullout load level (P_C/P_R) was examined by adopting different values for these parameters (P_C=0.4 P_R and 0.7 P_R, and A_C=0.2 P_R and 0.4 P_R). To assess whether or not the cyclic loading is susceptible to affect the peak pullout resistance mobilised in the subsequent test stage, load-controlled monotonic tests (benchmark tests) were also performed by adopting a constant loading rate of 0.2 kN/min. A relatively low vertical pressure was imposed in all of the tests ($\sigma = 25$ kPa at the reinforcement level).

Figure 1.7 illustrates the impact of the cyclic load amplitude (A_C) on the pullout behaviour of the reinforcements when subjected to P_C=0.4 P_R. Figure 1.7a and b plots the results related to the geocomposite reinforcement, specifically the evolution of pullout load with frontal displacement from monotonic and multistage tests, as well as the displacement profiles along the specimens at the onset and at the end of the cyclic loading phase, respectively. In turn, Figure 1.7c and d presents similar results obtained when the geogrid was used.

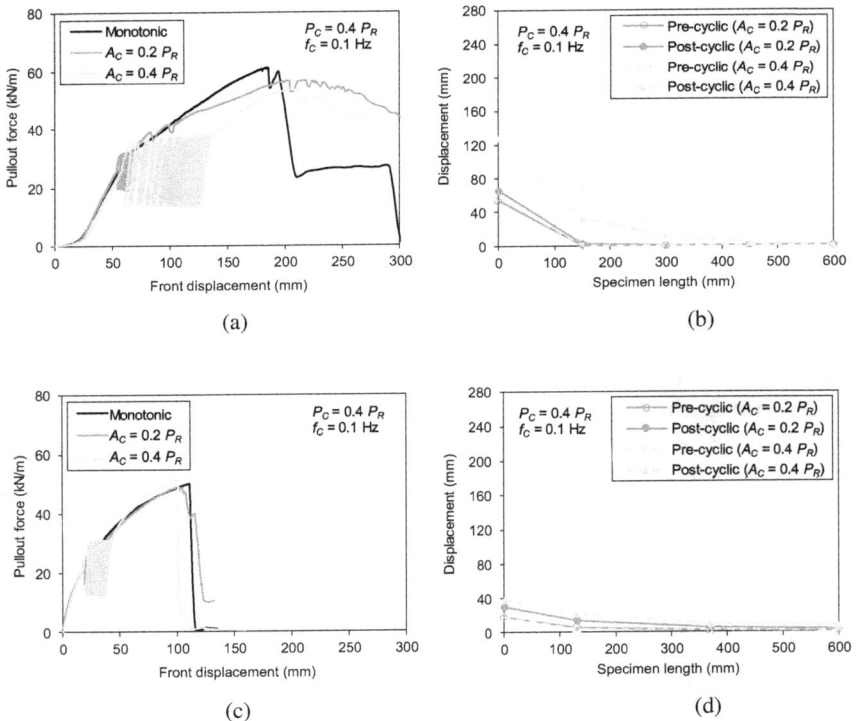

Figure 1.7 Effect of cyclic load amplitude (A_C) on the pullout response of (a) and (b) Geocomposite reinforcement. (c) and (d) HDPE uniaxial geogrid.

Source: Modified after Vieira et al. (2020a).

As shown in Figure 1.7a, regardless of the amplitude level, the maximum pullout force reached in the post-cyclic stage decreased comparatively with that attained under monotonic loading conditions. Such reduction was slightly greater for the highest amplitude, $A_C=0.4\ P_R$ ($\approx 13\%$). The displacements induced by cyclic loading throughout the length of the geocomposite were significantly affected by the A_C value (Figure 1.7b). When the lower amplitude was imposed ($A_C=0.2\ P_R$), only the front reinforcement segment (i.e. located closer to the clamp) exhibited considerable deformation. However, with the amplitude increase, additional segments located towards the free end of the reinforcement were mobilised during cyclic loading and the deformations experienced at the front section were substantially higher. Regardless of the amplitude, the displacements at the rear segment were negligible, indicating that this section was not mobilised and no sliding occurred during the cyclic stage.

The effect of the cyclic load amplitude on the geogrid pullout response is shown in Figure 1.7c and d. As opposed to what was observed for the geocomposite, the cyclic load did not significantly affect the geogrid peak pullout load. Similar values of ultimate pullout resistance were reached in the monotonic and multistage tests under different amplitude levels (Figure 1.7c). The displacements induced by cyclic loading over the length of the geogrid were slightly greater for the amplitude of $0.4\ P_R$ (Figure 1.7d).

Figure 1.8 illustrates how the pre-cyclic pullout load (P_C) influenced the behaviour of the geosynthetics for $A_C=0.4\ P_R$. Figure 1.8a and b shows the results for the geocomposite, whereas Figure 1.8c and d is related to the geogrid.

The increase in P_C in the pullout tests of the geocomposite led to the reinforcement rupture (i.e. tensile failure) during cyclic loading at $n=47$ cycles upon the accumulation of large strains, particularly along the front segment (Figure 1.8a and b). It can be observed that the pullout force-displacement curve from the monotonic test is steeper up to a pullout force of about 0.55 P_R, beyond which it shows a lower slope (reduced interface stiffness). This has a significant effect on the interface response under pullout loads exceeding the transition pullout force. For the highest value of P_C, the aforementioned transition force had already been exceeded and the interface was unable to withstand the predefined number of loading cycles (Figure 1.8a). However, for the geogrid, even when the highest P_C value was applied, the failure took place during the third stage of the test. Regardless of the value of P_C, the peak load capacity was not significantly affected by cyclic loading (Figure 1.8c). As observed for the geocomposite, the increase in P_C resulted in significantly higher displacements induced by cyclic loading over the length of the specimens (Figure 1.8d).

To gain further insight into the long-term response of recycled C&D waste-geosynthetic interfaces under pullout conditions, a large-scale pullout test programme involving sustained loads is currently under development and will be reported in future publications.

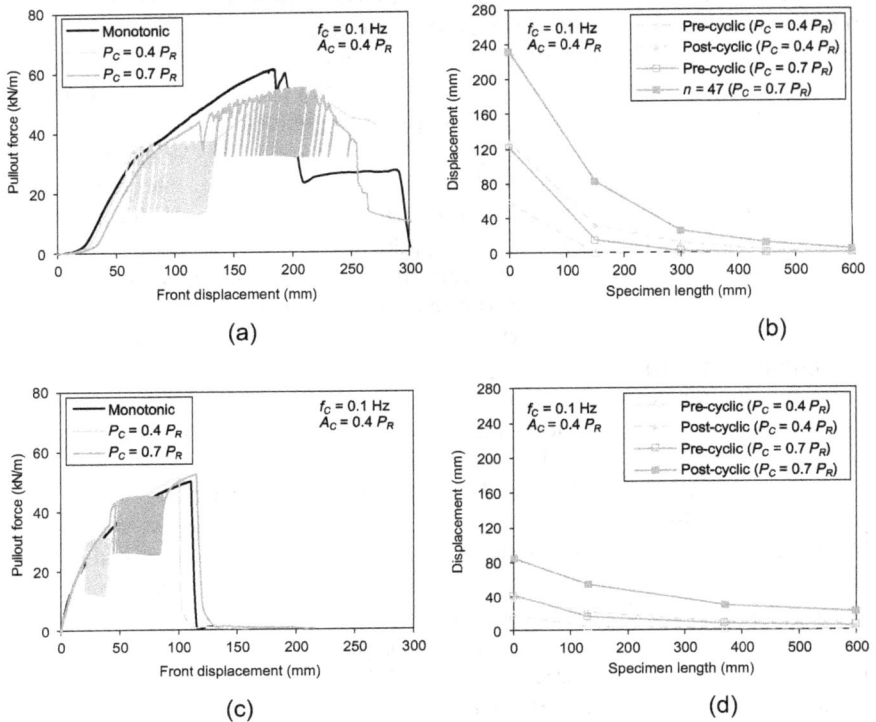

Figure 1.8 Effect of pre-cyclic pullout load level (P_C) on the pullout response of (a) and (b) geocomposite reinforcement. (c) and (d) HDPE uniaxial geogrid.

Source: Modified after Vieira et al. (2020a).

1.3.2 Direct shear behaviour

1.3.2.1 Conventional direct shear tests

The direct shear behaviour of soil-geosynthetic interfaces is typically examined by large-scale direct shear tests (Liu et al. 2009; Vieira et al. 2013; Anubhav and Wu 2015; Ferreira et al. 2015, 2016b, c). This test is considered as the most suitable test method to mimic the interaction between the soil (or other backfill material) and the geosynthetic when sliding of the backfill mass on the surface of the geosynthetic is susceptible to take place (Ferreira et al. 2015, 2023).

To study the interaction characteristics at the interface between a recycled C&D waste and a HDPE uniaxial geogrid (GGR) in direct shear mode, a large-scale direct shear test programme was conducted. The recycled C&D material was compacted to a dry density of 16.1 kN/m³ (i.e. about 80% of the maximum dry density) under two distinct moisture conditions: air-dried (hereinafter referred to as "dry") and at the optimum water content

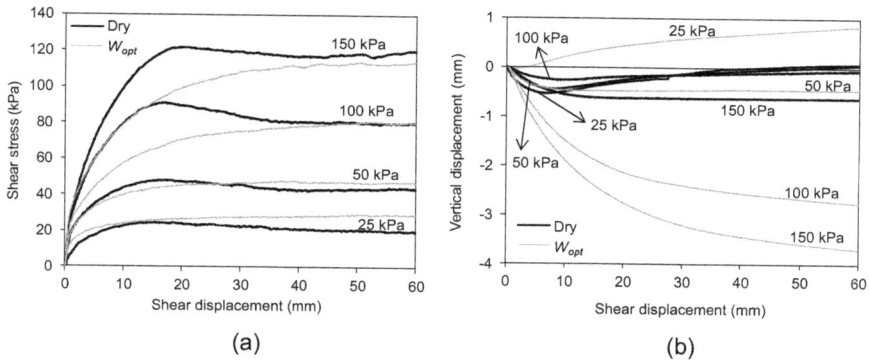

Figure 1.9 Influence of soil moisture content on the direct shear behaviour of a recycled C&D material-HDPE geogrid interface. (a) Shear stress plotted against shear displacement. (b) Vertical displacement plotted against shear displacement.

Source: (a) Modified after Vieira and Pereira (2022).

(w_{opt}=9%) as per the Modified Proctor test (EN 13286-2:2010). The direct shear tests were conducted at a constant rate of displacement of 1 mm/min under normal stresses varying between 25 and 150 kPa.

Figure 1.9 compares the evolution of shear stresses with shear displacement (Figure 1.9a) as well as the deformation behaviour of the test specimen during shearing (Figure 1.9b) for different moisture conditions of the recycled C&D material (dry and w_{opt}). As expected, the interface shear strength was affected by the moisture conditions of the backfill material. For the optimum water content, the peak of strength was no longer observed and the maximum stresses tended to be reached at larger shear displacements. Furthermore, under higher normal stress values the interface strength decreased significantly when compared to that obtained under dry conditions (Figure 1.9a).

The vertical displacement curves of the loading plate, presented in Figure 1.9b, show that when the recycled material was tested at the optimum water content, no dilation was observed during shearing, except for the lowest normal stress (25 kPa). Under higher normal stresses (100 and 150 kPa), the vertical settlement of the specimens was substantially higher when the recycled material was tested under moist conditions.

The coefficient of interaction (also termed as "interface shear strength coefficient") can be determined as the ratio of the maximum shear stress reached in the direct shear test on the backfill material-geosynthetic interface to the maximum shear stress obtained in the direct shear test on the backfill material under the same normal stress. The values of the coefficient of interaction for the studied interface ranged from 0.71 to 0.77 and 0.82 to 0.91 for dry and moist conditions, respectively. It is noteworthy that these values are within the typical range reported in previous studies for interfaces between geogrids and natural soils (Liu et al. 2009; Ferreira et al.

2015), thereby supporting the feasibility of using recycled C&D materials in lieu of traditional backfill materials in geosynthetic-reinforced structures.

1.3.2.2 Effect of stress relaxation

The time-dependent behaviour of backfill material-geosynthetic interfaces is an essential parameter concerning the long-term response of geosynthetic-reinforced soil systems. In this study, the time-dependent stress-strain behaviour of the interface between a recycled C&D material and a geocomposite reinforcement (GCR2) was investigated by means of multistage large-scale direct shear tests involving three different stages. Initially, the displacement rate was constant and equal to 1 mm/min. When a predefined shear displacement (μ) was attained, the test box was kept stationary for a given time interval (t); in the third stage, a constant rate of shear displacement of 1 mm/min was again imposed up to a total displacement of 60 mm. The shear displacement levels (μ) varied between $\mu = 1/3\ d_{max}$ and $\mu = 2/3\ d_{max}$ (where d_{max} is the displacement at which the maximum stress was reached in the benchmark test carried out under identical normal pressure). In addition, two different time intervals ($t = 30$ min and $t = 120$ min) were imposed. Conventional (benchmark) direct shear tests were also performed by adopting a constant rate of displacement of 1 mm/min.

The geocomposite reinforcement used in these tests (GCR2) is characterised by a nominal tensile strength of 40 kN/m with 10% elongation (Table 1.1). The recycled C&D material was compacted to a dry density of 17.1 kN/m³ (90% of the maximum dry density) and at the optimum water content ($w_{opt} = 11\%$), according to the Proctor test (EN 13286-2:2010). Both the multistage and benchmark tests were performed under normal stresses of 50, 100 and 150 kPa. The tests for the intermediate normal stress were repeated to analyse the repeatability of results.

Figure 1.10a and b presents the shear stress-shear displacement curves and the evolution of the vertical displacement of the rigid plate centre, respectively, obtained from the multistage tests under $\mu = 2/3\ d_{max}$ and $t = 120$ min. Also shown in these figures are the results from the benchmark tests. In general, the shear stress and vertical displacement curves from the multistage tests followed similar trends to those of the benchmark tests under the same normal stresses. As shown in Figure 1.10a, a significant shear stress reduction occurred in the multistage tests when the test box was kept stationary, which can be attributed to stress relaxation. At the beginning of the third stage (i.e. when the shear displacement was resumed), the interface showed high stiffness, with the shear stresses quickly reaching those recorded before the stress relaxation phase (Figure 1.10a).

Figure 1.10b indicates that the test specimens showed essentially a contractive response during shearing, irrespective of the test procedure and applied normal stress. During the second stage of the multistage tests (i.e. stress relaxation stage), the specimens underwent some additional vertical deformation.

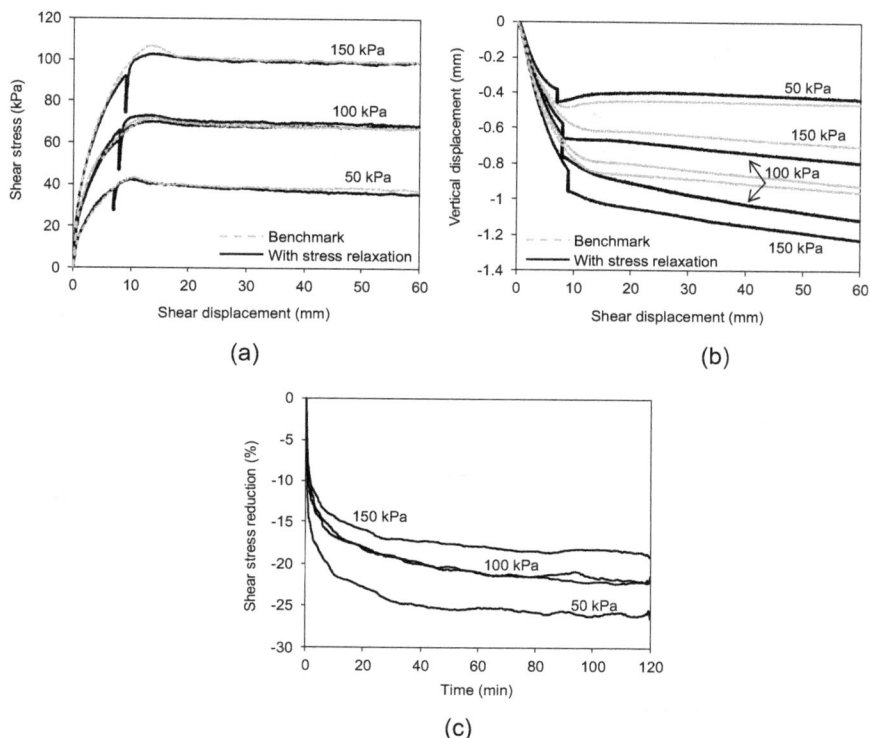

(a)

(b)

(c)

Figure 1.10 Influence of stress relaxation on the direct shear behaviour of a recycled C&D material-geocomposite reinforcement interface. (a) Shear stress plotted against shear displacement. (b) Vertical displacement plotted against shear displacement. (c) Percent reduction of shear stress with time.

Source: Modified after Ferreira et al. (2021b).

The maximum vertical displacements measured towards the end of the multistage and benchmark tests varied from 0.44 to 1.23 mm and 0.47 to 0.97 mm, respectively.

The shear stress reduction measured in the aforementioned multistage tests when the box was kept stationary (second stage) is plotted in Figure 1.10c. It can be seen that the percent reduction of shear stress increased with decreasing normal stress, implying that the time-dependent phenomena occurring at the interface were more significant under lower normal stresses. The highest percent reduction of shear stress (25.7%) measured at the end of the stress relaxation phase occurred under the normal pressure of 50 kPa. Irrespective of normal stress, the shear stress decay was more pronounced at the start of the stress relaxation phase and became marginal after the initial 30 min.

Figure 1.11a and b compares the peak and residual strength envelopes (Mohr-Coulomb envelopes) obtained on the basis of the conventional

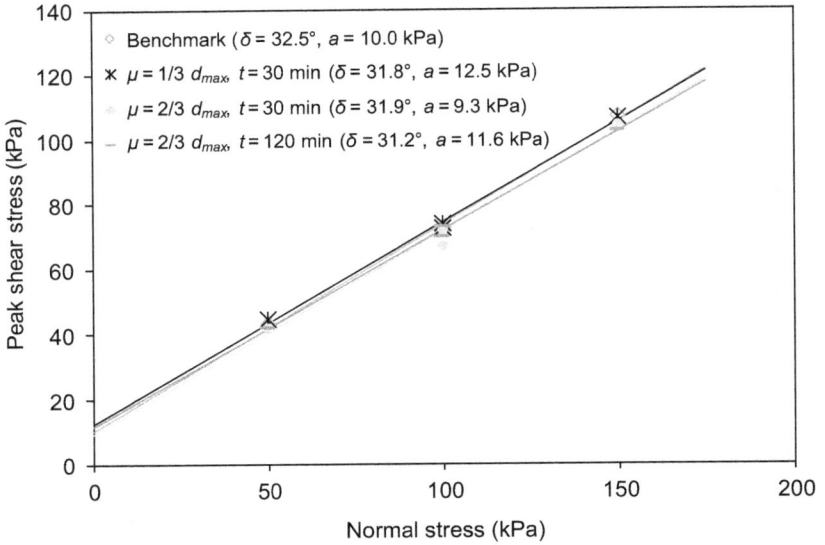

(a)

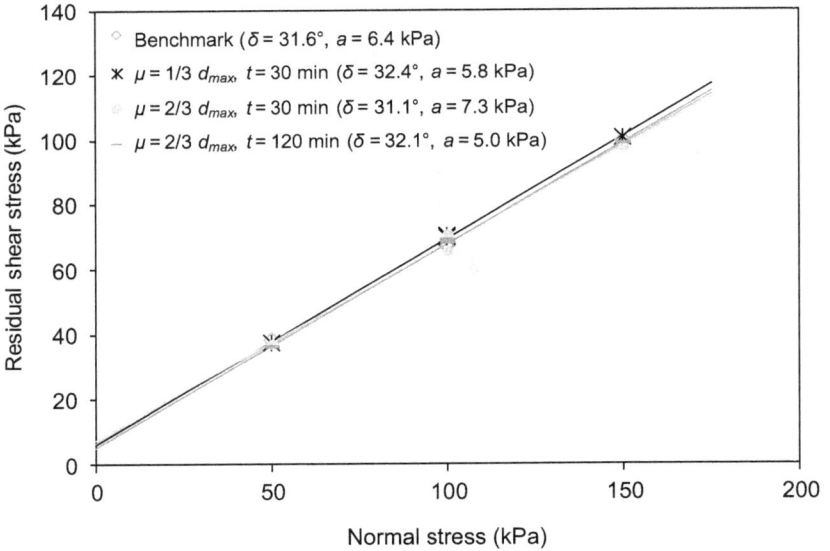

(b)

Figure 1.11 Shear strength envelopes of the interface for different test conditions. (a) Peak strength envelopes. (b) Residual strength envelopes.

Source: Modified after Ferreira et al. (2021b).

and multistage direct shear test results, considering different displacement levels (μ) and time intervals (t). The interface shear strength parameters, specifically the friction angle (δ) and adhesion (a), are also shown in these figures.

For the conditions examined, the impact of stress relaxation on the interface strength properties was almost negligible. Moreover, the displacement level or period of stress relaxation did not significantly influence the interface peak and residual shear strength. These findings suggest that the conventional (i.e. short-term) interface direct shear tests may be considered suitable to assess the interface shear strength parameters subjected to preliminary time-dependent phenomena. Nevertheless, additional tests involving sustained loads and different geosynthetic types are currently under development by the authors to further validate the above assumption.

1.4 CONCLUSIONS

This chapter reviewed recent research work conducted at the University of Porto to investigate the feasibility of using fine-grained recycled C&D waste in the construction of geosynthetic-reinforced systems, with special reference to the long-term behaviour.

- The short-term tensile response and the creep rupture behaviour of geosynthetics after exposure to recycled C&D waste and a natural soil under real environmental conditions were evaluated and compared with those of virgin specimens. The exposure of the geocomposite reinforcement to the backfill materials caused a significant degradation in its short-term tensile strength (by up to 27%). In contrast, the ultimate tensile strength of the HDPE geogrid was not significantly affected.
- For long-term conditions, the exhumed geocomposite specimens led to more optimistic predictions of the retained tensile strength, when compared to the results for virgin specimens. This implies that the use of virgin specimens for creep rupture testing can be regarded as a conservative (i.e. safe) approach.
- The pullout response of geosynthetics embedded in recycled C&D waste was characterised by large-scale pullout tests under monotonic and cyclic loading conditions. The monotonic tests showed that the measured peak pullout resistance of the HDPE geogrid increases with the displacement rate and specimen size. The pullout interaction coefficients for the studied interface ranged from 1.13 to 1.57, thereby representing identical performance to that generally achieved at soil-geogrid interfaces.
- The pullout tests involving cyclic loading revealed that depending on the geosynthetic type, the maximum pullout resistance attained in the post-cyclic stage may considerably decrease (by up to 13%)

comparatively with that reached under monotonic loading conditions. The increase in the cyclic loading amplitude and pre-cyclic pullout load resulted in greater displacements and strains along the specimens, regardless of geosynthetic type.

- The direct shear behaviour of recycled C&D waste-geosynthetic interfaces was assessed through conventional (short-term) and multistage large-scale direct shear tests. The conventional tests showed that when the backfill material is tested under moist conditions (i.e. $w=w_{opt}$), the interface strength decreases in comparison with that achieved under dry conditions, and no peak of strength is observed. The values of the coefficient of interaction ranged from 0.71 to 0.91, which is consistent with those typically reported for interfaces between geogrids and natural soils.

- The multistage direct shear tests showed a significant shear stress decay when the test box was kept restrained from any horizontal displacement (by up to 26%), which is attributed to stress relaxation. However, for the conditions examined, the effect of stress relaxation on the interface strength parameters was almost negligible.

The research reported herein contributes to a better understanding of the short- and long-term behaviours of geosynthetics and recycled C&D waste-geosynthetic interfaces under different conditions prevailing in field applications. While the results suggest that recycled C&D waste can be regarded as a feasible alternative to conventional fill materials of geosynthetic-reinforced systems, long-term field studies would be useful to provide further insight following the above conclusions. A full-scale geosynthetic-reinforced embankment constructed with recycled C&D waste as backfill material is currently being monitored by the authors. The outcomes of such study will be presented in future publications.

ACKNOWLEDGEMENTS

This work was financially supported by Project PTDC/ECI-EGC/30452/2017-POCI-01–0145-FEDER-030452 – funded by FEDER funds through COMPETE 2020-Programa Operacional Competitividade e Internacionalização (POCI) and by national funds (PIDDAC) through FCT/MCTES; Programmatic funding – UIDP/04708/2020 of the CONSTRUCT – Instituto de I&D em Estruturas e Construções – funded by national funds through the FCT/MCTES (PIDDAC). The first author wishes to acknowledge Grant No. 2021.03625. CEECIND from the Stimulus of Scientific Employment, Individual Support (CEECIND) – 4th Edition provided by Fundação para a Ciência e Tecnologia (FCT). The third author would also like to thank FCT for his research grant (SFRH/BD/147838/2019).

REFERENCES

Allen, T. M., and R. J. Bathurst. 1996. "Combined allowable strength reduction factor for geosynthetic creep and installation damage." *Geosynthetics International.* 3 (3): 407–439.

Allen, T. M., and R. J. Bathurst. 2002. "Observed long-term performance of geosynthetic walls and implications for design." *Geosynthetics International.* 9 (5–6): 567–606.

Allen, T. M., R. J. Bathurst, and R. R. Berg. 2002. "Global level of safety and performance of geosynthetic walls: an historical perspective." *Geosynthetics International.* 9 (5–6): 395–450.

Anubhav, and H. Wu. 2015. "Modelling of non-linear shear displacement behaviour of soil–geotextile interface." *International Journal of Geosynthetics and Ground Engineering.* 1 (2): 19.

Arulrajah, A., E. Yaghoubi, Y. C. Wong, and S. Horpibulsuk. 2017. "Recycled plastic granules and demolition wastes as construction materials: resilient moduli and strength characteristics." *Construction and Building Materials.* 147: 639–647.

Arulrajah, A., J. Piratheepan, M. M. Disfani, and M. W. Bo. 2013. "Geotechnical and geoenvironmental properties of recycled construction and demolition materials in pavement subbase applications." *Journal of Materials in Civil Engineering.* 25 (8): 1077–1088.

Arulrajah, A., M. A. Rahman, J. Piratheepan, M. W. Bo, and M. A. Imteaz. 2014. "Evaluation of interface shear strength properties of geogrid-reinforced construction and demolition materials using a modified large-scale direct shear testing apparatus." *Journal of Materials in Civil Engineering.* 26 (5): 974–982.

Arulrajah, A., M. Naeini, A. Mohammadinia, S. Horpibulsuk, and M. Leong. 2020. "Recovered plastic and demolition waste blends as railway capping materials." *Transportation Geotechnics.* 22: 100320.

ASTM D6706-01:2013. *Standard Test Method for Measuring Geosynthetic Pullout Resistance in Soil.* ASTM International, West Conshohocken, PA.

Bathurst, R. J., B. Q. Huang, and T. M. Allen. 2012. "Interpretation of laboratory creep testing for reliability-based analysis and load and resistance factor design (LRFD) calibration." *Geosynthetics International.* 19 (1): 39–53.

Benjamim, C. V. S., B. S. Bueno, and J. G. Zornberg. 2007. "Field monitoring evaluation of geotextile-reinforced soil-retaining walls." *Geosynthetics International.* 14 (2): 100–118.

Bueno, B. S., M. A. Costanzi, and J. G. Zornberg. 2005. "Conventional and accelerated creep tests on nonwoven needle-punched geotextiles." *Geosynthetics International.* 12 (6): 276–287.

Cardile, G., M. Pisano, and N. Moraci. 2019. "The influence of a cyclic loading history on soil-geogrid interaction under pullout condition." *Geotextiles and Geomembranes.* 47 (4): 552–565.

Cardile, G., M. Pisano, P. Recalcati, and N. Moraci. 2021. "A new apparatus for the study of pullout behaviour of soil-geosynthetic interfaces under sustained load over time." *Geotextiles and Geomembranes.* 49 (6): 1519–1528.

Deng, A., and Z. Huangfu. 2021. "Limit state and creep behaviour of high-density polyethylene geocell." *International Journal of Geosynthetics and Ground Engineering.* 7 (2): 28.

EN 13286-2:2010. *Unbound and Hydraulically Bound Mixtures - Part 2: Test Methods for Laboratory Reference Density and Water Content - Proctor Compaction.* CEN, Brussels, Belgium.

EN 13738:2004. *Geotextiles and Geotextile-Related Products - Determination of Pullout Resistance in Soil.* CEN, Brussels, Belgium.

EN ISO 10319:2015. *Geosynthetics - Wide-Width Tensile Test.* CEN, Brussels, Belgium.

EN ISO 13431:1999. *Geotextiles and Geotextile-Related Products - Determination of Tensile Creep and Creep Rupture Behaviour.* CEN (Brussels, Belgium) in collaboration with ISO (Geneva, Switzerland).

European Commission. 2020. "EU Circular Economy Action Plan." Accessed February 2021. https://ec.europa.eu/environment/circular-economy/.

Ferreira, F. B., A. T. Gomes, C. S. Vieira, and M. L. Lopes. 2016a. "Reliability analysis of geosynthetic-reinforced steep slopes." *Geosynthetics International.* 23 (4): 301–315.

Ferreira, F. B., C. S. Vieira, and M. L. Lopes. 2013. "Analysis of soil-geosynthetic interfaces shear strength through direct shear tests." In *Proc., International Symposium on Design and Practice of Geosynthetic-Reinforced Soil Structures,* 44–53. Bologna, Italy.

Ferreira, F. B., C. S. Vieira, and M. L. Lopes. 2016b. "Cyclic and post-cyclic shear behaviour of a granite residual soil-geogrid interface." In *Proc., Procedia Engineering, Vol. 143, Advances in Transportation Geotechnics III,* 379–386.

Ferreira, F. B., C. S. Vieira, and M. L. Lopes. 2016c. "Soil-geosynthetic interface strength properties from inclined plane and direct shear tests - A comparative analysis." In *Proc., 6th Asian Regional Conference on Geosynthetics: Geosynthetics for Infrastructure Development,* 925–937. New Delhi, India.

Ferreira, F. B., C. S. Vieira, and M. L. Lopes. 2020a. "Pullout behavior of different geosynthetics-Influence of soil density and moisture content." *Frontiers in Built Environment.* 6 (12): 1–13.

Ferreira, F. B., C. S. Vieira, M. L. Lopes, and D. M. Carlos. 2016d. "Experimental investigation on the pullout behaviour of geosynthetics embedded in a granite residual soil." *European Journal of Environmental and Civil Engineering.* 20 (9): 1147–1180.

Ferreira, F. B., C. S. Vieira, M. L. Lopes, and P. G. Ferreira. 2020b. "HDPE geogrid-residual soil interaction under monotonic and cyclic pullout loading." *Geosynthetics International.* 27 (1): 79–96.

Ferreira, F. B., P. M. Pereira, C. S. Vieira, and M. L. Lopes. 2021a. "Long-term tensile behavior of a high-strength geotextile after exposure to recycled construction and demolition materials." *Journal of Materials in Civil Engineering.* 34(5): 04022046.

Ferreira, F. B., P. M. Pereira, C. S. Vieira, and M. L. Lopes. 2021b. "Time-dependent response of a recycled C&D material-geotextile interface under direct shear mode." *Materials.* 14 (11): 3070.

Ferreira, F. B., C. S. Vieira, and M. L. Lopes. 2015. "Direct shear behaviour of residual soil–geosynthetic interfaces–influence of soil moisture content, soil density and geosynthetic type." *Geosynthetics International.* 22 (3): 257–272.

Ferreira, F.B., C. S. Vieira, G. Mendonça, and M. L. Lopes. 2023. "Effect of sustained loading on the direct shear behaviour of recycled C&D material–geosynthetic interfaces." *Materials.* 16 (4): 1722.

Goodhue, M. J., T. B. Edil, and C. H. Benson. 2001. "Interaction of foundry sands with geosynthetics." *Journal of Geotechnical and Geoenvironmental Engineering.* 127 (4): 353–362.

Greenway, D., J. R. Bell, and B. Vandre. 1999. "Snailback wall - First fabric wall revisited at 25 year milestone." In *Proc., Geosynthetics '99: Specifying Geosynthetics and Developing Design Details,* 905–919. Boston, MA.

Hegde, A., and R. Roy. 2018. "A comparative numerical study on soil–geosynthetic interactions using large scale direct shear test and pullout test." *International Journal of Geosynthetics and Ground Engineering.* 4 (1): 2.

Henzinger, C., and D. Heyer. 2018. "Soil improvement using recycled aggregates from demolition waste." *Proceedings of the Institution of Civil Engineers: Ground Improvement.* 171 (2): 74–81.

Hirakawa, D., W. Kongkitkul, F. Tatsuoka, and T. Uchimura. 2003. "Time-dependent stress-strain behaviour due to viscous properties of geogrid reinforcement." *Geosynthetics International.* 10 (6): 176–199.

Hsieh, C. W., G. H. Chen, and J. H. Wu. 2011. "The shear behavior obtained from the direct shear and pullout tests for different poor graded soil-geosynthetic systems." *Journal of GeoEngineering.* 6 (1): 15–26.

Karnam Prabhakara, B. K., U. Balunaini, and A. Arulrajah. 2021a. "Development of a unique test apparatus to conduct axial and transverse pullout testing on geogrid reinforcements." *Journal of Materials in Civil Engineering.* 33 (1): 04020406.

Karnamprabhakara, B. K., U. Balunaini, A. Arulrajah, and R. Evans. 2021b. "Axial pullout resistance and interface direct shear properties of geogrids in pond ash." *International Journal of Geosynthetics and Ground Engineering.* 7 (2): 22.

Kongkitkul, W., F. Tatsuoka, and D. Hirakawa. 2007. "Creep rupture curve for simultaneous creep deformation and degradation of geosynthetic reinforcement." *Geosynthetics International.* 14 (4): 189–200.

Leite, F. D., R. D. Motta, K. L. Vasconcelos, and L. Bernucci. 2011. "Laboratory evaluation of recycled construction and demolition waste for pavements." *Construction and Building Materials.* 25 (6): 2972–2979.

Leshchinsky, D., M. Dechasakulsom, V. N. Kaliakin, and H. I. Ling. 1997. "Creep and stress relaxation of geogrids." *Geosynthetics International.* 4 (5): 463–479.

Li, A. L., and R. K. Rowe. 2001. "Influence of creep and stress-relaxation of geosynthetic reinforcement on embankment behaviour." *Geosynthetics International.* 8 (3): 233–270.

Liu, C.-N., Y.-H. Ho, and J.-W. Huang. 2009. "Large scale direct shear tests of soil/PET-yarn geogrid interfaces." *Geotextiles and Geomembranes.* 27 (1): 19–30.

Liu, F. Y., P. Wang, X. Geng, J. Wang, and X. Lin. 2016. "Cyclic and post-cyclic behaviour from sand–geogrid interface large-scale direct shear tests." *Geosynthetics International.* 23 (2): 129–139.

Lopes, M. L. (2012). Soil-geosynthetic interaction. *Handbook of Geosynthetic Engineering.* Ice Publishing, Thomas Telford Ltd., London, UK. 45–66.

Lopes, M. L., A. S. Cardoso, and K. C. Yeo. 1994. "Modelling performance of a sloped reinforced soil wall using creep function." *Geotextiles and Geomembranes.* 13 (3): 181–197.

Lopes, M. L., and M. Ladeira. 1996. "Influence of the confinement, soil density and displacement rate on soil-geogrid interaction." *Geotextiles and Geomembranes.* 14 (10): 543–554.

Lu, C., J. Chen, C. Gu, J. Wang, Y. Cai, T. Zhang, and G. Lin. 2021. "Resilient and permanent deformation behaviors of construction and demolition wastes in unbound pavement base and subbase applications." *Transportation Geotechnics*. 28: 100541.

Maghool, F., A. Arulrajah, M. Mirzababaei, C. Suksiripattanapong, and S. Horpibulsuk. 2020. "Interface shear strength properties of geogrid-reinforced steel slags using a large-scale direct shear testing apparatus." *Geotextiles and Geomembranes*. 48 (5): 625–633.

Mahigir, A., A. Ardakani, and M. Hassanlourad. 2021. "Comparison between monotonic, cyclic and post-cyclic pullout behavior of a PET geogrid embedded in clean sand and clayey sand." *International Journal of Geosynthetics and Ground Engineering*. 7 (1): 10.

Min, Y., D. Leshchinsky, H. Ling, and V. Kaliakin. 1995. "Effects of sustained and repeated tensile loads on geogrid embedded in sand." *Geotechnical Testing Journal*. 18 (2): 204–225.

Miyata, Y., R. J. Bathurst, and T. M. Allen. 2014. "Reliability analysis of geogrid creep data in Japan." *Soils and Foundations*. 54 (4): 608–620.

Mohiuddin, A. (2003). Analysis of laboratory and field pullout tests of geosynthetics in clayey soils, Faculty of the Louisiana State University and Agricultural and Mechanical College, Osmania University.

Moraci, N., G. Cardile, D. Gioffrè, M. C. Mandaglio, L. S. Calvarano, and L. Carbone. 2014. "Soil geosynthetic interaction: design parameters from experimental and theoretical analysis." *Transportation Infrastructure Geotechnology*. 1 (2): 165–227.

Naeini, M., A. Mohammadinia, A. Arulrajah, and S. Horpibulsuk. 2021. "Recycled glass blends with recycled concrete aggregates in sustainable railway geotechnics." *Sustainability (Switzerland)*. 13 (5): 1–18.

Palmeira, E. M. 2009. "Soil-geosynthetic interaction: modelling and analysis." *Geotextiles and Geomembranes*. 27 (5): 368–390.

Pereira, P., F. Ferreira, C. Vieira, and M. Lopes. 2019. "Use of recycled C&D wastes in unpaved rural and forest roads - feasibility analysis." In *Proc., 5th International Conference WASTES - Solutions, Treatments and Opportunities*, 161–167. Lisbon.

Pinho-Lopes, M., A. M. Paula, and M. L. Lopes. 2018. "Long-term response and design of two geosynthetics: effect of field installation damage." *Geosynthetics International*. 25 (1): 98–117.

Pourkhorshidi, S., C. Sangiorgi, D. Torreggiani, and P. Tassinari. 2020. "Using recycled aggregates from construction and demolition waste in unbound layers of pavements." *Sustainability (Switzerland)*. 12 (22): 1–20.

Prashanth, V., A. M. Krishna, and S. K. Dash. 2016. "Pullout tests using modified direct shear test setup for measuring soil–geosynthetic interaction parameters." *International Journal of Geosynthetics and Ground Engineering*. 2 (2): 10.

Rahman, M. A., A. Arulrajah, J. Piratheepan, M. W. Bo, and M. A. Imteaz. 2014. "Resilient modulus and permanent deformation responses of geogrid-reinforced construction and demolition materials." *Journal of Materials in Civil Engineering*. 26 (3): 512–519.

Santos, E. C. G., E. M. Palmeira, and R. J. Bathurst. 2013. "Behaviour of a geogrid reinforced wall built with recycled construction and demolition waste backfill on a collapsible foundation." *Geotextiles and Geomembranes*. 39: 9–19.

Santos, E., E. Palmeira, and R. Bathurst. 2014. "Performance of two geosynthetic reinforced walls with recycled construction waste backfill and constructed on collapsible ground." *Geosynthetics International*. 21 (4): 256–269.

Silva, R. V., J. De Brito, and R. K. Dhir. 2014. "Properties and composition of recycled aggregates from construction and demolition waste suitable for concrete production." *Construction and Building Materials*. 65: 201–217.

Tang, X., G. R. Chehab, and A. Palomino. 2008. "Evaluation of geogrids for stabilising weak pavement subgrade." *International Journal of Pavement Engineering*. 9 (6): 413–429.

Vieira, C. S. 2020. "Valorization of fine-grain Construction and Demolition (C&D) waste in geosynthetic reinforced structures." *Waste and Biomass Valorization*. 11 (4): 1615–1626.

Vieira, C. S., and P. M. Pereira (2022). Geotechnical and geoenvironmental characterization of fine-grained construction and demolition recycled materials reinforced with geogrids. *Advances in Transportation Geotechnics IV*. Springer, Germany. 605–617.

Vieira, C. S., and P. M. Pereira. 2015a. "Damage induced by recycled Construction and Demolition Wastes on the short-term tensile behaviour of two geosynthetics." *Transportation Geotechnics*. 4: 64–75.

Vieira, C. S., and P. M. Pereira. 2015b. "Use of recycled construction and demolition materials in geotechnical applications: a review." *Resources, Conservation and Recycling*. 103: 192–204.

Vieira, C. S., and P. M. Pereira. 2021. "Short-term tensile behaviour of three geosynthetics after exposure to recycled construction and demolition materials." *Construction and Building Materials*. 273: 122031.

Vieira, C. S., F. B. Ferreira, P. M. Pereira, and M. L. Lopes. 2020a. "Pullout behaviour of geosynthetics in a recycled construction and demolition material - effects of cyclic loading." *Transportation Geotechnics*. 23: 100346.

Vieira, C. S., M. L. Lopes, and L. M. Caldeira. 2013. "Sand-geotextile interface characterisation through monotonic and cyclic direct shear tests." *Geosynthetics International*. 20 (1): 26–38.

Vieira, C. S., P. M. Pereira, and M. L. Lopes. 2016. "Recycled construction and demolition wastes as filling material for geosynthetic reinforced structures. Interface properties." *Journal of Cleaner Production*. 124: 299–311.

Vieira, C. S., P. M. Pereira, F. B. Ferreira, and M. L. Lopes. 2020b. "Pullout behaviour of geogrids embedded in a recycled construction and demolition material. Effects of specimen size and displacement rate." *Sustainability*. 12 (9): 3825.

Yaghoubi, E., N. Sudarsanan, and A. Arulrajah. 2021. "Stress-strain response analysis of demolition wastes as aggregate base course of pavements." *Transportation Geotechnics*. 30: 100599.

Zornberg, J. G., B. R. Byler, and J. W. Knudsen. 2004. "Creep of geotextiles using time-temperature superposition methods." *Journal of Geotechnical and Geoenvironmental Engineering*. 130 (11): 1158–1168.

Chapter 2

Investigation of economic feasibility between bamboo and conventional construction materials in non-residential projects

Vijayalaxmi J. and Aishwarya Sudheer Vadukkumchery
School of Planning and Architecture

CONTENTS

DOI: 10.1201/9781003368335-2

2.1 INTRODUCTION

Sustainability in building structures is achieved by employing resource-efficient and environmentally responsible methods throughout the building's life cycle. But currently, the construction industry heavily relies on non-renewable resources and non-green materials such as sand, cement, steel, and aggregates for building structures. Also, buildings contribute towards nearly half of India's carbon emissions, half of the country's water consumption, and about one-third of the landfill waste (Vadera et al., 2008). Sustainable buildings are an urgent requirement considering the footprint and economic impact the industry leaves behind.

There is a general misconception that sustainable alternatives are more expensive than the conventional ones due to the concern over the availability of natural building materials, the precautions or treatments involved in preserving these materials, the need for skilled workmanship, the initial capital cost, etc.

This is where materials like bamboo play a pronounced role. Bamboo is a hollow woody plant belonging to the grass family (Gramineae) and subfamily Bambusoideae. Around the world, there are approximately 70 genera and approximately 1,500 species of bamboo (Abdul Khalil et al., 2012). They thrive in a variety of climates, ranging from cold mountains to hot tropical regions. Asia is home to roughly 65% of all bamboo-growing areas (Paridah, 2013).

China is the world's most important producer of bamboo. It is the world leader in bamboo species, area, volume, and output. Bamboo has a profound and one-of-a-kind influence on life, and China has a one-of-a-kind ethnic bamboo culture (Liu et al., 2018).

In India, there are 11 exotic and 125 indigenous bamboo species (FSI, 2011). India is the world's second-richest country when it comes to bamboo genetic resources. Bamboo covers approximately 13.96 million hectares of land worldwide (FSI, 2011). *Dendrocalamus strictus, Bambusa arundinacea, Melocanna baccifera, Bambusa bambos, Ochlandra travancorica, Dendrocalamus hamiltonii, Bambusa tulda*, etc. are few common bamboo species available in India. Bamboo, an easily procurable, strong, durable, and environmentally friendly material with carbon sequestration capacity (Jyoti Nath et al., 2009), is one of the cost-effective alternatives among the marginalized sectors of the country's society (Manjunath, 2015). It is normally used in buildings as part of construction, decoration, or as a building component. Bamboo is not currently used as a primary structural material in design and construction practice (Habibi, 2019).

The usage of bamboo-based techniques in India is mostly limited to residential projects, mainly rural housing (Shah et al., 2012). There is a dearth in studies addressing or assessing the economic advantages of using bamboo in place of conventional building materials for non-residential buildings. Thereby, this chapter intends to explore bamboo's potential in replacing the commonly used materials in construction by studying the economic feasibility of employing bamboo as the major component in non-residential projects.

2.2 LITERATURE REVIEW

The basic engineering properties of bamboo, along with the processing of bamboo into various composite items, are being extensively studied in recent days. Its flexural and tensile strength and durability when properly treated have significantly promoted the usage among the construction industry. Bamboo has many other distinct advantages, including the ability to grow quickly and produce a high yield, as well as the ability to mature quickly. Bamboo can also be produced in bulk quantities at a reduced cost, thereby making it more cost-effective.

Bamboo's moisture content is the deciding factor in its use as a structural element, and all physicomechanical properties are functions of it (Wakchaure and Kute, 2012). The moisture content also attracts fungi and borer insects, thereby affecting its life span. Moisture content in bamboo varies according to its species (Anokye et al., 2016).

Bamboo's growth rate and high flexibility, as well as its low weight-to-height ratio, give it a diverse range of applications as a building material. Bamboo species have densities ranging from 700 to 800 kg/m³. The failure bending stress is 0.14 times the mass per unit volume. Bamboo has greater bending strength at failure due to its high mass per unit volume (Jit Kaur, 2018).

The density-strength relationship varies depending on properties and species. The density of bamboo material (whole stem weight/green bamboo volume) is between 400 and 900 kg/m³, and this is primarily determined by the density and composition of the vascular bundles (Anokye et al., 2016).

The tensile strength of Dagdi and Manga bamboo, if compared with the strength of steel bar, is half. The minimum requirement for flexural strength, which is 3.13 N/mm² for M20 grade of concrete, is easily available in bamboo-reinforced concrete. In comparison with steel-reinforced concrete, bamboo-reinforced concrete can be executed at least to some extent for G+1 structures or low-cost G+1 housing projects (as composite reinforcement), in that way the cost of construction can be reduced as well (Jayagond et al., 2020).

Bamboo has a tensile strength of 193.05 MPa, as compared to that of steel, which is 158.57 MPa. Bamboo requires about 50 times less energy to generate unit volume per unit stress as compared to steel or concrete. Therefore, bamboo is a viable alternative to steel in load-bearing applications (Jit Kaur, 2018).

Main and distribution reinforcement is possible with bamboo reinforcement technique like steel reinforcement, and also in comparison with the same, the bamboo reinforcement technique is cheaper by three times (Dange and Pataskar, 2017).

The bamboo fibre has a tensile strength of 650 MPa, whereas the tensile strength of steel is between 500 and 1,000 MPa. Also, the flexibility of bamboo fibre is higher (50 GPa) as compared to steel. Other properties of bamboo include an average weight of 0.625 kg/m, modulus of rupture between

610 and 1,600 kg/cm^2, modulus of elasticity between 1.5 and 2×105 kg/cm^2, ultimate compressive stress between 794 and 864 kg/cm^2, and safe working stress in compression about 105 kg/cm^2 (Hasan et al., 2015).

Dendrocalamus strictus has advantageous mechanical properties, and it can be a good eco-friendly and sustainable material for green building (Bhonde et al., 2014).

A study of bamboo's mechanical properties in concrete was conducted, in which the ultimate load of a bamboo-reinforced concrete beam proved to be 400% higher than un-reinforced concrete. It was observed that in comparison with steel, the bamboo and concrete bonding was weaker, and the modulus of elasticity of bamboo was 1/15 that of steel. Bamboo's compressive strength was found to be much lower than its tensile strength, and it was observed that along the fibres, there was high strength, whereas perpendicular to the fibres, the strength was low (Ghavami, 1995).

The Table 2.1 below summarizes the value range of Bamboo properties as observed in the review of literature.

From these studies, it is evident that bamboo's engineering properties are sufficient enough to replace the conventional materials if employed using a suitable technique.

Sustainable construction necessitates a long-term perspective where the initial capital cost is weighed against the structure's operating cost. Here, the primary economic benefits are lower operating and utility costs, lower maintenance costs, and overall improvements in building performance and efficiency (Lowe and Zhao, 2003).

Substituting conventional materials like steel, brick, timber, and cement with bamboo, even partially, for various building components can reduce the cost of construction of a residence by 40%. Such houses can sustain for a long period of time and requires low maintenance (Manjunath, 2015).

In a comparison study between bamboo and conventional building materials employed in the construction of a rural school building prototype conducted in Nigeria, it was found that the cost of the classroom block built using bamboo was 40.76% of the total cost of constructing the same block using conventional materials. Here, the bamboo techniques employed were

Table 2.1 Properties of bamboo

Property	Value	Reference
Density	0.2–0.85 kg/cm^2	Kaur et al. (2016)
Specific gravity	0.575–0.655 kg/m^3	Kaur et al. (2016)
Modulus of rupture	610–1,600 kg/cm^2	Dange and Pataskar (2017)
Modulus of elasticity	1.5–2.0×10^5 kg/cm^2	Dange and Pataskar (2017)
Ultimate compressive stress	794–864 kg/cm^2	Dange and Pataskar (2017)
Safe working stress in compression	105 kg/cm^2	Dange and Pataskar (2017)
Safe working stress in tension	160–350 kg/cm^2	Kaur et al. (2016)
Safe working stress in shear	115–180 kg/cm^2	Kaur et al. (2016)

bamboo laminate board for flooring and wall panelling, bamboo window, whole bamboo culm for roof support, and bamboo mat-corrugated sheet for roofing (Adetunji et al., 2015).

The cost of constructing a bamboo pole in a wooden batten frame partition wall was found to be 53% lesser than that of a brick wall, which was drawn when a comparative cost analysis between a conventional reinforced concrete structure and RC building involving bamboo partitions of same number of floors in Bangladesh was done (Hasan et al., 2015).

The economic advantages of using bamboo in place of conventional building materials for non-residential buildings are less explored in India. Here, the intention is to assess the economic prospects of replacing conventional materials with bamboo with the help of two field studies.

2.3 STUDY AREA

Kerala is the southwest state of India lying between 8°17′30″ and 12°47′ East Longitudes and 74°7″47″ and 77°37′12″ North Latitudes (Government of Kerala). The state has a total land area of 38,863 sq.km and a population of 3,33,87,677 people (Census, 2011).

2.3.1 Climate in Kerala

Kerala has a warm humid climate with its average maximum temperature as 36°C and average minimum temperature as 18°C. The relative humidity is also high and ranges between an annual high of 85% and an annual low of 16%. The average annual rainfall is 2,250 mm, but can go as high as 3,600 mm during certain periods (Kerenvis, 2021).

The most common bamboo species in Kerala is *Bambusa bamboos*, which grows in moist deciduous forests with an average rainfall of 1,200–2,000 mm and a minimum temperature range of 18°C–33°C (Kerala Forest Research Institute).

2.3.2 Bamboo in Kerala

Kerala is one of the country's major bamboo diversity centres. According to Kerala State Bamboo Mission, there are 28 varieties of bamboo found in the state – a few of them being *Bambusa bambos*, *Bambusa vulgaris*, *Dendrocalamus strictus*, *Ochlandra*, *Thyrsostachys Oliveri*, and *Pseudoxytenanthera* (Kerala Bamboo Corporation Ltd, 2011). Among these varieties, *Bambusa bambos*, *Ochlandra travancorica* (OT), and *Dendrocalamus strictus* are commonly used in the construction field for various purposes, including furniture making. In Kerala, bamboo is cultivated by both forest departments and homesteads. The fact that 67.3% of the bamboo extraction comes from home gardens rather than forests distinguishes the Kerala bamboo scene (Kerala State Bamboo Mission). In Kerala, the forest department allows the procurement of 1,000 tonne of bamboo a year for each establishment.

2.3.3 Preservation techniques

Bamboo, if not treated, is non-durable as it is prone to attack by insects, fungi, and even sap-stain. Thus, bamboo in all of its forms such as round, split, sticks, and slivers requires preservative treatment for enhanced durability. Previously, in Kerala, being a state having plenty of ponds, soaking in water for 10–15 days followed by air drying was commonly used as a local preservation technique of bamboo, which is still in use for small-scale construction works. Applying diesel or waste oil or even pesticides on bamboo followed by air drying would be another locally adopted technique.

In recent days, the following treatment methods are used:

Dip diffusion treatment: Soaking bamboo in a chemical solution of boric acid (5 kg) and borax (7.5 kg) in water (100 L) for a particular duration.

Vacuum-pressure impregnation treatment: A commercial pressure treatment cylinder facility is used to alternate between vacuum pressurization and chemical solution pressurization (Dhamodaran et al., 2020).

2.3.4 Joinery

Because of the round shape and reduction in diameter along the length, jointing is very cumbersome in bamboo. In Kerala, the commonly practised joinery technique is traditional joinery, which involves tying and using nuts and bolts. The possibilities of orthogonal joints are yet to be explored in the state due to the lack of skilled labour.

2.3.5 Construction methods using bamboo

Bamboo-based construction techniques are commonly employed in the construction of residential projects. Bamboo can be used on its own or in combination with other materials in the foundation, floor slabs, flooring, roof structure, walls, columns, beams, doors, windows, scaffolding, landscape components, and water treatment systems.

2.3.6 Bamboo products

Kerala State Bamboo Corporation Ltd has recently expanded into the production of bamboo mat veneer composites (BMVC) and bamboo mat high-density panel boards, which are the future of the home interior markets and are regarded as lifestyle products by the younger generation. Composites made of bamboo, such as bamboo fibre boards and bamboo ply, are exceptionally durable, resistant to fungal decay, insects and termite attacks, and they are extremely flexible as well, making them ideal for flooring, walls, roofing, concrete reinforcement, and scaffolding (Pavithra and Jacob, 2018).

2.3.7 Limitations

In Kerala, the most popular joinery techniques are tying and using nuts and bolts to secure bamboo parts. Bolted connections necessitate raw bamboo with high shear and splitting resistance. There is a possibility for cracking or local deformation of culm during drilling the hole. Also, the nuts and bolts don't fully fit the round-shaped body of bamboo pole, gradually leading to gaps and loose connections. Because each joint is a pin joint, the use of bolts necessitates a thorough understanding of structural design principles and the principle of triangulation. It is difficult to remove and replace poles once the structure has been built (Hong et al., 2019). The possibility of applying orthogonal joints or steel joints like hub, clamps, etc. in the state is yet to be explored and requires extensive study and skilled labour.

Bamboo, if exposed to extreme weather conditions, requires yearly maintenance in the form of varnish or some protective coat to protect it from fading and weakening.

Even though bamboo grows abundantly in Kerala, the forest department allows the procurement of only 1,000 tonne of bamboo a year for each establishment.

2.4 METHOD AND MATERIALS

2.4.1 Methodology

This chapter studies the engineering properties of bamboo and its capability of replacing conventional materials. The study area is limited to the district of Trivandrum, Kerala. Two non-residential buildings of similar scale and function are chosen for case study from the area and studied extensively in terms of cost incurred in constructing the two structures. A detailed study on a bamboo-based non-residential building is carried out to understand the various ways in which bamboo can be incorporated as a structural and non-structural element. Further to draw inference over the economic feasibility of bamboo-based buildings, a comparative study on the cost of construction of the bamboo-based building and conventional building has been done. The data is collected through unstructured personal interviews after obtaining informed consent from COSTFORD Trivandrum. The cost analysis includes the cost of procuring, treating, and transporting the materials along with the cost of labour. The primary data is collected from the respective designers of both the projects after obtaining informed consent. The secondary data is collected from Kerala State Bamboo Mission and Kerala State Bamboo Corporation Ltd. For estimating the cost incurred for construction, CPWD document, Delhi Scheduled Rates 2019, and Kerala PWD documents were referred as primary source. The market rates were obtained from COSTFORD Trivandrum through personal interviews.

2.4.2 Case study on Library and Documentation Centre, Trivandrum

The 2,200 sq. ft structure is situated at Laurie Baker Centre for Habitat Studies (LBC) in Vilappilsala village of Trivandrum District (8°53′0″ N and 77°03′0″ E). The three-storey building is the first public bamboo building in Kerala, designed and executed on the lines of Laurie Baker's initial sketches by COSTFORD Trivandrum. The majority of the building is constructed using various bamboo-based techniques.

2.4.2.1 Bambusa Bambos

The bamboo variety that is used in this building is *Bambusa bambos* (Indian thorny bamboo), locally known as "kallan mula". It typically grows in all types of soil with good drainage, but slightly acidic soils, sandy loam, and alluvial soils show the best growth. It grows abundantly in tropical and tropical-to-subtropical conditions, and it can be found up to 1,200 m above sea level on the plains. It accounts for approximately 28% of the total bamboo area in the country (Kerala State Bamboo Mission).

2.4.2.2 Source

The bamboo (660 bamboo poles in number) for the construction was procured from the Palode Range of the Forest Department at a rate of $1.28 per pole for cutting and loading in the year 2018.

2.4.2.3 Preservation technique

The pierced bamboo poles were treated using the dip diffusion treatment technique. The poles were soaked in a chemical solution of boric acid (2 kg) and borax (3 kg) in water (45 L) for 48 h in a pit of size $12 \times 6 \times 0.9$ m made of masonry block and cement plastering. Laurie Baker Centre for Habitat Studies has preservation set-up on-site, so the poles were treated within the premise.

2.4.2.4 Protective coat

A protective coat of cashew nut sap was applied on the exterior surface of exposed bamboos to guard it against harsh weather conditions prevailing in Kerala. The poles were also polished using bee wax to prolong its natural shine.

2.4.2.5 Bamboo-based techniques employed in the building

1. Wattle and daub walls: Bamboo pole frame with split bamboo wattle and 2-inch daub covering the wattle. The exterior side of the wall is finished with split bamboo, and the interior is plastered with lime mortar (1-lime, 4-coconut fibre, 6-mud).
2. Bamboo beams and columns: For 2.5–3 m spans, 4 bamboo poles are bolted together (as two layers) as beam. For 4–5 m spans, 8 bamboo poles (as two layers) are used. Bamboo poles are fixed vertically in between brick walls as well to transfer the load from the beam to these poles.
3. Bamboo-mud slab/roof: Bamboo split into six parts are laid over bamboo beam grid. Jute sacks dipped in cement slurry are laid over this, and further 2-inch-thick mud mix (stabilized with lime) is placed above the sacks. For exposed roofs, in addition to this, a 1-inch-thick Ferro cement slab is cast to protect the mud slab from rain.
4. Bamboo ventilator: The gap between lintel beam and roof band is lined with bamboo poles.
5. Panelling: Exterior panelling of fenestration with 4-mm-thick bamboo ply over 25–40 mm salvaged wood is done.

2.4.2.6 Unit rate of bamboo

Rate of bamboo pole = $0.81/m

Rate for cutting and loading bamboo pole = $1.28 per bamboo (a pole is 30 m in length)

Rate of treating bamboo pole = $2.47/m

The unit rate for the above project as specified in Annexure I is obtained from COSTFORD Trivandrum as per the market rate during the construction period (2018–2019).

Rate of the non-residential bamboo building (civil) = $18.36 per sq. ft (Source: COSTFORD Trivandrum)

The total cost of construction of the bamboo building excluding services is $40,298.52 (Source: COSTFORD Trivandrum).

2.4.3 Case study on a commercial building, Kerala

A commercial building of 2,615.63 sq. ft built-up area is situated in Kerala. It is a RCC-framed structure constructed using conventional building materials like cement, steel, and autoclave aerated cement blocks.

The unit rate for the above project as specified in Annexure II was obtained from DSR 2019 and market rate.

Rate of the non-residential conventional building (civil) = $19.06 per sq. ft.

The total cost of construction of the conventional excluding the services is $54,404.70.

2.5 FINDINGS

2.5.1 Bamboo building

The walls and slab contribute the most to the total cost of construction in the bamboo building (Figure 2.1). Wall, slab, beam, and column are the components in which bamboo is involved. But these components are not entirely composed of bamboo-based techniques except for beam. Figure 2.2 shows how much each technique contributes towards the total cost of the respective component.

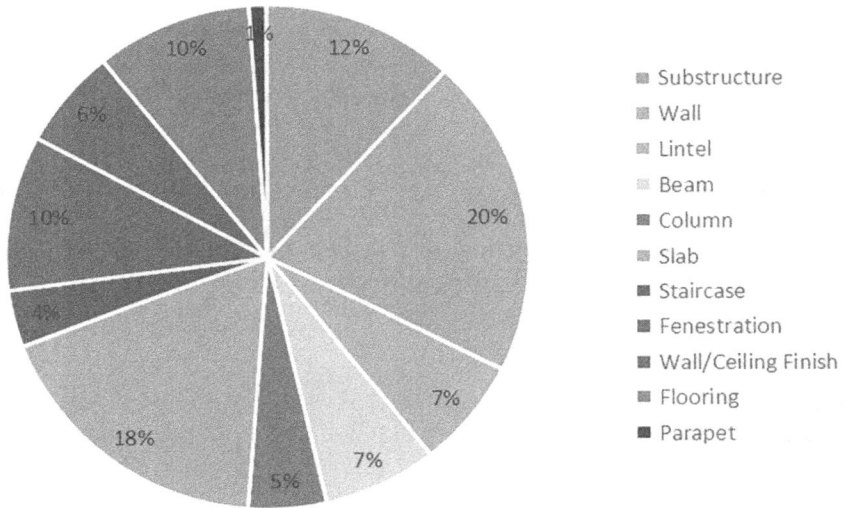

Figure 2.1 Cost percentage of components in bamboo building.

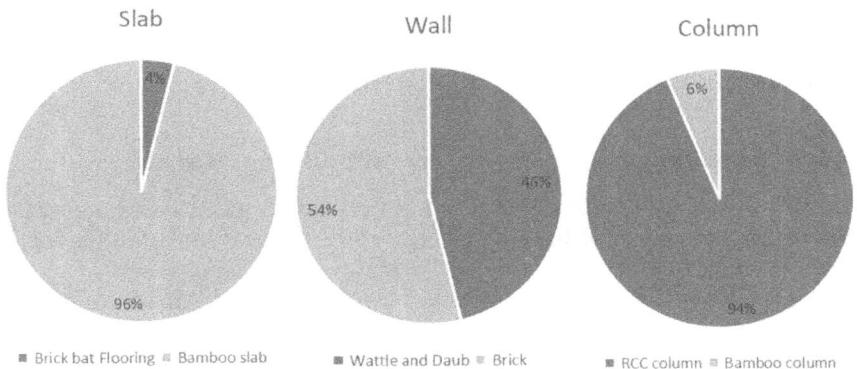

Figure 2.2 Internal cost split of slab, wall, and column.

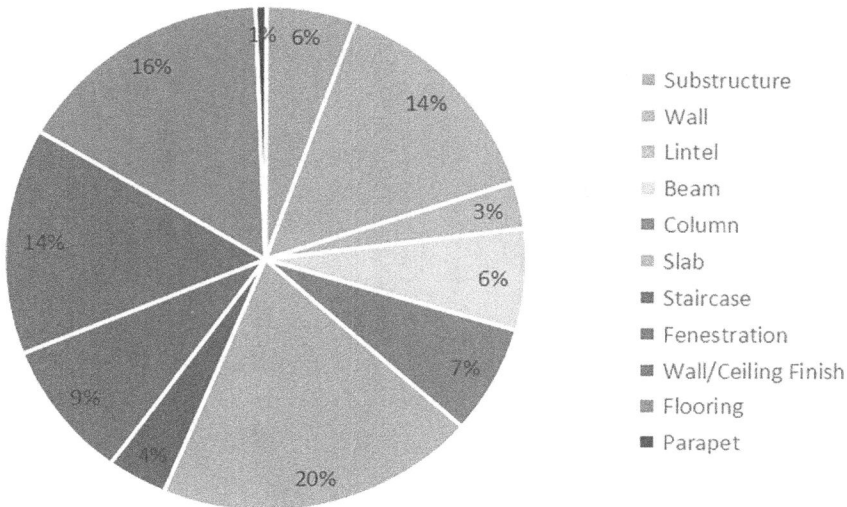

Figure 2.3 Cost percentage of components in conventional building.

2.5.2 Conventional building

The components slab and flooring contribute the most towards the total cost of construction in the above building (Figure 2.3). All the components are constructed using conventional materials.

2.6 DISCUSSION

2.6.1 Cost comparison between bamboo building and conventional building in Kerala

It is observed that the unit rate of the column and slab components of bamboo building which include bamboo material is lesser than that of the conventional building (Figure 2.4). At the same time, the unit rates of wall and beam components which again involve bamboo material are reflecting a difference in the previously observed pattern.

The rate of components in Figure 2.4 will be affected by the difference of total built-up area (2,615.63 − 2,200 = 415.63 sq. ft) between the two buildings. The quantity of masonry work, beam work, and column work involved is more for the bamboo building.

Also as seen in Table 2.2, the rate of these four components of bamboo building will be influenced by the conventional material-based techniques employed along with bamboo-based techniques.

Thus, the cost per same unit of work of the four components segregating bamboo and conventional materials is analysed and summarized in

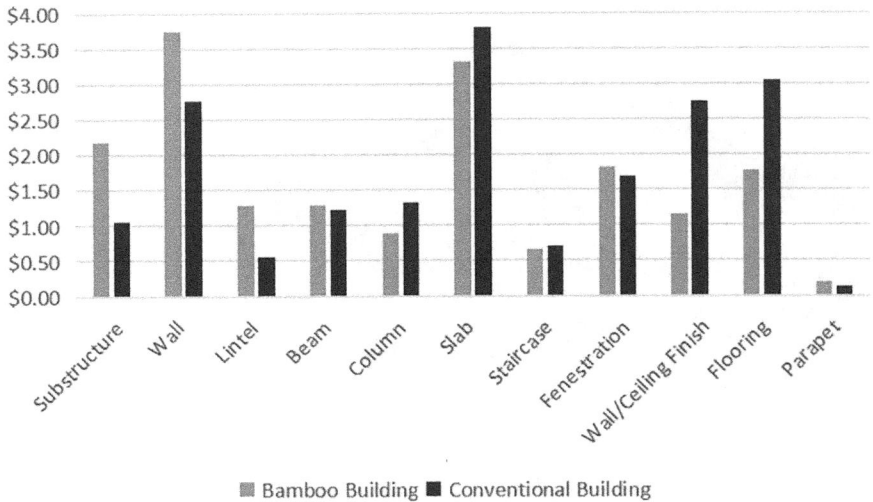

Figure 2.4 Rate of components per square feet area of the respective buildings.

Table 2.2 Cost comparison

Component	AAC (Conventional building)	Bamboo (Bamboo building)
Wall	$19.65/m²	$42.09/m²
Component	RCC (Conventional building)	Bamboo (Bamboo building)
Beam	$21.05/m	$4.58/m
Column	$28.72/m	$4.58/m
Slab	$259.48/m³	$56.71/m³

Table 2.3. Conventional cost per unit is converted for the ease of comparison with the bamboo components' unit.

From Table 2.4, it is observed that the construction cost of bamboo beam is 78.24% lesser than that of RCC. Similarly, the cost of construction of bamboo column and bamboo slab is 84.05% and 78.14% lesser than that of RCC, respectively. In these components, the bamboo poles are either fastened using nut and bolt or the poles are split and then laid in combination with mud layer. Here, the work involved in putting together the beam and column is minimal and does not involve heavy machinery. In case of the bamboo-based wattle and daub wall, the cost of construction is 53.31% higher than that of AAC wall. The reason for the increase in the cost of construction of a wattle and daub wall is the requirement for highly skilled labour to perform the task. The labour cost for constructing a wattle and daub wall using bamboo is 9.20% higher than that of AAC masonry. In general, bamboo subjected to extreme sun exposure must be applied with

Table 2.3 Cost details of bamboo building

Sub-structure

PCC Bed	1 cement: 4 coarse sand: 8 graded stone aggregate 40 mm nominal size	$82.61/m³
Foundation/plinth/ retaining wall	Random rubble masonry with hard stone in CM 1:6	$75.25/m³
Shuttering for foundation	Formwork	$6.38/m²
Plinth belt	RCC (1 cement: 1.5 coarse sand: 3 graded stone aggregate 20 mm nominal size)	$119.66/ m³

Wall

Wattle and daub wall	Bamboo pole (full and split)	$39.52/m²
Wattle and daub wall	Mud and lime plaster 1:4:6 (1-lime, 4-coconut fibre, 6-mud)	$51.42/m³
Bamboo exterior finish	Spar varnish	$2.23/m²
Brick masonry wall	Common burnt clay F.P.S. bricks in cement mortar 1:6	$100.91/ m³

Wall finish

Plaster	Mud plaster with 1:6 lime	$14.47/m²
Paint finish	Mud painting	$0.33/m²

Lintel

Lintel (exposed)	RCC (1 cement: 1.5 coarse sand: 3 graded stone aggregate 20 mm nominal size)	$139.87/ m³
Shuttering for lintel	Formwork	$6.38/m²

Beam

Bamboo beam	Pole, nut, and bolt	$4.58/m

Column

Bracing/columns	RCC (1 cement: 1.5 coarse sand: 3 graded stone aggregate 20 mm nominal size)	$135.87/ m³
Shuttering for column	Formwork	$8.71/m²
Bamboo column	Pole, nut, and bolt	$4.58/m

Staircase

Thread support	RCC (1 cement: 1.5 coarse sand: 3 graded stone aggregate 20 mm nominal size)	$139.87/ m³
Thread	4 cm Anjili wood	$80.06/m²

Reinforcement

Reinforcement	TMT steel bars	$1.31/kg

Slab

Brick bat flooring	Brick bat in 1:8 cement mortar	$9.15/m²
Bamboo-mud slab	Bamboo split pole, jute sack, cement slurry, mud mix, and Ferro cement	$56.71/m²

(Continued)

Table 2.3 (Continued) Cost details of bamboo building

Fenestration

Panelling	4-mm-thick bamboo ply, salvaged wood	$25.19/m²
Fenestration	Timber (excluding fixing)	$1,403.66/m³
Fenestration fixing	30-mm-thick salvaged wood shutters	$33.74/m²
Fenestration fixing	Glazed shutters	$38.32/m²

Flooring

Tiles	Terracotta tiles	$18.05/m²
Tiles	Ceramic tiles (wall)	$16.07/m²
Tiles	Ceramic tiles (floor)	$15.08/m²
Cement floor	12 mm (1:4), red oxide	$8.20/m²
Red oxide polish	Red oxide with wax polish	$0.66/m²

Others

Parapet	Half brick wall	$39.52/m²

Table 2.4 Cost details of conventional building

Components	Materials	Unit rate
Sub-structure		
PCC	PCC in 1:4:8	$91.27/m³
Foundation/plinth	Concrete M20	$112.92/m³
Shuttering for foundation	Formwork	$3.83/m²
Plinth beam	Concrete M20	$112.92/m³
Shuttering for plinth beam	Formwork	$8.19/m²
Wall		
Wall	AAC blocks in cement mortar 1:4 (230 mm)	$89.23/m³
Half wall	AAC blocks in cement mortar 1:4 (100 mm)	$110.61/m³
Wall/ceiling finish		
Plastering	15 mm 1:4 cement plaster	$4.14/m²
Putty	1 mm cement-based putty	$1.55/m²
Paint	Acrylic paint	$1.53/m²
Ceiling finish	15 mm 1:4 cement plaster	$4.14/m²
Lintel		
Lintel	Concrete M20	$131.79/m³
Shuttering for lintel	Formwork	$7.42/m²

(Continued)

Table 2.4 (Continued) Cost details of conventional building

Components	Materials	Unit rate
Beam		
Beam	Concrete M20	$131.79/m³
Shuttering for beam	Formwork	$7.42/m²
Column		
Column	Concrete M20	$131.79/m³
Shuttering for column	Formwork	$9.85/m²
Slab		
Slab	Concrete M20	$131.79/m³
Shuttering for slab	Formwork	$9.32/m²
Reinforcement		
Reinforcement	TMT steel bars	$1.12/kg
Staircase		
Staircase	Concrete M20	$131.79/m³
Shuttering for stairs	Formwork	$8.37/m²
Handrail	SS (1,150 mm)	$7.74/kg
Fenestration		
Door frame	Purple heart wood	$10.02 per 10 cu. Dm
Door shutter	Purple heart wood (2.5 cm)	$4.18 per 10 sq. dm
Window	Casement window	$97.37/m²
Ventilator	Casement ventilator	$87.10/m²
Grill work	12 kg/m² area of shutter	$103.64 per quintal
Toilet door frame	uPVC	$2.98/m
Toilet door shutter	uPVC	$23.66/m²
Flooring		
Polished granite	18-mm-thick slabs over 20 mm screed	$47.41/m²
Ceramic tiles	300×300 mm with 12 mm 1:3 cement mortar	$15.74/m²
Cement floor finish	1:2:4	$6.70/m²
Others		
Parapet	AAC blocks in cement mortar 1:4 (100 mm)	$110.61/m³

varnish like cashew nut sap yearly to maintain the shine and colour. Apart from this cost, bamboo does not incur any other maintenance cost. With the help of the above field studies, it is observed that the overall total cost of construction of a bamboo-based non-residential building is 3.75% lesser than a conventional material-based non-residential building in the context of Kerala.

2.7 CONCLUSIONS

The intention of this chapter is to assess if bamboo can replace conventional materials in the construction of non-residential buildings with respect to cost. In the case of Kerala, bamboo is a readily available material given its favourable climatic conditions for its cultivation. In Kerala, bamboo is widely used in housing projects as a sustainable and economical alternative but that is not the same scenario in the case of non-residential buildings.

With the help of the above field studies,

1. It is observed that employing bamboo as a major material in building components will significantly reduce the total cost of construction.
2. For simple joinery-based bamboo components, the labour as well as the equipment cost is minimal; therefore, the total cost of construction reduces substantially. The decrease percentage ranges between 78% and 84%.
3. The only hurdle in the process is the availability of skilled labour for carrying out various bamboo-based techniques. Techniques like wattle and daub can be tedious, and it also requires highly skilled and specialized labour; thus, the cost of labour for the same increases in comparison with wall masonry using conventional materials.
4. But an overall drop of 3.75% in the rate per square feet is observed in the studied bamboo-based building as compared to the conventional building. This will serve as an encouragement factor to pursue bamboo as a viable alternative in small-scale non-residential buildings in the context of Kerala.
5. To curb the overutilization of bamboo components (mass usage), the Forest Department of Kerala allows the procurement of only 1,000 tonnes of bamboo a year for each establishment. This step will help to maintain a sustainable use of bamboo construction for small-scale projects.

REFERENCES

Abdul Khalil, H., Bhat, I., Jawaid, M., Zaidon, A., Hermawan, D., & Hadi, Y. (2012), Bamboo fibre reinforced biocomposites: a review. *Materials & Design*, 42, 353–368, Doi: 10.1016/j.matdes.2012.06.015.

Adetunji, O., Moyanga, D., & Bayegun, A. (2015), Comparison of bamboo and conventional building materials for low-cost classroom construction in Isarun, Nigeria. *Proceedings of the 2nd Nigerian Institute of Quantity Surveyors Research Conference Federal University Of Technology*, Akure, 807–818.

Anokye, R., Bakar, E. S., Ratnansingam J., & Awang, K. B. (2016), Bamboo properties and suitability as a replacement for wood. *Pertanika Journal of Scholarly Research Reviews*, 2(1), 64–80.

Bhonde, D., Nagarnaik, P. B., Parbat, D. K., & Waghe, U. P. (2014), Physical and mechanical properties of bamboo (*Dendrocalmus Strictus*). *International Journal of Scientific & Engineering Research*, 5(1), 455–459.

Dange, S., & V. Pataskar, S. (2017), Cost and design analysis of steel and bamboo reinforcement. *International Journal of Innovative Research in Science, Engineering and Technology*, 6(12), 22464–22477.

Dhamodaran, T.K., Johny, J., & Ganesh Gopal, T.M. (2020). *A Manual for Preservative Treatment of Bamboo*, KFRI Handbook No. 18, Published by Kerala Forest Research Institute, Peechi 2020.

Ghavami, K. (1995), Ultimate load behaviour of bamboo-reinforced lightweight concrete beams. *Cement and Concrete Composites*, 17(4), 281–288, Doi: 10.1016/0958-9465(95)00018-8.

Habibi, S. (2019), Design concepts for the integration of bamboo in contemporary vernacular architecture. *Architectural Engineering and Design Management*, 15(6), 475–489.

Hasan, E., Karim, M. R., Shill, S. K., Mia, M. S., & Uddin, M. S. (2015), Utilization of bamboo as a construction material for low cost housing and resorts in Bangladesh. *International Conference on Recent Innovation in Civil Engineering for Sustainable Development*, 81–86.

Hong, C., Li, H., Lorenzo, R., Wu, G., Corbi, I., & Corbi, O. et al. (2019), Review on connections for original bamboo structures. *Journal of Renewable Materials*, 7(8), 713–730, Doi: 10.32604/jrm.2019.07647.

Jayagond, S., Pathak, S., & Tatikonde, M., (2020), Bamboo reinforced concrete: experimental investigation on Manga & Dagdi Bamboo. *International Research Journal of Engineering and Technology*, 7(4), 1510–1518.

Jit Kaur, P. (2018), Bamboo availability and utilization potential as a building material. *Forestry Research and Engineering: International Journal*, 2(5), Doi: 10.15406/freij.2018.02.00056.

Jyoti Nath, A., Das, G., & Das, A. (2009), Above ground standing biomass and carbon storage in village bamboos in North East India. *Biomass and Bioenergy*, 33(9), 1188–1196, Doi: 10.1016/j.biombioe.2009.05.020.

Kaur, P., Kardam, V., Pant, K., Naik, S., & Satya, S. (2016), Characterization of commercially important Asian bamboo species. *European Journal of Wood and Wood Products*, 74(1), 137–139, Doi: 10.1007/s00107-015-0977-y.

Kerala Bamboo Corporation Ltd. (2011), Retrieved from http://www.bambooworld-india.com/global_index.php?fname=bambooinkerala.

Kerenvis.nic.in. (2021), Administrative Profile: Climate – Status of Environment related issues: Kerala ENVIS Centre, Ministry of Environment and Forests, Govt. of India. Retrieved from http://www.kerenvis.nic.in/Database/Climate_829.aspx.

Liu, W., Hui, C., Wang, F., Wang, M., & Liu, G. (2018), Review of the resources and utilization of bamboo in China. *Bamboo - Current and Future Prospects*, doi: 10.5772/intechopen.76485.

Lowe, D. J. & Zhou, L. (2003), Economic challenges of sustainable construction. *RICS COBRA Foundation Construction and Building Research Conference.* University of Wolverhampton 1st–2nd September 2003. London: The RICS Foundation, pp. 113–126.

Manjunath, N., (2015), Contemporary Bamboo Architecture in India and Its Acceptability. *10th World Bamboo Congress*, Korea, 2015.

Paridah, M.T. (2013), *Bonding with Natural Fibres*, 1st edn. Universiti Putra Malaysia Press, Serdang.

Pavithra, G. M., & Jacob, K. J. (2018), Building a successful Bamboo based community: a case study of Kerala State Bamboo Corporation Limited, Kerala, India. *Journal of Bamboo & Rattan*, 17(2), 26–35.

Shah, R., Bambhava, H., & Pitroda, J. (2012), Bamboo: eco-friendly building material in Indian context. *International Journal of Scientific Research*, 2(3), 129–133, Doi: 10.15373/22778179/mar2013/41.

State of Forest Report (FSI Report). (2011), State of bamboo sector in Kerala. *Proceedings of National Workshop on Global Warming and its Implications on Kerala*, Published by KFRI, 2011.

Vadera, S., Woolas, P., Flint, C., Pearson, I., Hodge, M., Jordan, W., & Davies, M. (2008), Strategy for Sustainable Construction. Retrieved from http://www.berr.gov.uk/files/file46535.pdf.

Wakchaure, M.R., & Kute, S.Y. (2012), Effect of moisture content on physical and mechanical properties of bamboo. *Asian Journal of Civil Engineering (Building and Housing)*, 13(6), 753–763.

Chapter 3

An insight into sustainability in construction of shallow foundations

Muhammad Nouman Amjad Raja
University of Management and Technology

Sanjay Kumar Shukla
Edith Cowan University

CONTENTS

3.1 INTRODUCTION

In 2009, the United Nations Environment Programme (UNEP) reported that building development was responsible for one-third of world greenhouse gas (GHG) emissions, which account for almost 40% of global energy demand (UNEP, 2009). Population growth will raise the demand for new construction. The life-cycle (LC) impact of the building can broadly be categorized into two phases: (i) operational phase and (ii) embodied phase. The former is defined as the period of time in which building is in operation/working after construction and commissioning, whereas the latter consists of construction, material manufacturing, maintenance, or demolition. The operational phase of the building is significantly higher than the embodied

DOI: 10.1201/9781003368335-3

phase of the building, mainly due to the heating, ventilation, and air conditioning (HVAC) operation. However, the life-cycle assessment (LCA) for residential buildings revealed that for low-energy buildings, the embodied phase may account for more than 50% of the total LC impact (Ghattas et al., 2013). This is due to the lower impact of operation and the high usage of energy-intensive materials. The building is broadly divided into two components, namely superstructure and substructure. The superstructure is the part of the building above the ground and mainly comprises floor, roof, parapet, beams, columns, lintel, etc., while the substructure is the part of building beneath the ground and usually consists of foundation and plinth of a building. This chapter will focus on the design of shallow foundations by considering the strategies and measures that account for the sustainable construction of such foundation systems.

The lowermost part of the structure that transfers the load from the building or structure to the underneath soil is generally referred as *foundation*. Terzaghi and Peck (1948) defined the shallow foundation having embedment depth (D_f) lower than or equal to the width of the footing (B) (i.e., $D_f/B \leq 1$). However, many studies have shown that D_f/B can be greater than 1, even up to 3 to 4 (Shukla, 2014, 2015). The main parameters governing the foundation design are bearing capacity and settlement. An adequately designed foundation should be able to support the load of the structure and transfer it safely throughout the soil strata without overstressing the soil. According to the study conducted by Gonzalez and Navarro (2006) on low-terraced buildings, underground works such as foundation construction, plumbing/pipe work, and ground movements account for 60% of carbon dioxide (CO_2) emissions. In general, CO_2 emissions released during the foundation works are limited over a short span of time when compared to the construction of the whole building. Therefore, these emissions are usually ignored during foundation design; hence, there is a need for optimizing strategies that will help in reducing the environmental impact of foundation construction.

3.2 SHALLOW FOUNDATION CONSTRUCTION

Shallow foundations are usually classified into two types: (i) isolated concrete foundations (ICFs) and (ii) mat/raft foundations. ICFs are usually used to support individual columns, whereas mat or raft foundation is generally used to support the entire area of a building. Figure 3.1 describes the geometry and parameters of the individual ICF of width b. ICF is rigid when the column-to-edge length (s) is less than or equals two times of depth (d) $(s \leq 2d)$ and flexible when the column-to-edge length s is greater than $2d$ $(s > 2d)$. The thickness of the footing is shown by d, and y represents the distance of steel reinforcement from the top of footing. However, c_1 and c_2 represent the clear cover.

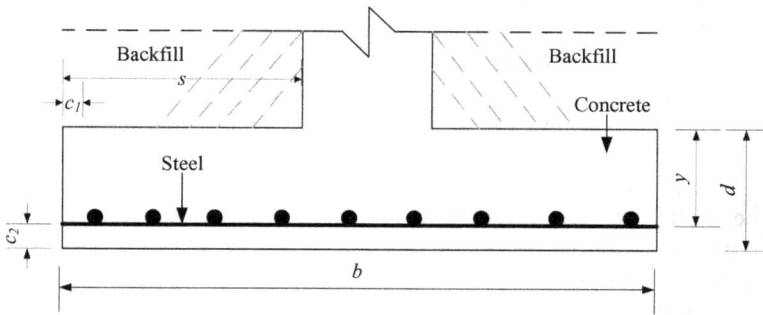

Figure 3.1 Annotations and geometry of ICF.

The construction of shallow foundations generally involves the following steps:

1. Excavation of ground and preparation of framework
2. Pouring of lean concrete and levelling
3. Adequate steel reinforcement
4. Concreting the foundation
5. Backfilling

Generally, the cast-in-situ foundations need 28 days to acquire the required strength. Thereafter, the foundations are ready for load bearing.

3.3 COMMON PRACTICES AROUND THE WORLD

The design of shallow foundations around the world is more or less similar. The design is divided into two parts: (i) geotechnical design and (ii) structural design. The geotechnical design deals with measurements related to the properties and type of soil(s), while the structural component is related to reinforced concrete design. In various parts of the world, different design codes are utilized to design a shallow foundation, for example, Eurocode 7 (Driscoll and Simpson, 2001), Spanish codes (CTE-SE-C, 2008), United States (Kimmerling, 2002), and India (IS:1904, 1986). These codes address the shallow foundation design with respect to the geotechnical and structural design of the foundations.

The major issue with the current practices around the world is the non-attendance to the environmental impacts of the foundations. According to the study conducted by Ondova and Estokova (2016), reinforced concrete foundations are the most commonly used type all around the world. The construction of these concrete foundations results in the release of more embodied energy. A typical raft foundation for the high-rise structures can result in the GHG emissions of approximately 67% just due to the amount of utilized material (Pujadas-Gispert et al., 2018). The replacement of reinforced concrete foundations with other types of foundations such as earthbag foundations, or replacement of materials, for example, bricks with rubble masonry, will result in low embodied energy, thus having a considerable positive impact on the environment with low emissions. An optimized concrete reinforcement can also reduce the amount of GHG emissions during the shallow foundation construction. Similarly, good soil support will result into sustainable foundation construction. The earth reinforcement techniques, especially geosynthetics or fiber reinforcement, may be utilized to reinforce the weak soil, thus eliminating the need of less environmental-friendly methods like deep foundations or other soil stabilization techniques such as blasting, high-impact compaction, or soil replacement.

3.4 ENVIRONMENTAL INDICATORS IN SHALLOW FOUNDATION DESIGN

As discussed in the previous section, in the foundation design, the GHG emissions are primarily related to the amount of concrete and steel, type of foundation (cast-in-situ or precast), optimization of reinforcement materials, and condition of the soil (weak or strong). The following section describes the environmental indicators used to estimate the impact of overall construction and, in particular, building foundations.

3.4.1 Primary energy intensity (PEI)

In the evaluation of LC impact of a building, primary energy or embodied energy intensity is defined as the energy resources depleted during the construction phase and usually measured in megaJoule (MJ). Primary energy includes but not restricted to non-renewable fossil fuels consumption, feedstock biomass/fossils, and wind/solar/water/geothermal energy. Mathematically, it is expressed as follows:

$$PEI = \sum_i w_i.PEI_i \tag{3.1}$$

where w_i is the weight of a specific material and PEI_i is the unit measurement of primary energy (MJkg^{-1}). PEI for building foundation and whole building is evaluated from the LCA data for a unit weight (kg) of a material.

3.4.2 Global warming potential (GWP)

Global warming potential (GWP) indicator is utilized to evaluate the amount of GHG emissions during the LC of the building. Amongst all the anthropogenic processes, the leading cause of GHG emissions is the burning of fossil fuels. According to the Environmental Protection Agency (EPA), the main GHGs are carbon dioxide (CO_2), methane, nitrous oxide (NO), and fluorinated gases. In the USA alone, in 2019, the amounts of these gases emitted were 80%, 10%, 7%, and 3%, respectively. GWP is usually measured in terms of CO_2 equivalent emissions and measured in $kgCO_2eq$. If the effect of other GHG is ignored, then GWP is confined to the CO_2 emissions, which are sometimes called "embodied carbon emissions". Mathematically, it is expressed as follows:

$$GWP = \sum_i w_i.GWP_i \tag{3.2}$$

where w_i is the weight of a specific material and PEI_i is the unit measurement of GWP ($kgCO_2$). The amount of GWP is related to the time horizon for which it is assessed. For long-term effects, the time span is usually 100 years for buildings materials.

3.4.3 Acidification potential (AP)

Acidification potential (AP) is estimated based on the amount of sulphur dioxide (SO_2), nitrogen oxides (NO_x), hydrogen fluoride (HF), hydrochloric acid (HCl), and ammonia (NH_3). All these substances are responsible for potential acid deposition and consequently decrease the pH value of the environment. AP is generally calculated in terms of SO_2 equivalent. For building materials, the SO_2eq is usually measured as follows:

$$AP = \sum_i w_i.AP_i \tag{3.3}$$

where w_i is the weight of a specific material and AP_i is the unit measurement of material's acidification potential measured in $kgSO_2$.

3.4.4 Environmental impact assessment of building foundations

The typical building materials used for construction of residential/family houses are bulk materials (soil and gravels), reinforced concrete, hollow concrete blocks or bricks for wall footings, ceramics (bricks and tiles), insulation materials (polyvinyl (PVC), damp-proof course (DPC) with bitumen, expanded polystyrene (EPS), or extruded polystyrene (X-EPS), etc.), wood, plasters, glass, and laminate materials (floor covering). Among these materials, the typical foundation materials are gravels, bricks, concrete, bitumen, and PVC. Table 3.1 summarizes the estimated environmental impact of foundations from the previous studies.

Table 3.1 Environmental impact indicators' values of house foundations

Foundation material	Embodied energy (PEI) (MJ)	Embodied carbon emissions (GWP) (kg CO_2eq)	Acidification potential (AP) (kg SO_2eq)	Study
Reinforced concrete	1.76×10^5	0.14×10^5	46.88	Ondova and Estokova (2016)
Reinforced concrete	2.92×10^5	0.1×10^5 (mud-brick) 0.14×10^5 (concrete block)	Not reported	Abanda et al. (2014)
Earthbag foundations	1.67×10^4	0.015×10^5	Not reported	Magwood and Feigin (2014)
Rammed earth tire foundations	2.92×10^4	0.03×10^5	Not reported	Magwood and Feigin (2014)
Dry stone	3.58×10^4	0.027×10^5	Not reported	Magwood and Feigin (2014)

Note: These are reference values obtained by normalizing the values by floor area, built-up volume, and weight of foundation material.

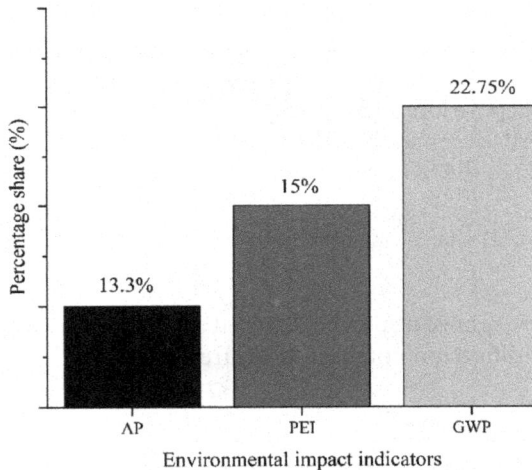

Figure 3.2 Average percentage share of building environmental impact indicators.

These indicators (PEI, GWP, and AP) are calculated for unit amount of material confined to system boundaries. The environmental performance of the materials is estimated through the IBO database. Figure 3.2 represents the average of all the environmental impact indicators of building houses. The values are derived from the detailed study conducted by Ondova and Estokova (2016) by comparing the eight types of family houses building foundations in Slovakia.

3.4.5 Optimizing strategies for shallow foundation design

The most common strategies for minimizing the environmental impact of building foundations are as follows:

1. Optimizing the amount of steel utilized to reinforcement
2. Optimizing the amount of concrete and steel reinforcement simultaneously
3. Opting for flexible foundation instead of rigid foundation
4. Precast foundations instead of cast-in-situ foundations

All these scenarios are summarized in Table 3.2. It may be noted that the comparison here is only presented for Eurocode 2 (British Standard, 2004) for structural design of shallow foundation. The characteristics for the foundation are as follows:

a. Isolated concrete square footing of 0.4 m width.
b. Total dead and live/imposed load of 400 and 150 kN, respectively.
c. Bearing capacity of silty soil is 150 kPa.
d. No presence of groundwater table, seismic activities, or chemical reactions.
e. Design life of 50 years.

It can be observed that for the rigid concrete foundation with same amount of concrete, precast foundations (R-PC-S) consumed approximately 37.34% more steel than cast-in-situ foundation (R-CI-S). For flexible foundations, the difference is not significant between the precast and cast-in-situ

Table 3.2 Concrete shallow foundations characteristics (Pujadas-Gispert et al., 2018)

Scenarios and variables	Foundation type	Abbreviation	b(m)	h(m)	Volume of concrete (m³)	Steel bar (∅)	Steel weight (kg)	c_1 (m)	c_2 (m)
Same amount of concrete	Rigid	R-CI-S	2.02	0.60	2.45	14	46.45	0.075	0.030
		R-PC-S	2.02	0.60	2.45	21	74.14	0.015	0.015
	Flexible	F-CI-S	1.97	0.30	1.16	13	41.86	0.075	0.030
		F-PC-S	1.97	0.30	1.16	12	41.19	0.015	0.015
Reduced amount of concrete	Rigid	R-CI-Ru	1.98	0.40	1.57	10	32.48	0.075	0.030
		R-PC-Ru	1.98	0.40	0.81	12	41.00	0.015	0.015
	Flexible	F-CI-Ru	1.96	0.25	0.95	18	57.64	0.075	0.003
		F-PC-Ru	1.96	0.20	0.75	19	64.58	0.015	0.015

Note: b is the base of foundation; h is the depth of foundations; c_1 is the lateral concrete cover; c_2 is the bottom concrete cover; CI is the cast-in-situ; R is the rigid foundation; S is the same amount of concrete; Ru is the reduced amount of concrete; F is the flexible foundation; and P is the precast.

techniques. However, it is also important to note that precast techniques give better quality control and, hence, more concrete strength. For the reduced amount of concrete, precast rigid foundation (R-PC-Ru) tends to take much less volume of concrete (approximately 49% less) in comparison with cast-in-situ foundation (R-CI-Ru). Similarly, for flexible foundations, F-PC-Ru performed better than F-CI-Ru with 21% less concrete volume. Nevertheless, this reduction in concrete is compensated by increasing the amount of steel in the precast versions of the foundations.

The environmental impact of the first category (same amount of concrete) is depicted in Figure 3.3. Pujadas-Gispert et al. (2018) estimated these indicators' values using the database of the Institute of Construction Technology of Catalonia (ITeC, 2017) for building materials.

The rigid precast foundation is the worst option in this case as indicated by all the environmental impact indices. It may be noted that for each environmental category, indicators are reported relative to the minimum/worst option. The best option is the flexible foundation with cast-in-situ technique

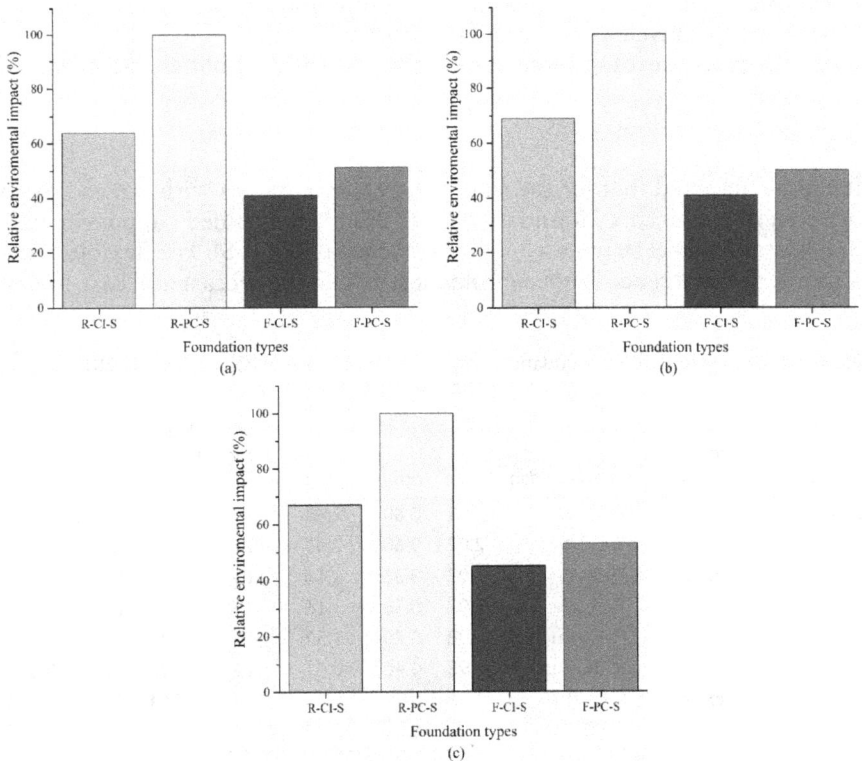

Figure 3.3 Environmental impact indicators for shallow foundations construction with same amount of concrete. (a) PEI. (b) GWP. (c) AP.

Table 3.3 Comparison of environmental impact
indicators for shallow foundations
construction with same amount of concrete

Foundation type	PEI	GWP
R-CI-Ru	90%	100%
R-PC-Ru	86%	85%
F-CI-Ru	91%	88%
F-PC-Ru	100%	94%

with the lowest impact indicator values of 41%, 41%, and 45% for PEI, GWP, and AP, respectively. This indicates that, in quantitative terms, on average, flexible cast-in-situ foundation is approximately 45% more environmentally friendly than its counterparts. It may also be noted that all the precast foundations have higher impact (35% more) in comparison with cast-in-situ versions. This is due to higher per cubic impact along with other factors like transportation, machinery for on-site installation, and higher steel requirement.

The comparison of all the environmental impact indicators for the foundations with reduced amount of concrete is presented in Table 3.3.

As concrete is the main contributor to the increase in the environmental impact, therefore, cast-in-situ foundations have the highest impact on the environment. Mostly, for single concrete-reinforced foundation design, precast concrete foundations are less environmentally friendly than its comparative cast-in-situ versions. However, if the concrete and steel are sufficiently reduced such is case of R-PC-Ru, precast foundation shows more promising results than cast-in-situ foundations.

3.5 RAFT CONCRETE FOUNDATION: CASE STUDY OF LOCAL COMMUNITY CENTRE IN PAKISTAN

The environmental impact assessment of raft foundation of a 3-story local community centre building situated in Lahore, Pakistan, is presented below. The key features of the building are provided in Table 3.4.

The building has no basement, and soil is strong enough to curtail the settlement. The groundwater table is at greater depth; therefore, its impact is negligible. The photographs from the construction site during the raft construction phase are presented in Figure 3.4. The main construction materials are concrete and steel. Some of the excavated soil is used in backfill, while the rest is used to level the empty plots located nearby, therefore minimizing the effect of transportation needed for soil dumping. An average one-way distance for concrete and steel transportation is less than 20 km. This also helps in neutralizing the emission effects due to transportation of materials.

Table 3.4 Key features of the building area and raft foundation

Building floor area	177 m²
Foundation type	Raft
Foundation depth	1.5 m
Foundation thickness	275 mm
Total load (dead and live)	~60 kPa
Total material for foundation (concrete and steel)	1.2 × 10⁵ kg
Grade of steel	Grade 40 (ASTM)

(a)

(b)

Figure 3.4 Construction site during the foundation construction phase. (a) Before concrete pouring. (b) After concrete pouring. Note: Local community centre in Lahore, Pakistan.

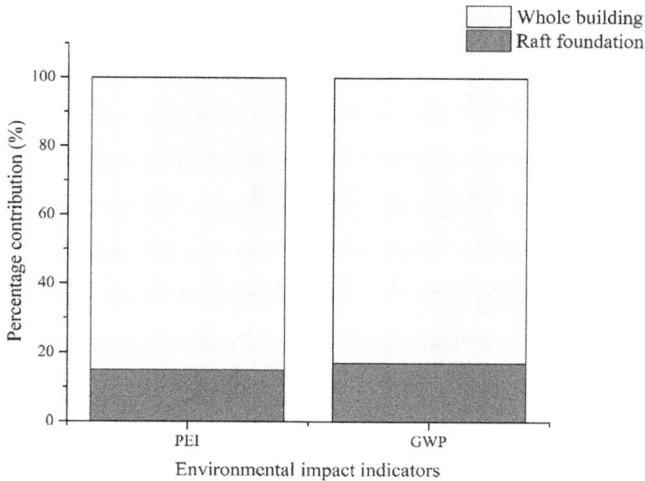

Figure 3.5 Contribution of raft foundation emissions in comparison with whole building.

For the raft foundation, the estimated environmental indicators, namely PEI and GWP, are estimated to be 1.37×10^5 MJ and 0.16×10^5 CO_2eq, respectively. The value of AP is not estimated as there is no significant harm from the acidification. In comparison with the whole building, the foundation emissions are presented in Figure 3.5. It can be observed that the emission footprints from foundation work range from 1/5 to 1/6 in comparison with whole building.

3.6 GEOSYNTHETIC-REINFORCED SOIL FOUNDATIONS (GRSF)

The earlier discussion was primarily concerned with the structural design of the foundation, considering the soil has enough strength to withstand the applied loads. However, this is not always the case in the field. Often, the underground strata are not strong enough to bear the load of the structure without causing any damage to the structure, such as excessive settlement and/or bearing capacity failure. In those cases, alternative treatments are required. Most of the times, these alternatives include soil replacement, deep (pile foundations), or soil reinforcement. In comparison with deep foundations, the geosynthetically reinforced soil is more ecologically friendly solution, which offers high performance and excellent bearing capacity (e.g., Hasthi et al., 2022; Phillips et al., 2016; Shukla, 2022; Raja and Shukla, 2021; Raja et al., 2022). The previous studies have shown that the GRSF not only increases the strength of the soil but also reduces the

GHG emissions. Chulski (2015) compared the cradle-to-factory-gate GHG and PEI of the retaining walls and found that the geotextile-reinforced retaining walls have much less GHG emissions and PEI in comparison with gravity retaining and mechanically stabilized earth (MSE) walls. Similarly, the geosynthetic-reinforced soil technology has shown a promising impact on the environment in road construction by reducing the energy demand by 5.4% (PEI) and GHG emissions by 32% (CO_2.eq) in comparison with traditional pavement foundation system (Frischknecht et al., 2012). On the same basis, it can be concurred that the GRSF will help in reducing the amount of GHG emissions and present more sustainability in the shallow foundation construction.

In GRSF, traditionally, soil is reinforced with one or more planar layer of geosynthetic reinforcement. Such foundations spread the load to the wider area, thus increasing the strength (bearing capacity) and reducing the deformation (settlement). For practical simplicity, the geosynthetic sheets are often laid horizontally. However, most recently, the wraparound reinforcement technique, as shown in Figure 3.6, shows that the wraparound ends/edges provide additional confinement to the soil, thus increasing its bearing capacity and reducing the settlement. This technique also helps in saving the land required for foundation, thus saving the project cost (Kazi et al., 2015, 2016; Raja et al., 2021). It may be noted that for shallow foundation,

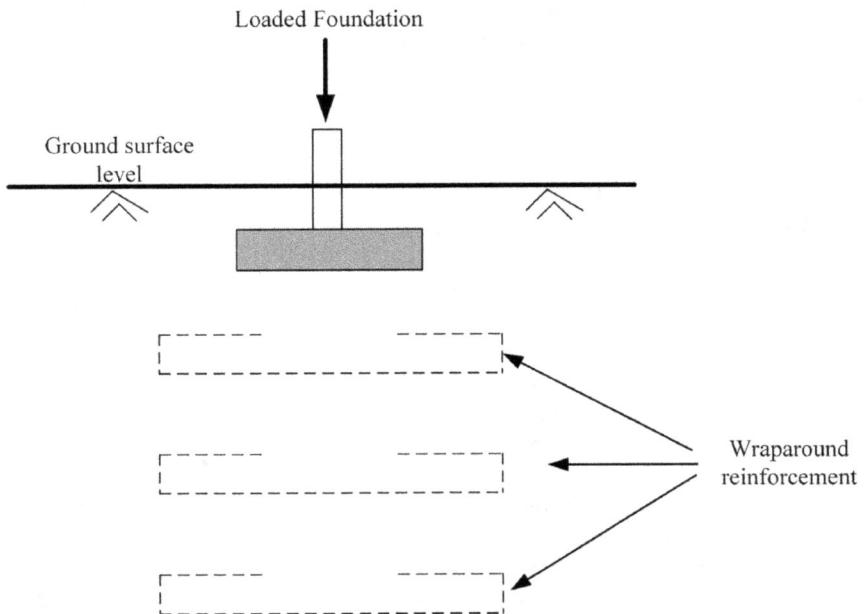

Figure 3.6 Footing resting on soil reinforced with geosynthetic layers having wrap-around ends.

this technique was implemented by the second author earlier in the field construction project in 2007–2008 (Shukla, 2016; Raja and Shukla, 2020).

Such a reinforcing mechanism can prove to be extremely useful in diminishing the impact of GHG emissions related to foundation works. However, it is important to estimate the cost-benefit ratio of such foundations, as the initial cost is likely to be on the higher end for GRSF when compared with unreinforced soil foundations.

There is a need of good communication between the structural, geotechnical, LC assessment practitioners, and all the other relevant stakeholders to collaborate during the design stages of the building foundations. The consideration of the environmental impacts can significantly help in reducing the GHGs and leads towards more sustainable construction practices in the foundation design.

3.7 CONCLUSIONS

The following general conclusions can be drawn from the discussion presented above:

- In shallow foundations, concrete and steel are the main contributors to the increase in the environmental impact indicators.
- Among the environmental impact indicators (PEI, GWP, and AP) discussed in the study, GWP is significant due to the GHG emissions during the foundation construction.
- The optimization of the reinforcement process (concrete and steel amount) in cast-in-situ foundations can help in diminishing the environmental impacts during the foundation construction process.
- If the soil is weak, GRSF can play a significant part in sustainable design of shallow foundations. The need for deep foundations, soil improvement/replacement, or other stabilization methods can be avoided by constructing GRSF.

REFERENCES

Abanda, F.H., Nkeng, G.E., Tah, J.H.M., Ohanjah, E.N.F., & Manijia, M.B., (2014). Embodied energy and CO$_2$ analysis of mud-brick and cement-block houses. *Energy*, 2, 18–40

Chulski, K.D., (2015). Life cycle assessment and costing of geosynthetics versus Earthen materials. *Doctoral dissertation*, University of Toledo.

CTE-SE-C, (2008). Structural Safety. Foundations. *Spanish Technical Building Code*, Madrid.

Driscoll, R., and Simpson, B., (2001). EN1997 Eurocode 7: geotechnical design. *Proceedings of the Institution of Civil Engineers-Civil Engineering*, 144(6), 49–54.

British Standard (2004). *Eurocode 2 (EN1992): Design of Concrete Structures – Part 573 1-1: General Rules and Rules for Buildings*. British Standards Institution, London.

Frischknecht, R., Stucki, M., Büsser, S., and Itten, R., (2012). Comparative life cycle assessment of geosynthetics versus conventional construction materials. *Ground Engineering*, 45(10), 24–28.

Ghattas, R., Gregory, J., Olivetti, E., and Greene, S., (2013). Life cycle assessment for residential buildings: a literature review and gap analysis. Massachusetts Institute of Technology. https://cshub.mit.edu/sites/default/files/documents/ LCAforResidentialBuildings.pdf (accessed 13.08.21).

Gonzalez, M.J. and García Navarro, J., (2006). Assessment of the decrease of CO_2 emissions in the construction field through the selection of materials: practical case study of three houses of low environmental impact. *Building and Environment*, 41, 902–909.

Hasthi, V., Raja, M. N. A., Hegde, A., and Shukla, S. K., (2022). Experimental and intelligent modelling for predicting the amplitude of footing resting on geocell-reinforced soil bed under vibratory load. *Transportation Geotechnics*, (35), 100783.

IS: 1904, (1986). *Code of Practice for Design and Construction of Foundation in Soils: General requirements*. Bureau of Indian Standards, New Delhi.

ITeC, (2017). Online ITeC Database: Prices, Technical Details, Companies, Certificates, Product Pictures and Environmental Data. https://itec.cat/.

Kazi, M., Shukla, S. K., and Habibi, D., (2015). Behavior of embedded strip footing on sand bed reinforced with multilayer geotextile with wraparound ends. *International Journal of Geotechnical Engineering*, 9(5), 437–452.

Kazi, M., Shukla, S. K., and Habibi, D., (2016). Behaviour of an embedded footing on geotextile-reinforced sand. *Proceedings of the Institution of Civil Engineers-Ground Improvement*, 169(2), 120–133.

Kimmerling, R., (2002). *Geotechnical Engineering Circular No. 6 Shallow Foundations (No. FHWA-SA-02–054)*. United States. Federal Highway Administration. Office of Bridge Technology.

Magwood, C. and Feigin, J., (2014). *Making Better Buildings: A Comparative Guide to Sustainable Construction for Homeowners and Contractors*. 1st Edition, 441 p. New Society Publishers, Gabriola Island, Canada.

Ondova, M. and Estokova, A., (2016). Environmental impact assessment of building foundation in masonry family houses related to the total used building materials. *Environmental Progress & Sustainable Energy*, 35(4), 1113–1120.

Phillips, E. K., Shillaber, C. M., Mitchell, J. K., Dove, J. E., and Filz, G. M. (2016). Sustainability comparison of a geosynthetic-reinforced soil abutment and a traditionally-founded abutment: a case history. In *Geotechnical and Structural Engineering Congress* 2016. pp. 699–711.

Pujadas-Gispert, E., Sanjuan-Delmás, D., and Josa, A., (2018). Environmental analysis of building shallow foundations: the influence of prefabrication, typology, and structural design codes. *Journal of Cleaner Production*, 186, 407–417.

Raja, M. N. A. and Shukla, S. K., (2020). Ultimate bearing capacity of strip footing resting on soil bed strengthened by wraparound geosynthetic reinforcement technique. *Geotextiles and Geomembranes*, 48(6), 867–874.

Raja, M. N. A. and Shukla, S.K., (2021a). Experimental study on repeatedly loaded foundation soil strengthened by wraparound geosynthetic reinforcement technique. *Journal of Rock Mechanics and Geotechnical Engineering*, 13(4), 899–911.

Raja, M. N. A. and Shukla, S.K., (2021b). Predicting the settlement of geosynthetic-reinforced soil foundations using evolutionary artificial intelligence technique. *Geotextiles and Geomembranes*, 49(5), 1280–1293.

Raja, M. N. A., Jaffar, S. T. A., Bardhan, A., and Shukla, S. K., (2022). Predicting and validating the load-settlement behavior of large-scale geosynthetic-reinforced soil abutments using hybrid intelligent modeling. *Journal of Rock Mechanics and Geotechnical Engineering*. DOI: 10.1016/j.jrmge.2022.04.012

Shukla, S. K., (2014). *Core Principles of Soil Mechanics*, ICE Publishing, London.

Shukla, S. K., (2015). *Core Concepts of Geotechnical Engineering*, ICE Publishing, London.

Shukla, S. K., (2016). *An Introduction to Geosynthetic Engineering*. CRC Press, Taylor and Francis, London.

Shukla, S. K., (2022). *Handbook of Geosynthetic Engineering. Geosynthetics and Their Applications*. 3rd Edition. ICE Publishing, London.

Terzaghi, K. and Peck, R.B., (1948). *Soil Mechanics in Engineering Practice*. John Wiley & Sons, New York.

UNEP (2009). Buildings and climate change. Summary for decision makers. https://europa.eu/capacity4dev/unep/document/buildings-and-climate-changesummary-decision-makers/ (accessed 17.08.21).

Chapter 4

Enhancement of bearing capacity of cohesive and granular soils using sustainable Kenaf fibre geotextile

Ahmad Safuan A Rashid
Universiti Teknologi Malaysia

Roohollah Kalatehjari
Auckland University of Technology

Mohammad Gharehzadeh Shirazi
Universiti Teknologi Malaysia

CONTENTS

4.1 INTRODUCTION

Several researchers have studied the properties of bio-based materials, including Kenaf, sugarcane bagasse, and coir, as potential replacements for synthetic fibres used in commercially available geotextiles during the past two decades (Shirazi et al., 2020). The main reasons behind this effort are the more sustainable and less costly nature of bio-based geotextiles that have been proven effective in the reinforcement and stabilisation of soils due to their high tensile strength (Huber et al., 2012; Sarsby, 2007).

Kenaf fibre is abundant in tropical and subtropical regions, with the potential to be used as a sustainable bio-based material with the high tensile strength to replace commercial geotextiles (Yetimoglu & Salbas, 2003). However, this is still a young field of study, with only a few investigations done on the use of this geotextile for soil reinforcement (Chaiyaput et al., 2014; Meon et al., 2012; Rashid et al., 2017, 2020; Shirazi et al., 2020). Kenaf is an annual plant growing in tropical and subtropical climates to

DOI: 10.1201/9781003368335-4

300 m to 600 cm in less than half a year (Summerscales et al., 2010). Like other bio-based materials, this natural fibre is degradable when buried in soil and undergoes a progressive loss of strength (Shukla, 2017). This process causes the material to break down into its constituent elements in an irreversible process (Arshad et al., 2014). The degradation process is affected by several parameters of the host soil, such as particle size and shape, degree of acidity or alkalinity, presence of moisture, heavy metals, organic matter, and temperature. However, this problem can be overcome by chemical or physical treatment of the natural fibres, such as treatment by chemicals and using different coatings (Shukla, 2017).

This study initially investigates the effect of textile pattern, moisture content, and chemical treatment on the tensile strength and durability of the Kenaf geotextile. A 6% sodium hydroxide solution is used to treat Kenaf fibres, and the durability of treated and untreated samples is investigated by burying them for 1 year in natural soil. To investigate the performance of Kenaf geotextile as reinforcement on the bearing capacity of different types of soils, a set of small-scale physical modelling of a shallow foundation under unit gravity is carried out. Bearing capacity and deformation of unreinforced and Kenaf geotextile-reinforced clay and sandy soils are separately investigated under a shallow foundation with a constant axial penetration speed.

The required granular soil model was prepared by sand at medium compactness, and Kaolin clay was used to prepare the required cohesive soil. For reinforced soil models, the Kenaf geotextile layers were placed on the surface and three different depths of the models up to a depth equal to the width of the footing. All treated and untreated models were tested by a rigid shallow foundation on the surface of the soil to measure the bearing capacity and settlement under vertical loading.

4.2 TREATED AND UNTREATED KENAF GEOTEXTILE

The bio-fibres of Kenaf were sourced from the Malaysian Agriculture Research and Development Institute (MARDI), located in Selangor, Malaysia. Table 4.1 includes the properties of the Kenaf fibres used in this study.

Table 4.1 Properties of Kenaf fibres

Properties	Values
Density	0.75 g/cm^3
Holocellulose	80.9
Lignin	15.1
Alpha-cellulose	56.4
Moisture content	<6%

Source: After Shirazi et al. (2019).

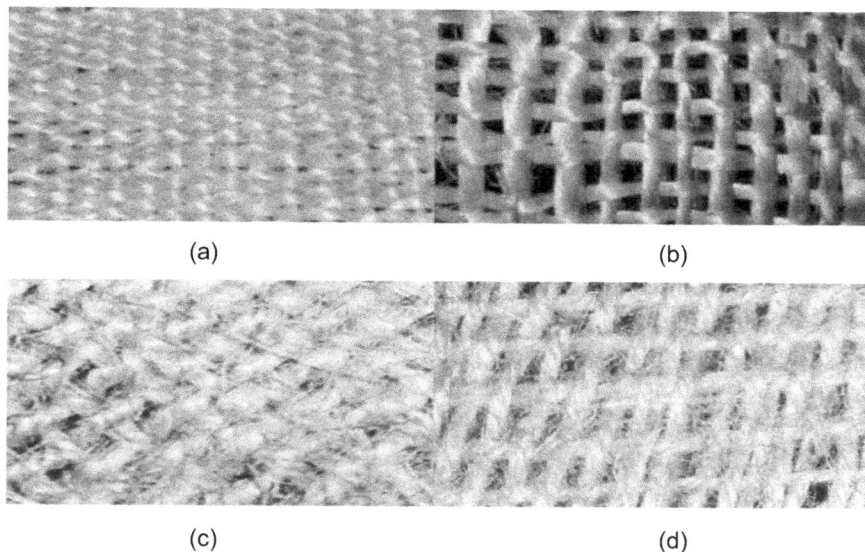

Figure 4.1 Kenaf geotextiles were produced using (a) plain pattern and 0 mm opening, (b) plain pattern and 2 mm spacing, (c) inclined pattern and 0 mm opening, and (d) inclined pattern and 2 mm opening. (After Shirazi et al., 2019.)

To prepare the geotextile, the Kenaf fibres needed to be woven by a weaving device into pieces with the required dimensions, such as 300 mm (length) by 150 mm (width). The weaving device allows various patterns and spacings between Kenaf fibres that will produce different Kenaf geotextiles, as shown in Figure 4.1. Two different weaving settings were used in the investigation for tensile strength to produce the woven Kenaf geotextiles with a plain pattern and an inclined pattern both with two opening sizes of 0 mm (0 mm×0 mm) – replicating the texture of commercially available synthetic geotextile – and 2 mm (2 mm×2 mm). The plain pattern produced a network of fibres intersecting at a 90° angle, and the included patterns created a texture with 45° of inclinations between fibres.

A total of four numbers were produced from each geotextile to be used in further tests. Half of the produced geotextiles were preserved in natural condition, and the other half were chemically treated by soaking in a 6% sodium hydroxide solution for 24 h before being washed with running water (Edeerozey et al., 2007; Li et al., 2009; Meon et al., 2012). All samples were then oven-dried for 24 h at 100°C, and half was submerged in water for 24 h to produce wet samples. The physical characteristics of the oven-dried woven Kenaf geotextiles are listed in Table 4.2.

A series of tensile tests were carried out on all Kenaf geotextiles to determine their ultimate tensile strength (UTS) using a universal testing machine (UTM) with 2.5 kN load capacity and equipped with fine clamps and

Table 4.2 Physical characteristics of oven-dried woven Kenaf geotextiles

Pattern	Opening (mm)	Mass per unit area[a] (g/m²)		Thickness[b] (mm)
		Untreated	Treated	
Plain	0	1,158	1,243	3
	2	603	628	3
Inclined	0	1,221	1,296	3
	2	627	650	3

Source: After Shirazi et al. (2019).
[a] According to ASTM D5261-10, 2010.
[b] According to ASTM D5199-12, 2012.

Figure 4.2 Universal testing machine equipped with fine clamps and rubber tabs. (After Shirazi et al., 2019.)

additional rubber tabs to ensure full-length engagement of the textiles with the clams and avoid slippage during the tests, as shown in Figure 4.2. The ASTM D4595-17 (D4595-17, 2017) standard was used for the tensile test with an axial speed of 0.1 mm/s. The test was conducted at room temperature (recorded as 23°C) on a 150 mm by 100 mm piece of each geotextile, where the longer side was clamped to the device to minimise the contraction effect under the tensile stress.

A linear variable displacement transformer (LVDT) was used to record the textile's axial deformation during the test. A load cell was also employed to record the applied tension against the elongation to produce a load-displacement graph for each sample. An observation of the trend of tensile force versus elongation during the tests indicated that all samples behave the same from the start of the loading until the breaking point, with a gradual increase in the tensile force by the increase of the elongation. However, irregular behaviour was observed when samples reached their peak tensile strength, which is believed to be caused by the breakage of the Kenaf fibres. Equation (4.1) presents the calculations required for determining the tensile strength of the samples, and the axial deformation of the sample is translated into the elongation percentage using Equation (4.2).

$$\alpha_f = \frac{F_f}{W_s} \tag{4.1}$$

$$\varepsilon_p = \frac{\Delta_L}{L_g} \times 100 \tag{4.2}$$

where α_f is the tensile strength in N/m, F_f is the force per unit width of the sample in N acquired at the time of failure, W_s is the width of the sample in m, ε_p is the percentage of elongation, Δ_L is the axial deformation in mm, and L_g is the original length of the sample in the axial direction in mm.

Table 4.3 presents the results of the UTS test on the textile samples under dry and wet conditions. The table also includes the rankings of textiles based on UTS results as well as the percentage of increase in UTS for treated samples compared with the untreated samples with the same pattern and moisture condition. The results prove the effectiveness of the applied chemical treatment in improving the UTS of all the textiles and extending the elongation before the sample break (lowering the stiffness). The highest UTS recorded was 47.1 kN/m, which belongs to a treated dry textile with a plain pattern and an opening size of 0 mm with 4.6% elongation before breakpoint. The second strongest sample has the same textile treatment, pattern, and opening but under the wet condition, resulting in a UTS of 41.6 kN/m and 4.8% elongation before breakpoint.

In both dry and wet conditions, the treated samples showed significantly higher UTS compared with natural samples of the same pattern and opening size, with an improvement of 41.7%–52% for dry samples and 27%–45.5% for wet samples. This trend also applied to the elongation percentage before the break, with an increment between 15.0% and 17.1% for dry samples and 12.2% and 14.3% for wet samples found in treated textiles. Therefore, it can be concluded that the applied chemical treatment significantly improved the UTS of the Kenaf geotextiles and extended their elongation before the break. The results also show that the elongation of the wet

Table 4.3 Mechanical characteristics of treated and untreated woven Kenaf geotextiles

Treatment condition	Textile pattern	Moisture condition	Textile mass per unit area (g/m²)	Ultimate tensile strength (UTS) (kN/m)	Textile tensile strength rankings	Textile elongation at break (%)	UTS increase compared to equivalent untreated sample (%)
Untreated (Natural) Kenaf textile	Inclined 2 mm	Dry	627	12.7	11	3.5	–
	Inclined 2 mm	Wet	1,408	11.8	16	3.8	–
	Inclined 0 mm	Dry	1,221	24.5	7	3.8	–
	Inclined 0 mm	Wet	1,998	22.2	9	4.1	–
	Plain 2 mm	Dry	603	16.8	12	3.7	–
	Plain 2 mm	Wet	1,387	15.5	13	3.9	–
	Plain 0 mm	Dry	1,158	31.2	5	4	–
	Plain 0 mm	Wet	2,111	28.6	6	4	–
Treated (6% NaOH) Kenaf textile	Inclined 2 mm	Dry	650	19.3	11	4	52
	Inclined 2 mm	Wet	1,436	15.0	14	4.3	27
	Inclined 0 mm	Dry	1,296	36.2	3	4.4	47.8
	Inclined 0 mm	Wet	2,207	32.3	4	4.6	45.5
	Plain 2 mm	Dry	628	23.8	8	4.3	41.7
	Plain 2 mm	Wet	1,411	20.1	10	4.4	29.7
	Plain 0 mm	Dry	1,243	47.1	1	4.6	51
	Plain 0 mm	Wet	2,168	41.6	2	4.8	45.5

Source: After Shirazi et al. (2019).

samples before break increased up to 8.6% for treated and up to 4.9% for untreated samples compared to similar textiles under dry conditions, which indicates that the presence of moisture contributes to the reduction of textile stiffness at the cost of reducing its UTS. The highest-ranked geotextiles in terms of UTS in both treated and untreated categories are identified as dry textiles with plain patterns and 0 mm spacing.

Further investigation was done on these samples in terms of the durability of the Kenaf geotextile under natural soil conditions by comparing their UTS and elongations at the break with fresh geotextiles. Samples of both natural and treated geotextile with plain patterns and an opening size of 0 mm were buried in natural soil conditions at a depth of 500 mm under the surface for 1 year. This approach was inspired by studies of Kim et al. (2005) and Yaacob et al. (2016) on determining the biodegradability properties of rice straw powder and polylactic acid, respectively. All samples were oven-dried before the test to provide the same moisture condition for unburied and buried textiles.

Figure 4.3 shows the results of the UTS test on buried and unburied samples. Both treated and untreated buried samples exhibit lower UTS and elongation at break compared with unburied samples, with a more noticeable decrease in the treated buried samples. This shows that degradation can significantly impact the durability of the Kenaf geotextile over time. Although the ratio of UTS reduction is more significant for the treated

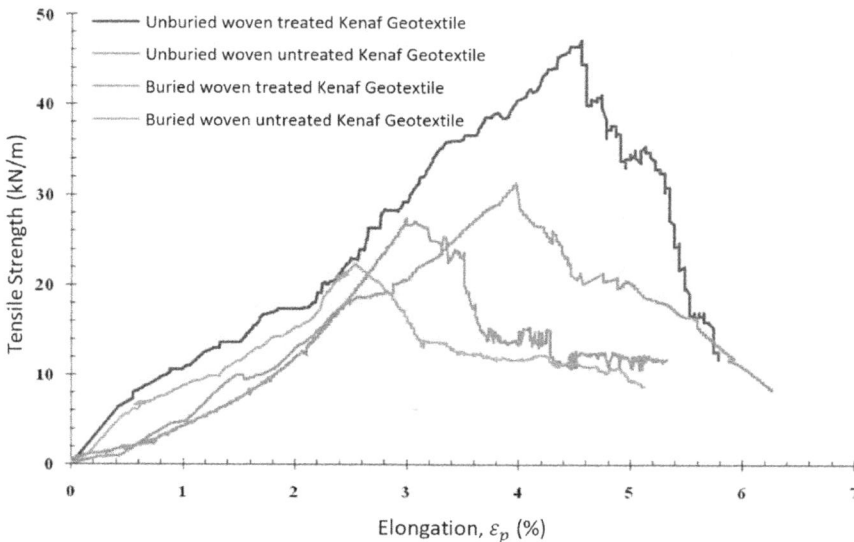

Figure 4.3 Tensile strength versus elongation for 1-year buried treated and untreated plain pattern geotextiles with 0 mm spacing and those of the unburied samples. (Shirazi et al., 2019.)

buried sample than for the untreated buried sample when compared with their corresponding unburied samples, the absolute UTS value for the treated buried sample is still larger than that for the untreated sample. This is evidence of the positive impact of the applied chemical treatment on the durability of the Kenaf geotextile in natural soil conditions.

The observed reduction in the UTS of the buried samples could result from the activity of living matter, especially microorganisms (Arshad et al., 2014). These living beings can consume bio-geotextile as their food (Zaikov, 2014). In addition, the absorbed water by the geotextile could lead to the swelling of the Kenaf fibres and eventually introduce microcracks or voids to their structure (Mohanty et al., 2005; Pickering, 2008). Domination of the dominant factors in degradation and their time-dependent impact on the performance of the geotextile cannot be easily done without extensive investigation. Therefore, further investigation would be needed to find the main factors involved in degradation and their long-term effect on the performance of the Kenaf geotextile.

4.3 SMALL-SCALE PHYSICAL MODEL TESTS

The performance of the Kenaf geotextile as soil reinforcement was investigated by designing and conducting a series of small-scale physical models to investigate the bearing capacity of a shallow foundation placed on reinforced and unreinforced granular and cohesive soils. Sandy soil was used as representative of granular soils, and cohesive soils were represented by Kaolin clay soil. A 430-mm-high rigid aluminium frame soil chamber equipped with a Perspex observation window was constructed with base dimensions of 400 mm (length) by 150 mm (width) as the test box. A 100-mm-wide rigid block exemplified the required shallow foundation with a length of 150 mm and thickness of 15 mm constructed with a scale factor of 1–10 to represent a 1-m-wide strip footing. The loading and data acquisition system included a driving unit, a load cell, and an LVDT to fulfil the aims of this investigation. Due to the well-known limitations of the plate load test, such as time limitation for ultimate settlement to occur, underestimating the bearing capacity of dense sandy soils, personal interpretation errors in estimating the failure load, etc. (Patel, 2019), the findings for bearing capacity and settlement should be used with caution.

The width of the soil chamber was designed based on the Prandtl (1921) failure mechanism assuming the width of the sliding block surrounding the footing is equal to the width of the footing and including an additional factor of 1.5 to avoid the boundary effect. By selecting a footing length equal to the width of the soil chamber, a plane strain condition was created for the tests. The contact friction between the chamber walls and sides of the footing, as well as between the soil and chamber walls, was minimised by

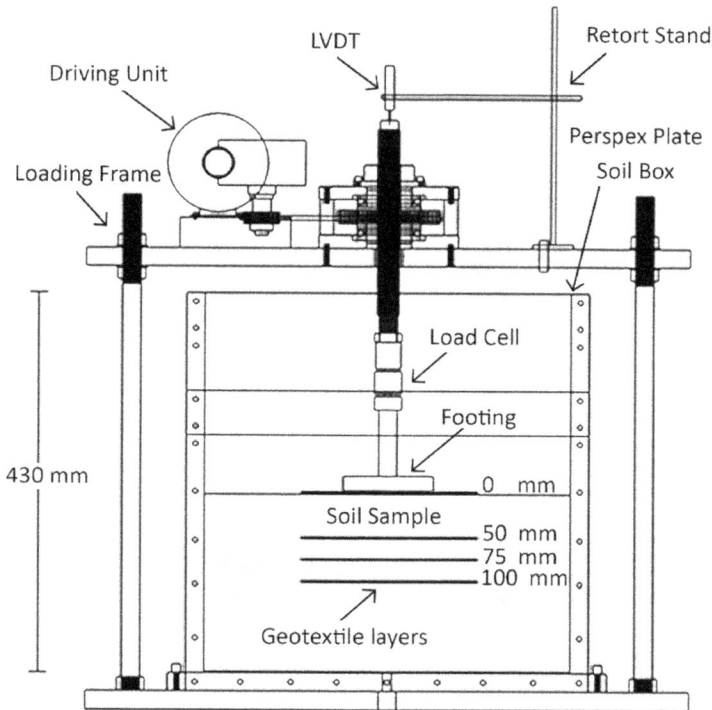

Figure 4.4 Small-scale physical model test set-up. (After Rashid et al., 2017.)

applying a layer of transparent grease on the inside of the chamber walls before constructing the models (Rashid & Noor, 2012). The parametric study concept was used in the design of the tests where the model itself is a prototype (Wood, 2017). In addition, due to the applied unit gravity scale test, there was no need to consider the scale effect in this study. Figure 4.4 shows the details of the small-scale physical model test set-up.

Several scenarios were created for physical model tests by placing the Kenaf geotextile at different depths below the footing and inside the model ground. The results of these tests were compared with the results of the tests on unreinforced model ground to investigate the impact of the application and depth of the Kenaf geotextile on the bearing capacity of the shallow foundation. The Kenaf geotextile used in the physical test was woven to a rectangular 190 mm by 150 mm shape with a thickness of 10 mm using a plain pattern and spacing of 5 mm and was used in natural and dry conditions. The load-deformation behaviour of this geotextile was determined using the UTM, as shown in Figure 4.5, which indicated a UTS of 5.53 kN/m and deformation of approximately 90 mm at the breakpoint.

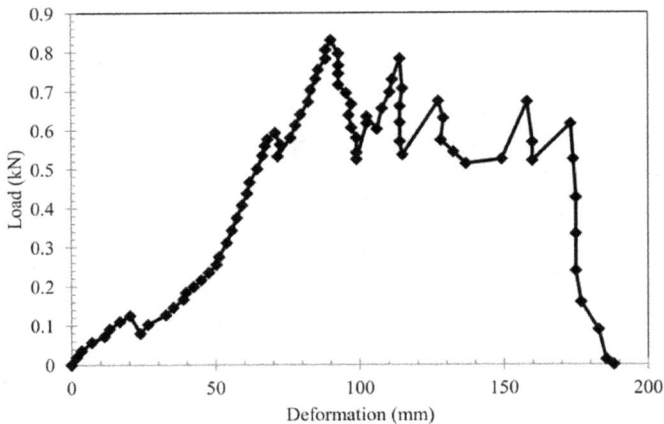

Figure 4.5 Tensile strength test for Kenaf geotextile of physical model tests. (Rashid et al., 2017.)

4.4 EFFECT OF KENAF GEOTEXTILE ON BEARING CAPACITY OF COHESIVE SOIL

A slurry deposition method was used for the constitution of the Kaolin clay ground model by applying a moisture content of two times the liquid limit of the soil. Using two drainage lines at the bottom of the soil chamber and applying incremental consolidation stresses up to 50 KPa by a pneumatic piston and a plate on top of the slurry, a cohesive soil ground model with 200 mm height was constructed inside the chamber (Braim et al., 2016; Moradi et al., 2018).

Four tests were planned with the geotextile installed as reinforcement at specific depths after the completion of each model ground. This was done by carefully cutting the soil to the target depth and laying back the soil, followed by ground surface levelling after installing the geotextile. A 50 kPa surcharge was applied on top of the soil surface to ensure the replaced soil was consolidated and fully in contact with the geotextile. After that, an overconsolidated ratio of 10 was attained by reducing the surcharge to 5 kPa. To investigate the short-term bearing capacity of clay, an undrained situation was created using a strain-controlled driving system capable of applying a normalised velocity by using a penetration rate of 6.28 mm/min (Lehane et al., 2008). The axial displacement and axial compressive stress were captured and recorded using the LVDT and load cell, respectively. The termination criteria for the test were set to observable shear failure or a maximum 20 mm penetration, equal to 20% of the footing width. It was assumed that the footing has fully mobilised the soil strength by reaching this depth. Further, two tests with the same setting were carried out for each scenario to ensure the consistency of the results.

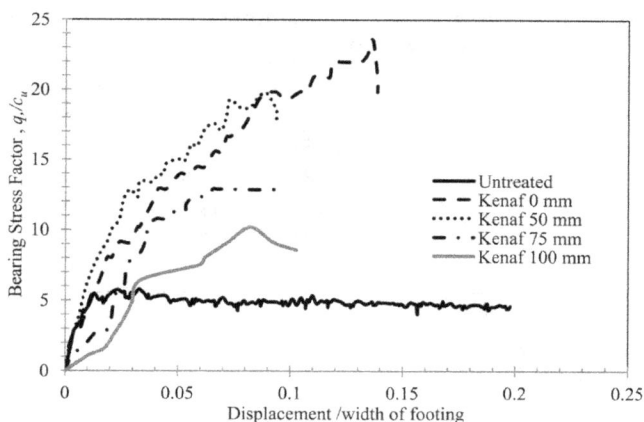

Figure 4.6 Bearing capacity factor versus displacement/footing width for cohesive soil. (Rashid et al., 2017.)

Figure 4.6 shows the curves of the bearing stress factor against displacement per width of the footing for the tested ground models. It is observed that a progressive settlement occurs for the unreinforced model with no sudden failure. The visual observation of the model also indicated a slight heaving around the perimeter of the footing with no sign of rupture surface on the ground. Therefore, a local shear failure must have occurred by reaching the plateau at a settlement per footing width ratio of about 0.01–0.03. A completely different failure mechanism, however, is observed for all cases with Kenaf geotextile reinforcement, where there is a sharp increase in the bearing stress factor by the progress of settlement until reading the peak stress. The bearing stress factor is then dropped or levelled off as a sign of shear failure.

Table 4.4 summarises the strength parameters of cohesive soil model ground under different test conditions. A vane shear test was used to estimate the ground model's undrained shear strength surrounding the foundation after the completion of each test. The slightly higher value of undrained shear strength for the reinforced models compared to the unreinforced model is believed to be caused by reapplying the surcharge after installation of the geotextile. The impact of this parameter, however, was offset in the calculation of the bearing capacity factor.

The value of the bearing capacity factor obtained from the test for unreinforced soft clay is found to be greater than the proposed theoretical value of 5.14 for strip footing by Meyerhof (1963), which is believed to be caused by slight irregularity in the ground surface resulting in friction between the footing ends and the soil chamber walls during the downward movement (Rashid & Norhazilan, 2012). The bearing capacity factor increased in all reinforced models compared with the unreinforced model. Also, by

Table 4.4 Strength parameters of cohesive soil ground model under different test conditions

Reinforcement depth (mm)	$C_u(kPa)^a$	$q_{ult}(kPa)^b$	N_c^c	Improvement[d] (%)
Unreinforced	5.7	33.57	5.9	-
0	7	156.8	22.5	281.4
50	7	138.6	19.8	235.6
75	7	90.3	12.9	118.6
100	7	71.4	10.2	72.9

Source: After Rashid et al. (2020).

[a] Undrained shear strength.
[b] Ultimate bearing capacity.
[c] Bearing capacity factor $\left(\dfrac{q_{ult}}{C_u}\right)$.
[d] Compared to unreinforced condition.

increasing the depth of the reinforcement, the bearing capacity factors registered higher numbers, and the magnitude of settlement under the footing was reduced. The best performance of reinforcement in terms of improvement in the bearing capacity factor was achieved by installation at the ground surface with a 281.4% increase compared with the unreinforced model. This finding is aligned with the Prandtl (1921) failure mechanism, indicating the upper part of the failure block is where the highest strain occurs. With an agreement with the same concept, a 72.9% increase in the bearing capacity factor was achieved as the minimum improvement compared with the unreinforced model, where the geotextile was installed at a depth of 100 mm below the base of the footing. Being equal to the width of the foundation, this depth is where the failure wedge becomes effective.

4.5 EFFECT OF KENAF GEOTEXTILE ON BEARING CAPACITY OF GRANULAR SOIL

A 50% relative density was targeted to prepare medium sandy soil using well-graded sand through the dry pluviation method to thickness the same 200 mm of soil as the clay model. The characteristics of the sand used in this study are detailed in Rashid et al. (2017). Using the minimum and maximum dry unit weight values, the 50% of relative density was targeted at a unit weight of 16.53 kN/mm^3. Five physical model grounds, namely one unreinforced and four reinforced, were constructed like those of cohesive ground models. After the constitution of each model and placing the footing on the surface, the same strain-controlled loading system and data

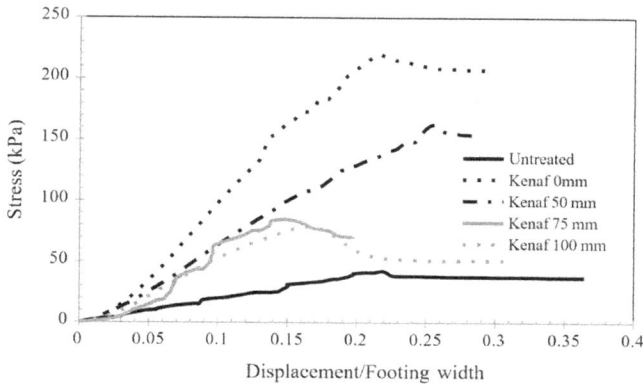

Figure 4.7 Vertical stress versus displacement/footing width for granular soil. (After Rashid et al., 2020.)

acquisition used in the clay model were employed to conduct the test and record the stress and displacement during the test. The same termination criteria as the clay model were also applied to this test.

The curves of vertical stress versus settlement per width of the footing for all the sand model grounds are shown in Figure 4.7. A slight increase in vertical stress by the progress of settlement is evident in unreinforced sand, where the soil continues to be compacted under axial compression. However, this behaviour changes by introducing the reinforcement where the stress reaches a plateau and is then levelled off by the strain-hardening behaviour of the soil.

The strength parameters of granular soil model ground for reinforcements installed at depths of 0, 50, 75, and 100 mm show an improvement of 414.9%, 277.0%, 97.9%, and 64.0%, respectively, compared to unreinforced condition (Rashid et al., 2017). The bearing capacity of the unreinforced sand obtained from the physical model test can be compared with the theoretical ultimate bearing capacity of Terzaghi and Peck (1968) using Equation (4.3).

$$q_u = \left(1 + 0.3\frac{B}{L}\right)cN_c + qN_q + \frac{1}{2}\left(1 - 0.2\frac{B}{L}\right)\gamma BN_\gamma \qquad (4.3)$$

where B is the width, L is the length of the footing, N_c, N_q, and N_γ are the bearing capacity factors, c is the soil cohesion, and q is the overburden pressure at the base of the foundation. The theoretical q_u is simplified by setting q and c equal to zero and is calculated as 38.93 kPa, which is almost 10% less than the result of the physical test. The slightly higher bearing capacity obtained at the physical test is believed to be caused by the same factors discussed for clay models.

An improvement of between 64.0% and 414.9% in the bearing capacity of the sand models is evident as the result of reinforcement, where the most significant improvement occurred by installing the geotextile on the surface, and the lowest improvement was achieved at the reinforcement depth of 100 mm. The main contributing factors to this improvement are believed to be the interaction between the sandy soil grains and Kenaf geotextile fibres (Artidteang et al., 2012; Tanchaisawat et al., 2013). The overall pattern of bearing capacity improvement for sand models is aligned with the results obtained for clay models.

4.6 CONCLUDING REMARKS

- An investigation made in this study revealed a significant degradation in the quality of the Kenaf geotextile when it was buried in natural soil for 1 year. A significant reduction was observed in the UTS and elongation before the breakpoint of the geotextile recovered from the ground compared to the unburied one.
- The time-dependent decay of bio-geotextile would be a favourable factor for limited-life projects with minimum impact on the in situ soil in terms of contamination or environmental issues where higher bearing capacity is only needed for a limited time during the project.
- Further investigation on the geotextiles showed that chemical treatment of this bio-geotextile with 6% sodium sulphate solution results in higher tensile strength and elongation before the break compared to untreated geotextile. However, it does not prevent the degradation of the materials in natural soil conditions.
- Although researchers identified some agents, such as micro-biomechanisms and moisture, to be responsible for the degradation of biomaterials in soil, further studies are needed to examine the behaviour of the Kenaf geotextiles in longer periods, determining the main contributing factors in their durability, their effective life, required chemical treatments to slow down their degradation process, and the environmental effects of releasing their absorbed chemicals into the soil.
- A series of small-scale physical model tests on Kaolin clay and sandy soils confirmed a significant improvement of soil's bearing capacity in Kenaf geotextile-reinforced soils compared to the unreinforced soils. The magnitude of the improvement was dependent on the depth of installation of the geotextile, where the maximum improvement occurs at surface installation and the lowest improvement at a depth equal to the width of the foundation.
- The enhancement in bearing capacity for reinforced soil versus unreinforced soil was between 72.9% and 281.4% for clay and between 46.0% and 414.9% for sand. This strongly supports the idea of using Kenaf geotextile as a sustainable and degradable biomaterial

for limited-life soil reinforcement to improve the short-term bearing capacity of clay and the ultimate bearing capacity of sand. Kenaf geotextile can effectively replace synthetic geotextiles in limited-life applications such as platform reinforcement for prefabricated vertical drains (PVD) installation and piling.

REFERENCES

Arshad, K., Skrifvars, M., Vivod, V., Valh, J., & Voncina, B. (2014). Biodegradation of natural textile materials in soil. *Tekstilec, 57*(2), 118–132.

Artidteang, S., Bergado, D., Tanchaisawat, T., & Saowapakpiboon, J. (2012). Investigation of tensile and soil-geotextile interface strength of kenaf woven limited life geotextiles (LLGs). *Lowland Technology International, 14*(2), 1–8.

Braim, K. S., Ahmad, S., Rashid, A. S. A., & Mohamad, H. (2016). Strip footing settlement on sandy soil due to eccentricity load. *International Journal of GEOMATE, 11*(5), 2741–2746.

Chaiyaput, S., Bergado, D., & Artidteang, S. (2014). Measured and simulated results of a Kenaf Limited Life Geosynthetics (LLGs) reinforced test embankment on soft clay. *Geotextiles and Geomembranes, 42*(1), 39–47.

D4595-17, A. (2017). *Standard Test Method for Tensile Properties of Geotextiles by the Wide-Width Strip Method*. ASTM International.

Edeerozey, A. M., Akil, H. M., Azhar, A., & Ariffin, M. Z. (2007). Chemical modification of kenaf fibers. *Materials Letters, 61*(10), 2023–2025.

Huber, T., Müssig, J., Curnow, O., Pang, S., Bickerton, S., & Staiger, M. P. (2012). A critical review of all-cellulose composites. *Journal of Materials Science, 47*(3), 1171–1186.

Kim, H. S., Yang, H. S., & Kim, H. J. (2005). Biodegradability and mechanical properties of agro-flour–filled polybutylene succinate biocomposites. *Journal of Applied Polymer Science, 97*(4), 1513–1521.

Lehane, B., Gaudin, C., Richards, D., & Rattley, M. (2008). Rate effects on the vertical uplift capacity of footings founded in clay. *Géotechnique, 58*(1), 13–21.

Li, X., Panigrahi, S., & Tabil, L. (2009). A study on flax fiber-reinforced polyethylene biocomposites. *Applied Engineering in Agriculture, 25*(4), 525–531.

Meon, M. S., Othman, M. F., Husain, H., Remeli, M. F., & Syawal, M. S. M. (2012). Improving tensile properties of kenaf fibers treated with sodium hydroxide. *Procedia Engineering, 41*, 1587–1592.

Meyerhof, G. G. (1963). Some recent research on the bearing capacity of foundations. *Canadian Geotechnical Journal, 1*(1), 16–26.

Mohanty, A. K., Misra, M., & Drzal, L. T. (2005). *Natural Fibers, Biopolymers, and Biocomposites*. CRC Press.

Moradi, R., Marto, A., Rashid, A. S. A., Moradi, M. M., Ganiyu, A. A., & Horpibulsuk, S. (2018). Bearing capacity of soft soil model treated with end-bearing bottom ash columns. *Environmental Earth Sciences, 77*(3), 1–9.

Patel, A. (2019). *Geotechnical Investigations and Improvement of Ground Conditions*. Woodhead Publishing.

Pickering, K. (2008). *Properties and Performance of Natural-Fibre Composites*. Elsevier.

Prandtl, L. (1921). Hauptaufsätze: Über die Eindringungsfestigkeit (Härte) plastischer Baustoffe und die Festigkeit von Schneiden. *ZAMM - Journal of Applied Mathematics and Mechanics / Zeitschrift für Angewandte Mathematik und Mechanik, 1*(1), 15–20. doi: 10.1002/zamm.19210010102.

Rashid, A. S. A., & Noor, N. M. (2012). Estimation of interface resistance between testing chamber and soil model using shear box test. *European Journal of Scientific Research, 80*(4), 472–478.

Rashid, A. S. A., & Norhazilan, M. N. (2012). Estimation of wall friction of chamber box using consolidation characteristic. *Paper Presented at the Applied Mechanics and Materials.*

Rashid, A. S. A., Shirazi, M. G., Mohamad, H., & Sahdi, F. (2017). Bearing capacity of sandy soil treated by Kenaf fibre geotextile. *Environmental Earth Sciences, 76*(12), 1–6.

Rashid, A. S. A., Shirazi, M. G., Nazir, R., Mohamad, H., Sahdi, F., & Horpibulsuk, S. (2020). Bearing capacity performance of soft cohesive soil treated by kenaf limited life geotextile. *Marine Georesources & Geotechnology, 38*(6), 755–760.

Sarsby, R. W. (2007). Use of 'Limited Life Geotextiles' (LLGs) for basal reinforcement of embankments built on soft clay. *Geotextiles and Geomembranes, 25*(4–5), 302–310.

Shirazi, M. G., Rashid, A. S. B. A., Nazir, R. B., Rashid, A. H. B. A., Moayedi, H., Horpibulsuk, S., & Samingthong, W. (2020). Sustainable soil bearing capacity improvement using natural limited life geotextile reinforcement—a review. *Minerals, 10*(5), 479.

Shukla, S. K. (2017). *Fundamentals of Fibre-Reinforced Soil Engineering.* Springer.

Summerscales, J., Dissanayake, N. P., Virk, A. S., & Hall, W. (2010). A review of bast fibres and their composites. Part 1–Fibres as reinforcements. *Composites Part A: Applied Science and Manufacturing, 41*(10), 1329–1335.

Tanchaisawat, T., Bergado, D., & Artidteang, S. (2013). Measured and simulated interactions between Kenaf geogrid limited life geosynthetics (LLGs) and silty sand backfill.

Terzaghi, K., & Peck, R. B. (1968). *Soil Mechanics in Engineering Practice: 2d Ed.* John Wiley.

Wood, D. M. (2017). *Geotechnical Modelling.* CRC Press.

Yaacob, N. D., Ismail, H., & Ting, S. S. (2016). Soil burial of polylactic acid/paddy straw powder biocomposite. *BioResources, 11*(1), 1255–1269.

Yetimoglu, T., & Salbas, O. (2003). A study on shear strength of sands reinforced with randomly distributed discrete fibers. *Geotextiles and Geomembranes, 21*(2), 103–110.

Zaikov, G. (2014). *Biodegradation and Durability of Materials under the Effect of Microorganisms.* CRC Press.

Chapter 5

Energy-efficient buildings for sustainable development

Sateesh Kumar Pisini and Swetha Priya Thammadi
Fiji National University

Suzanne Wilkinson
Massey University

CONTENTS

5.1 INTRODUCTION

With 0.7% population growth per year, coupled with rapid urbanization and climate change, world energy consumption is projected to rise nearly 50% by 2050 (IEO, 2020). The global energy-related CO_2 emissions in 2020 were 31.5 Gt (gigatonnes), causing the CO_2 concentration in atmosphere to reach 412.5 ppm (parts per million), which is the highest ever average annual concentration. It is estimated that by 2035, global energy-related CO_2 emissions will exceed today's levels by 20% (IEA, 2019). About 33% of the total global CO_2 emissions is linked to energy use in buildings (Price et al., 2006). IPCC estimated rate of growth of building-related greenhouse gas (GHG) emissions to be 2.5% per year for commercial buildings and 1.7% per year for residential buildings (Levine et al., 2007), thus emphasizing the need for energy and technological transformations. Electricity consumed is expected to increase in the industrial, transportation, and building sectors (REEEP, 2011), with approximately 40% of the total energy consumed by the residential and commercial building sectors used to heat or cool

DOI: 10.1201/9781003368335-5

buildings, this being mainly due to the long periods people spend indoors as well as inadequate building insulation. This percentage varies based on the degree of electrification, the level of urbanization, the amount of building area per capita, the prevailing climate, as well as national and local policies to promote efficiency. Energy consumption in buildings is > 40% of total energy consumption in European Union countries, while it is 15%–20% in the Philippines[6], 42% in Brazil[7], 47% in Florida, USA (Laar and Grimme, 2002), and 66% in California[9]. The International Energy Agency (IEA) statistics estimate that globally, the building sector is responsible for 42% more electricity consumption than any other sector (IEA, 2004b). In order to help lower GHG emissions across the building and construction sector, there are two types of emissions mitigation frameworks: (i) whole-of-life embodied carbon reduction, i.e., whole-of-life carbon emissions, including from the materials used in construction, the construction process, construction waste, and the disposal of a building at the end of its life, and (ii) transforming operational efficiency, i.e., efficiency for energy use, water use, and minimum indoor environmental quality measures for buildings. The reduction of embodied carbon emissions can be achieved by tackling the emissions (illustrated in Figure 5.3) during the lifecycle of a built asset, which is usually evaluated using lifecycle assessment (LCA). LCA methodology assesses the resource consumption and environmental impacts at each stage of the building's lifecycle, including an assessment of the potential benefits from the reuse or recycling of components after the end of a building's useful life. The application of recycled products like recycled aggregates, and using waste materials like fly ash and ground granulated blast furnace slag are some of the notable examples of material efficiency in reducing the embodied carbon emissions of buildings. However, the current study focuses on operational efficiency aspect of buildings to mitigate GHG emissions in the context of zero energy building (ZEB).

According to US Energy Information Administration IEO (International Energy Outlook)-2019, the global energy-related CO_2 emissions will increase by 0.6% per year during the period 2018–2050. Figure 5.1 shows the history and projection of global energy-related CO_2 emissions from 1990 to 2050. It is to be noted that the projection of these emissions is not uniformly distributed across the world. OECD (Organisation for Economic Cooperation and Development) countries account for nearly 20% of world's population. Projection of energy-related CO_2 emissions from OECD countries display a significant contribution from non-OECD countries compared to OECD countries. Furthermore, as shown in Figure 5.2, building sector energy consumption is driven by non-OECD countries due to their rising income, urbanization, and increase in access to electricity, thus leading to increase in energy demand. Energy consumption from building sector in non-OECD countries increases at about 2% per year, which is nearly five times faster than OECD countries. It is to be noted that energy consumption of building sector from non-OECD countries surpasses that of

Energy-related carbon dioxide emissions
billion metric tons

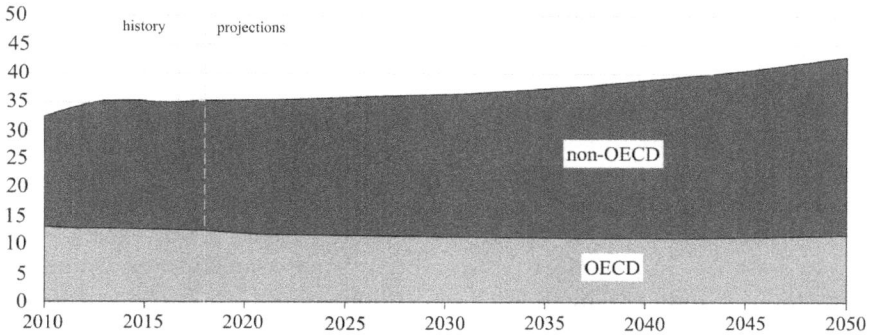

Figure 5.1 The projection of global energy-related CO_2 emissions. (IEO 2019.)

Building sector energy consumption
quadrillion British thermal units

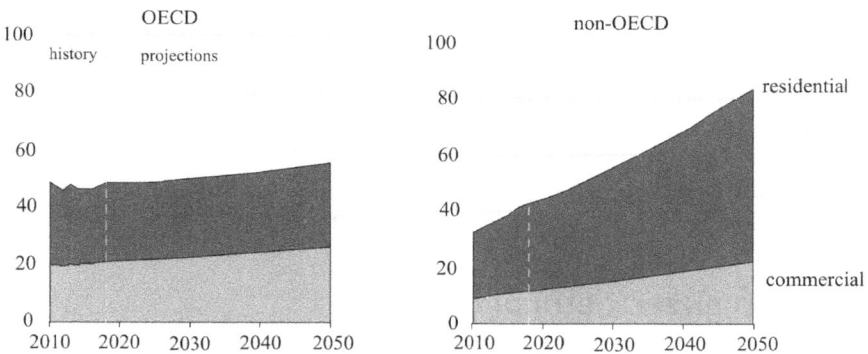

Figure 5.2 Projected energy consumption in building sector. (IEO 2019.)

OECD countries by 2025. The growth in building sector energy demand of developing countries is due to improved access to energy, use of energy-consuming devices, greater ownership, and drastic increase in floor area of buildings globally. On the other hand, although the residential energy consumption per person increases in non-OECD countries, it is lower than average of residential energy consumption per person increases in OECD countries due to the increased access to energy sources and energy-efficient equipment and appliances.

The energy efficiency of a building is the extent to which the energy consumption per square metre of floor area of the building measures up to established energy consumption benchmarks for that particular type of building under defined climatic conditions. Benchmarks are applied mainly to heating, cooling, air conditioning, ventilation, lighting, fans, pumps and controls, office or other electrical equipment, and electricity consumption for external lighting. These benchmarks vary with the country and type of building.

In industrial countries, buildings account for 25%–40% of total energy consumption. Major energy consumption is during the building's operational phase, for heating, cooling, and lighting purposes, which contribute towards substantial amount of CO_2 emissions (UNEP, 2007). In today's technological society, the main activities of living and working take place in an enclosed space in which people spend more than 90% of their time (Jenkins et al., 1990), and in more than 40% of the enclosed space, people suffer from health-, comfort-, and safety related complaints and illnesses (Dorgan, 1993) as a result of the "sick building syndrome". The emergence of the building-related sickness among building occupants can significantly reduce comfort and productivity (Dorgan, 1993; Bonnefoy et al., 2004).

UN Framework Convention on Climate Change and the Kyoto Protocol, associated with more than 100 countries with the Copenhagen Accords, set an objective to limit the increase in global temperature to below two degrees Celsius. This implies reducing global CO_2 emissions by 50% by 2050. The main objective is to promote sustainable development by implementing policies and measures to, among others, enhance energy efficiency, protect and enhance sinks and reservoirs of GHGs, and increase the usage of new and renewable forms of energy and of advanced and innovative environmentally sound technologies. Applying renewable energy sources as an alternative for summer air conditioning not only has an extensive potential to meet the sustainability commitments but also it will have a significant contribution to improve living standards of global population.

5.2 ZERO ENERGY BUILDINGS

Emissions generated over the lifecycle of a building are categorized into embodied emissions and operational emissions. Embodied carbon emissions are produced during the construction phase of a building, from extraction of materials, transportation, and construction. Operational carbon emissions are generated during the building's life after construction, from the energy sources utilized during the operation phase. Operational carbon emissions can be direct or indirect carbon emissions. Direct carbon emissions are produced through fossil fuel burning to provide electricity and heating water. Indirect carbon emissions are produced from electricity and water usage.

Whole-life thinking involves all the life stages of a built asset, buildings in this case, from material extraction, manufacturing, conveyance and installation, operation and maintenance, and eventually disposal as shown in Figure 5.3. However, the current study focuses on operational carbon

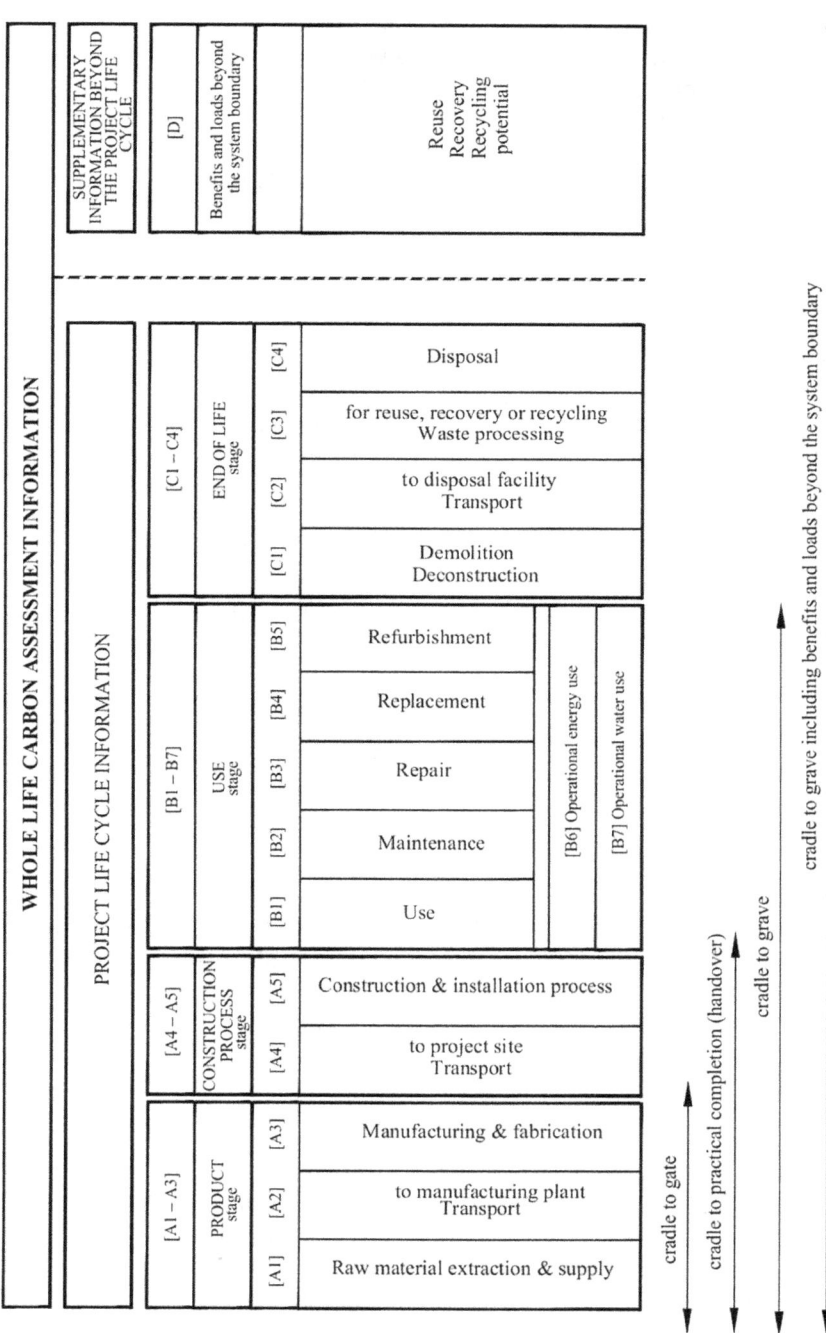

Figure 5.3 Stages in the lifecycle of a built asset. (RICS, 2017.)

emissions associated with operation of the building from handover to the end of service life. This includes operational carbon emissions due to energy and water use as well as embodied carbon emissions due to repair, maintenance, replacement, and refurbishment, i.e., B1–B5 in Figure 5.3.

B1 (in-use emissions) includes carbon emissions from components of the building during the entire lifecycle such as release of GHGs from insulation, paints, refrigerators, etc. B2 (maintenance emissions) accounts for carbon emissions from the maintenance activities associated with building envelope (walls, windows, roof, finishes, doors) and services, such as cleaning and use of other products including energy and water use for the activities. B3 (repair emissions) deals with carbon emissions arising from repair activities and associated products and can be taken as 25% of B2 in the absence of relevant data. B4 (replacement emissions) and B5 (refurbishment emissions) account for carbon emissions released from the replacement and refurbishment of components of building envelope and services, from production, conveyance to installation including losses associated with these activities, as well as carbon emissions associated with their removal. B6 (operational energy use) includes carbon emissions from building-integrated systems throughout the building lifecycle. These include emissions from building-regulated energy use as well as non-building-related systems. Building-regulated energy use accounts for domestic hot water supply, air conditioning, lighting, ventilation, heating and auxiliary systems, lifts, security, and communication setup. Non-building-related carbon emissions considered are for, e.g., cooking appliances, ICT equipment which represent a significant part of total operational CO_2 emissions. B7 (operational water use) includes carbon emissions pertaining to water (e.g., domestic hot water supply) and wastewater treatment excluding repair, maintenance, replacement, and refurbishment over the lifecycle of a building (RICS, 2017).

Net zero energy buildings (NZEB) are highly efficient buildings with ideally net zero energy consumption, i.e., minimizing energy consumption, thereby decreasing operational CO_2 emissions, maximizing energy generation using renewable energy produced onsite or nearby, and promoting energy conservation.

5.2.1 Envelope optimization

Building envelope separates the interior and exterior environment, i.e., roof, windows, walls, and foundation. Maintenance emissions (B2), repair emissions (B3), replacement emissions (B4), and refurbishment emissions (B5) play a significant role in optimizing the building envelope-related CO_2 emissions. B2 can be calculated based on properties and facilities management, lifecycle cost reports, and professional guidance. Data for B3 is extracted from properties/facilities maintenance and strategy reports, operation, and maintenance manuals apart from the data sources for calculation of B2. In the absence of any of the mentioned data sources, B3 can be assumed to be 25% of B2. As B4 is associated with replacement of building components, full replacement of items is assumed after the lifespan of the component. In the absence of specific data, it can be assumed that the life expectancy of roof and superstructure is 30 years, wall and floor finishes – 10–30 years, ceiling finishes – 10–20 years, fittings and heat equipment – 10–30 years, façade – 30–35 years, and services – 15–30 years depending on the building element/component (RICS, 2017). B5 is evaluated based on material additions as per the formula:

Carbon emissions = Material quantity × Material embodied carbon factor

The building envelope insulation properties along with construction quality affect the moisture and heat flows in the building. Cao et al. (2017) concluded that an optimal combination of wind turbine and photovoltaics (PV) generates net zero energy when generation from PV reaches 20% and 60% for Finnish and German climates, respectively. This study has investigated a hybrid zero energy system, i.e., ZEB and hydrogen-fuelled vehicle system. Experimental and numerical investigations conducted by Cellat et al. (2015) deduced that the use of phase change materials (PCMs) improves the sustainability of buildings by increasing the energy savings, reducing energy consumption, and improving the thermal comfort of humans. Different forms of PCMs such as usage of bio-based fatty acids as PCM in concrete, concrete with microencapsulated PCM, and butyl stearate have been studied under diverse conditions (Cellat et al., 2015; 2019a; Beyhan et al., 2017).

Efficient insulation is important for retrofitting due to limited space for additional insulation. Lightweight silica aerogel and vacuum insulation are new approaches for efficient insulation (Baetens, 2013). Several applications have been developed for this specific sector to improve the durability and enhanced performance of construction components, energy efficiency and safety of the buildings; to facilitate the ease of maintenance; and to provide increased comfort for living.

5.2.2 Operational CO_2 emissions

Indirect carbon emissions from electricity and water usage in buildings expressly contribute to the operational carbon emissions. The building sector accounts for 76% of electricity usage globally, which is eventually associated with GHG emissions (IEA, 2019). Therefore, it is essential to reduce the energy consumption in buildings to meet a nation's environmental and energy challenge. Major components of building energy consumption are heating, ventilation, and air conditioning (HVAC), which constitute 35% of total energy consumption in a building. Lighting consumes 11%, and other major appliances such as refrigerators, freezers, water heaters, and dryers consume 18% of total building energy. Remaining 36% of energy is consumed by miscellaneous appliances, including electronics (IEA, 2009). Therefore, the three components of operational CO_2 emissions, i.e., carbon emissions directly from combustion of fossil fuels, and indirect carbon emissions from electricity and water usage, contribute towards operational efficiency.

In-use emissions (B1) and operational energy emissions (B6) contribute towards the evaluation of operational CO_2 emissions. Reduction in operational CO_2 emissions is useful for climate change mitigation and contributes towards abatement of operational emissions. According to the World Energy Outlook 2009 (IEA, 2009), 52% energy efficiency is to be achieved in the short to medium term for end-use needs – and buildings must deliver a large part of this reduction.

As per 2020 MBIE report, the operational CO_2 emissions are expressed in kilograms of CO_2-equivalent per square metre per annum, i.e., kg CO_2-e/ $(m^2.a)$ as shown below (MBIE, 2020).

Operational CO_2 emissions $=$ kg CO_2-e/$(m^2.a)$

Combustion of fossil fuel $+$ kg CO_2-e/$(m^2.a)$	Electricity usage $\times$ Grid emissions factor $+$ kWh/$(m^2.a)$ kg CO_2 e/kWh	Water usage $\times$ Water emissions factor m^2/$(m^2.a)$ kg CO_2-e/m^2

5.2.3 Water usage

Potable water usage in buildings generates a remarkable amount wastewater. Electricity is used for treatment of potable water and wastewater as well as conveyance. This contributes towards indirect CO_2 emissions. Water consumption estimates for a nation are based on official reports from respective water authorities. Carbon conversion factors for water use and treatment as published by the local water supplier should be used. In the absence of relevant data, the relevant generic carbon conversion factors from available sources from different countries should be used.

Due to the changing climate patterns and the increasing risk of drought, countries that are highly dependent on electricity from hydro as their main source of electricity are losing much of their generation capacity, resulting in intensive power rationing, and it is anticipated to worsen further. The amount of water used in buildings impacts the demand on water resources. Although renewable sources of electricity such as hydro, geothermal, or wind provide electricity at a much lower cost than electricity generation from petroleum, their capital outlay is large, and they are complex but sustainable. Petroleum-based generation is usually brought in in the short term to meet this demand, which results in increased cost of electricity, overdependence on petroleum, and subsequently vulnerability to oil price fluctuations. Therefore, investment in adapting alternative sources of electricity today for improving water efficiency reduces operational CO_2 emissions and ensures uninterrupted supply of water for the future by building climate change resilience and adaptation.

5.2.4 Indoor air quality

In 2010, undesirable residential air leaks caused more than one half quads of space cooling energy loss and two quads of space heating energy loss along with one quad of commercial heating energy loss (EERE, 2014). Improper ventilation causes exposure to indoor pollutants such as mould, smoke, and other materials that eventually impact human health and well-being. Build-up of moisture in buildings can cause structural damage. Indoor air quality is the operational efficiency in terms of the amount of energy necessary to sustain indoor conditions. Energy lost in ventilation systems can be decreased using new technologies such as Acoustic Building Infiltration Measuring System apart from natural ventilation and minimizing leaks from ducts (Muehleisen, 2014). Sensors detect CO_2 concentrations and make necessary adjustments to the rate of ventilation. Good control systems can decrease the energy related to ventilation by nearly 40% (Walker, 2014). Efficient HVAC systems involve efficiency in heating or cooling air and advanced technology for effective removal of moisture from air. Operational efficiency adaptation helps move quickly with new buildings to achieve desired outcomes. As per United States Department of Environment (USDOE), efficient window, wall, and HVAC equipment currently available could reduce residential cooling by 61%, as shown in Figure 5.4.

Building sector plays a critical role in achieving the transition to a low-carbon economy. Energy needs of building sector can be reduced by approx. 30% by 2050 with energy renovations, efficiency in equipment and appliances, and using solar thermal energy; though there is a twofold increase in global floor area.

5.2.5 Climate and regional considerations

Adaptation of technologies is not uniform but varies with climate and region, which is challenging. Economics, efficiency, feasibility, resilience, and climate change are the five dimensions to be considered for optimizing the performance of a building, thus resulting in cost-effective operation and maintenance. High insulation is essential for colder and hotter weather due to the adversities caused by climate change. Also, space cooling/heating equipment needs to be climate specific. Efficient windows and space cooling/heating equipment serve better for climate change mitigation and adaptation. Climate zone affects the building design and orientation. Dryness and humidity affect the location of HVAC ducts and ventilation design.

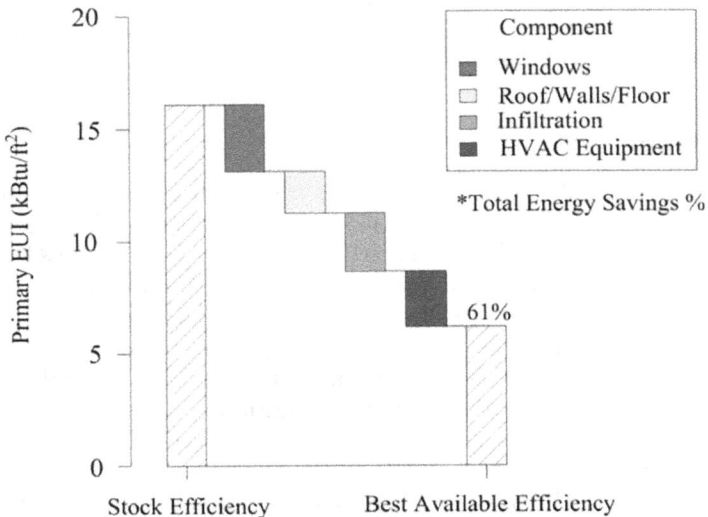

Figure 5.4 Energy-efficient building components. (USDOE.)

Geographic location characterizes the susceptibility to natural disasters, which are again aggravated due to climate change. It affects construction schedules and severity of disasters such as flooding, wild fires, earthquakes, termites, hurricanes, etc. Design and construction practices have to vary with location such as coastal zones, flood-prone areas, new and retrofitting buildings, and disaster-resilient attributes. Various tools can be used to measure the mitigation costs for individual and communities to reduce loss of life and property due to disasters.

5.3 EVALUATION TOOLS

Energy certification of buildings is a vital instrument to reduce the energy consumption and enhance the energy efficiency of new and existing buildings. Energy efficiency of buildings can be summarized using rating schemes, which has become a common practice globally. Certificates for energy efficiency provide information in two ways: (i) **comparative label** provides information about a building's performance compared to similar buildings and (ii) **positive** or **endorsement label** certifies a building that meets the standards.

European Union issues Energy Performance Certificates (EPCs) which provide vital information to the public for promoting energy efficiency.

The recommendation reports provide information on the potential to lower energy consumption, cost-effective alternatives, and paybacks for energy improvements. EPCs help EU reach the building sector's emissions reduction targets. Similarly, USA EPA (Environmental Protection Agency) and DOE (Department of Energy) run a program ENERGY STAR® to promote energy efficiency for products and devices using various standards. This label is used for many categories of products including houses, commercial buildings, and industries. RESNET/HERS are certification schemes in USA to calculate the energy efficiency of buildings using International Energy Conservation Code (IEEC) and the American Society of Heating, Refrigerating and Air-Conditioning Engineers (ASHRAE) standards. In Singapore, Energy Smart is the certification scheme to rate the energy performance of commercial buildings. In Singapore, BCA Green Mark, developed by Building and Construction Authority (BCA), supported by the National Environment Agency, is the positive label used for commercial and residential buildings. In India, LEED (Leadership in Environment and Energy Design) and GRIHA (Green Rating for Integrated Habitat Assessment) are the building environment assessment schemes based on environment, social, and economic pillars of sustainability. In New Zealand, energy modelling tools used are EnergyPlus, PHPP, IES, DesignBuilder, and AccuRateNZ for accessing operational carbon.

As discussed above, many countries have developed frameworks to achieve long-term sustainable building targets and immediate milestone targets for delivery. Denmark is among the countries that trails very low-energy buildings due to the strict energy requirements and a certification scheme in place since 1997, even before the EPBD (European Directive on Energy Performance of Buildings) was implemented. Operational energy certification linked to the granting of a building use permit has kept energy use constant over the years despite an ever-growing building stock (Thomsen, 2009). Irrespective of the EPBD requirements, increasingly low-energy buildings that go beyond building regulation standards are being certified.

MeetMED (Mitigation Enabling Energy Transition in the Mediterranean Region) focuses on policies related to "Energy Efficiency in Buildings" in the MeetMED target countries: Morocco, Algeria, Egypt, Palestine, Libya, Jordan, Lebanon, and Tunisia. Global Buildings Performance Network is a globally organized and regionally focused network whose mission is to advance best practice policies that can expressively reduce energy consumption and associated CO_2 emissions from buildings. International Renewable Energy Agency (IRENA) is an intergovernmental organization dedicated to renewable energy. In accordance with its Statutes, IRENA's objective is to "promote the widespread and increased adoption and the sustainable use of all forms of renewable energy". These member organizations are a few examples of collaborations between key institutions on clean energy and sustainable development of buildings.

5.4 CASE STUDIES

Implementation of certification schemes could be expensive, and resolution is based on cost-benefit analysis of reduction in CO_2 emissions, health, and well-being. Following case studies demonstrate successful implementation of ZEB concept in public arena. Also, innovative certification schemes from various nations around the world successfully demonstrated by dedicated financing, incentivized, cost-effective measures have been briefed.

Colorado-The National Renewable Energy Laboratory (NREL) is an award-winning energy research facility for establishing an energy-efficient NZEB. Consisting of an energy-efficient data centre with a combination of evaporative cooling, waste heat capture, highly efficient servers, and outside air ventilation, it reduces the energy use by 50% compared to conventional approaches. Renewable energy generated onsite is nearly 1.6 MW of PV. On the other hand, Wayne Aspinall Federal Building & U.S. Courthouse is an example of renovating an existing building into a highly energy-efficient and sustainable building. Objectives of NZEB are met through a combination of energy-efficient materials, HVAC with variable refrigerant flow, high-efficiency lighting systems, high thermal performance of building envelope, geo-exchange systems, and generation of renewable energy onsite, which provides for 100% of the annual energy requirements. With 385 PV roof panels, 123 kW of power is generated (equivalent to accommodation of 15 average American homes) and the building is now 50% energy efficient than a traditional office building (UNIDO, 2003).

In *India*, a NZEB was constructed by Central Public Works Department for the Ministry of Environment, Forest and Climate Change, first of its kind in India, generating onsite solar power to provide for total energy requirements. The building was constructed conforming to the highest norms and rated five star as per GRIHA ratings of India. Main features of the building include energy-efficient building envelope, HVAC system, lighting system, energy-efficient lifts, and geothermal heat exchange system for AC. Solar panels have been installed to cater to the annual power needs of the building (Soni et al., 2020).

Ireland has implemented EPBD requirements in spirit and law (IEA, 2020). It has a national grant scheme for energy retrofit and provides an additional certificate after the measure is completed. In *Portugal,* IFFRU 2020 is a successful financing tool, with a budget up to 1.4 million euros (1.65 million USD), in the Portuguese territory to support investments in urban rehabilitation focussing on buildings. Loans and guarantees are provided by funding agencies such as European Investment Bank, Council of the Europe Development Bank, European funds from PORTUGAL 2020 along with other commercial banking resources (MeetMED, 2020).

Loans and direct subsidies in *Spain* are provided for energy retrofitting of buildings across the nation using PAREER, a successful financial programme in Spain. PAREER is addressed towards enhancing the thermal envelope, i.e., insulation of roofs, windows and facades, as well as energy-efficient thermal and electric installations in existing buildings. EPC is used to check the energy efficiency of a building before and after retrofitting measures are taken. Nearly 42,000 dwellings have been improved in the past 8 years, and it is projected that 3% of public buildings will be retrofitted by 2030 as per the new National Energy and Climate Plans, Spain. In *Italy,* ECOBONUS system is a tax deduction mechanism to address the energy renovation of existing residential buildings in Italy. Over 3.3 million USD worth incentivized activities have been taken up since 2007, resulting in a saving of 1.87 Mton/yr (MeetMED, 2020).

5.5 CATALOGUE FOR ENERGY-EFFICIENT BUILDINGS

The below catalogue is a general guide for quick reference to assist various stakeholders to progress towards a ZEB.

5.6 CONCLUSIONS

- Globally, there is an increasing scrutiny of carbon emissions generated from building materials and products. There is a huge scope for reduction of operational carbon emissions from buildings overtime, as new buildings can become more efficient in use through improved design and technology.
- Emissions cap on new buildings and energy certification for new and retrofitting buildings should be made a regulatory requirement as practised in few advanced economies. Globally successful innovative financial schemes illustrated through case studies can be customized, making it appropriate to the regional needs by nations aiming towards ZEB.
- The building sector needs to focus on implementation of design and construction of ZEBs focussing on building envelope optimization, reduction of operational CO_2 emissions, maintenance of indoor air quality, and efficiency in water usage for new and existing buildings with insights demonstrated in the catalogue for alternative approaches to conventional building for energy-efficient buildings (Table 5.1).
- It is to be noted that energy-efficient buildings focus on using renewable energy to meet the energy requirements, meaning fewer emissions and less stress on the environment. Therefore, carbon emissions from building operations will significantly decrease by adapting ZEBs, and contribute towards achieving net zero carbon emissions by 2050.

Table 5.1 Catalogue for alternatives approaches to conventional building

	Component	Conventional building	Energy-efficient building
Building envelope	Windows and doors	No air/moisture sealing	U values < 0.35 for windows, < R-5 for doors
	Wall	Brick/concrete	Lightweight concrete and bricks
	Wall insulation	Fibreglass	PCMs
	Infrastructure	Cost criteria	Economics, efficiency, feasibility, renewable energy resilience
Operational CO_2 emissions	Solar power	No consideration	Installation of PV panels, window alignment, increase in solar glazing area
	Appliances	Standard appliances	Energy star ratings applied
	Lighting	Incandescent bulbs	Energy star-rated CFLs and LED bulbs, lighting controls
	Electricity	Fossil fuel	Renewable energy source onsite
	Water heating	Standard water heater	Solar water heater
	Heat island effect	No consideration	Usage of heat deflecting materials
	Space heating	Standard air furnace	Geo-exchange systems, thermostats
Indoor air quality	Passive solar cooling	No consideration	Ventilated windows, energy labelled equipment/products, architectural considerations for shade structures and landscaping
	Day-lighting	No strategies	Architectural design for day-lighting, tubular skylights, PV panels

REFERENCES

Baetens, R. (2013) High performance thermal insulation materials for buildings, *Nanotechnology in Eco-Efficient Construction, Materials, Processes and Applications*, Woodhead Publishing Series in Civil and Structural Engineering, pp. 188–206.

Beyhan, B., Cellat, K., Konuklu, Y., Güngör, C., Karahan, O., Dündar, C., Paksoy, H. (2017) Robust microencapsulated phase change materials in concrete mixes for sustainable buildings. *Int J Energy Res* 41:113–126. https://doi.org/10.1002/er.3603.

Bonnefoy, X.R., Annesi-Maesano, I., Aznar, L.M., Braubachi, M., Croxford, B. (2004) Review of evidence of housing and health, *Fourth Ministerial Conference on Environmental and Health*, Budapest, Hungary, 23–25 June 2004.

California Energy Commission (2005) Options for energy efficiency in existing buildings.

Cao, S., Klein, K., Herkel, S., Sirén, K. (2017) Approaches to enhance the energy performance of a zero-energy building integrated with a commercial-scale hydrogen fueled zero-energy vehicle under Finnish and German conditions. *Energy Convers Manag* 142:153–175. https://doi.org/10.1016/j.enconman.2017.03.037.

Cellat, K., Beyhan, B., Güngör, C., Konuklu, Y., Karahan, O., Dündar, C., Paksoy, H. (2015) Thermal enhancement of concrete by adding bio-based fatty acids as phase change materials. *Energy Build* 106:156–163 https://doi.org/10.1016/J.ENBUILD.2015.05.035.

Cellat, K., Beyhan, B., Konuklu, Y., Dündar, C., Karahan, O., Güngör, C., Paksoy, H. (2019a) 2 years of monitoring results from passive solar energy storage in test cabins with phase change materials. *Solar Energy* https://doi.org/10.1016/J.SOLENER.2019.01.045.

Directive 2002/91/ec of the European Parliament and of the Council on the Energy Performance of Buildings, 2002.

Dorgan Associated (1993) *Productivity and Indoor Environmental Quality Study*, Alexandria, VA, National Management Institute.

EERE (2014) *Energy Efficiency and Renewable Energy report*, US Department of Energy.

Energy Efficiency Division of the Philippines Department of Energy (DOE) (2002) Philippines. Guidelines for Energy Conserving Design of Buildings and Utility Systems.

IEA (2004b) International Energy Agency, Energy Balances for OECD Countries and Energy Balances for non-OECD Countries; Energy Statistics for OECD Countries and Energy Statistics for non-OECD Countries (2004 editions) Paris.

IEA (2009) International Energy Agency, World Energy Outlook.

IEA (2019) International Energy Agency, World Energy Outlook.

IEA (2020) International Energy Agency, World Energy Outlook.

IEO (2020) International Energy Outlook, US Energy Information Administration, October 2020.

Jenkins, P.L., Philips, T.J., Mulberg, E.J. (1990) Activity patterns of Californias: use of and proximity to indoor pollutant sources. *Proceeding of Indoor Air '90*, Toronto, Vol. 2, pp. 465–470.

Laar, M., Grimme, F.W. (2002). Sustainable Buildings in the Tropics. Institute of Technology in the Tropics ITT, University of Applied Sciences Cologne: Presented at RIO 02 – World Climate & Energy Event, January 6–11, 2002.

Levine, M., Urge-Vorsatz, D., Blok, K., Geng, L., Harvey, D., Land, S., Levermore, G., Mongameli Mehlwana, A., Mirasgedis, S., Novikova, A., Rlling, J., Yoshino, H. (2007) Residential and commercial buildings, Climate Change 2007: Mitigation, *Contribution of Working Group III to the Fourth Assessment Report of the Intergovernmental Panel on Climate Change* [B. Metz, O.R. Davidson, P.R. Bosch, R. Dave, L.A. Meyer (eds)], Cambridge University Press, Cambridge, U.K. & New York, NY.

MBIE (2020) *Ministry of Business, Innovation and Employment New Zealand report*, Transforming Operational Efficiency, Building for climate change programme.

MeetMED (2020), Report on Mitigation Enabling Energy Transition in the MEDiterranean region, funded by the European Union

Muehleisen, R. (2014) "Acoustic Building Infiltration Measurement System." Washington, DC: EERE/DOE. http://energy.gov/sites/prod/files/2014/07/f17/emt40_Muehleisen_042414.pdf.

Price, L., De la Rue du Can, S., Sinton, J., Worrell, E. (2006) Sectoral Trends in Global Energy Use and GHG Emissions. Lawrence Berkeley National Laboratory, California, USA.

REEEP (2011) Renewable Energy and Energy Efficiency Partnership Annual Report, 2011–12.

RICS (2017) Royal Institution of Chartered Surveyors professional statement, Whole life carbon assessment for the built environment, November 2017.

Soni K.M., Bhagat Singh, P. (2020) First onsite net zero energy green building of India, *International Journal of Environmental Science and Technology* 17:2197–2204 https://doi.org/10.1007/s13762-019-02514-0.

Thomsen, K. (2009) Denmark: Impact, Compliance and Control of Legislation, EC ASEIPI P175, EU, available at: www.aseipi.eu and www.buildup.eu.

UNIDO (2003) United Nations Industrial Development Organization, Clean Development Mechanism (CDM), Investor Guide, South Africa, Vienna 2003.

United Nations Environment Programme (UNEP) (2007) Buildings and Climate Change. Status, Challenges and Opportunities.

Walker, A. (2014) "Natural Ventilation." Washington, DC: National Institute of Building Sciences. Available at: http://www.wbdg.org/resources/naturalventilation.php.

Chapter 6

Development and application of FBG sensors for landslide monitoring

Hong-Hu Zhu and Xiao Ye
Nanjing University

Hua-Fu Pei
Dalian University of Technology

Dao-Yuan Tan
The Hong Kong Polytechnic University

Bin Shi
Nanjing University

CONTENTS

6.1 INTRODUCTION

Fiber Bragg grating (FBG), as one of the most promising intelligent sensing technologies, has increasingly become popular to measure strain or temperature changes in geotechnical and geohazard monitoring (Rao, 1999; Othonos et al., 2006; Zhu et al., 2017). Compared with other sensing technologies, FBG has the characteristics of small size, high accuracy,

DOI: 10.1201/9781003368335-6

conspicuous durability, and immunity to electromagnetic interference. These sensors can be adhered to the surface of structural elements or be embedded in boreholes for performance monitoring and early warning of potential geohazards (Peng et al., 2006; Wu et al., 2021). In addition, multiple FBG sensors can be connected in series into a cluster of sensing arrays, which is called quasi-distributed monitoring. The sensing signals can be transmitted over a long distance to a remote client unit. Therefore, FBG sensing technology offers the possibility of remote monitoring for massive sequential data in real time.

In 1978, Hill et al. made the world's first FBG by using the standing wave writing method in germanium-doped fiber (Hill et al., 1978). With the rapid development of manufacturing technologies (e.g., the phase mask method) (Meltz et al., 1989), there is an ever-increasing application scope of FBG in engineering monitoring. It has become one of the most mature and widely used fiber optic sensing technologies (Zhu et al., 2017). Particularly, as an advanced version of FBG, ultra-weak fiber Bragg grating (UWFBG) has made major efforts in improving the multiplexing quantity, which complements the deficiency of point-wise monitoring using conventional FBG sensors. It is possible to reach near fully distributed monitoring, almost comparable to Brillouin optical time domain reflectometer (BOTDR) and Brillouin optical time domain analysis (BOTDA). Thus, a series of new FBG/UWFBG sensors have been designed and developed in recent decades for measuring stress, displacement, moisture, and earth/water pressure, in addition to strain and temperature. It is expected that this technology will start a revolution in geoengineering monitoring (Li and Zhang, 2018; Jiang et al., 2021).

Landslides, one of the most common geohazards, are generally regarded as the result of the structural failure of geological bodies under the action of endogenous and exogeneous geological processes. Undoubtedly, detailed monitoring, whether at the surface or at the subsurface, is critical to understand the mechanisms and evolution stages of sliding occurrences (Chae et al., 2017; Song et al., 2018; Tang et al., 2019). In laboratory model tests, miniature FBG sensors have been employed to investigate the mechanism of slope failure and the initiation of debris flows (Zhu et al., 2014, 2015; Xu et al., 2019; Li et al., 2021). There are also many successful applications of FBG systems to in-situ monitoring of engineering slopes (Zhu et al., 2012; Wang et al., 2015; Zhang et al., 2021). Previous studies primarily concentrated on deformation measurements (i.e., strain, displacement, and tilt), but only a few have focused on multi-parameter monitoring using FBGs.

In this paper, the sensing principle of FBG/UWFBG technology for landslide monitoring is briefly introduced. New development of FBG sensors for measuring critical physical and mechanical parameters is subsequently summarized. Finally, a case study for FBG-based landslide monitoring is presented.

6.2 SENSING PRINCIPLE

An FBG is a reflection filter-passive sensitive component with high performance, forming a spatial phase grating inside the fiber core. As shown in Figure 6.1, when the broadband light source passes through the FBG, some part of the light will be reflected. The reflected wavelength depends on the effective refractive index of the grating period and reverse coupling mode. Through the physical or thermal elongation of the sensing section, as well as the change in the refractive index of the optical fiber due to photoelasticity and thermo-optical effects, the Bragg wavelength λ_B of an FBG sensor will change linearly with the applied strain $\Delta\varepsilon$ or temperature ΔT (Othonos and Kalli, 1999):

$$\frac{\Delta\lambda_B}{\lambda_B} = (1 - p^{eff})\Delta\varepsilon + (\alpha + \xi)\Delta T \tag{6.1}$$

where $\Delta\lambda_B$ is the change in the Bragg wavelength due to applied strain and temperature changes; λ_B is the original Bragg wavelength under strain-free and 0°C conditions; p^{eff} is the photo-elastic parameter; and α and ξ are the thermal expansion and thermo-optic coefficients, respectively.

The multiplexing technology of the FBG mainly includes wavelength division multiplexing (WDM), time-division multiplexing (TDM), hybrid multiplexing technology, etc. For the WDM technology, gratings with different central wavelengths occupy various frequency band sources so that the power of incident light at each frequency can be fully utilized. FBG sensors with different Bragg wavelengths can be multiplexed in series using the WDM technique. To avoid spectral overlap during measurement, there should be sufficient spacing of the Bragg wavelengths for the serially connected FBG

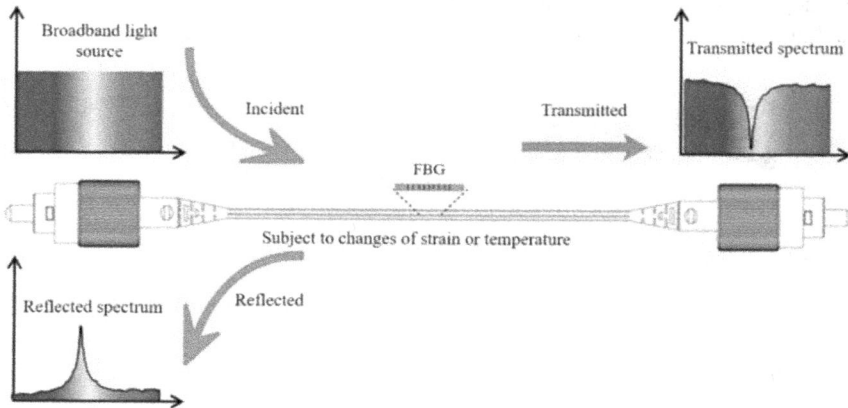

Figure 6.1 Conceptual graph of the FBG/UWFBG sensors.

sensors. In most cases, however, several gratings can often be connected in series, and the integrated number of gratings is no more than ten.

The TDM technology is different from WDM, in which the light source bandwidth will no longer restrict the number of multiplexing sensors. UWFBG enables the improvement of the multiplexing number and sensing distance of grating sensors and realizes high-precision dense distributed sensing due to the characteristics of narrow bandwidth and weak reflection. Therefore, this novel technology will take quasi-distributed fiber optic monitoring to a new level for real-time and accurate measurement.

6.3 DEVELOPMENT OF FBG SENSORS

6.3.1 FBG strain sensor

As mentioned earlier, the change in strain will directly induce the shift of the central wavelength of an FBG; thus, it can be used as a strain sensor. When the working environment is gentle or the structure measured has strict requirements on the size of the sensor, the FBG strain sensor can be installed on the surface of the structure with proper glue such as epoxy resin or be directly buried inside the object. At this time, attention should be given to the shear strength, long-term durability, creeping motion, hysteresis, and other properties of the selected glue. Because the bare FBG is very fragile with an outer diameter of 250 μm, it is easily damaged and fails in harsh environments. Thus, special packaging measures must be taken in most cases for FBG protection (Zhu et al., 2017).

As shown in Figure 6.2, in geological and geotechnical engineering practices, FBG strain sensors generally adopt four packaging methods: (i) metallization packaging, that is, a layer of a metal film is plated on the surface of an FBG by the vacuum evaporation method, ion beam deposition method, electrochemical deposition method, or chemical vapor deposition method; (ii) substrate packaging, mostly resin substrate and metal substrates such as steel, copper,

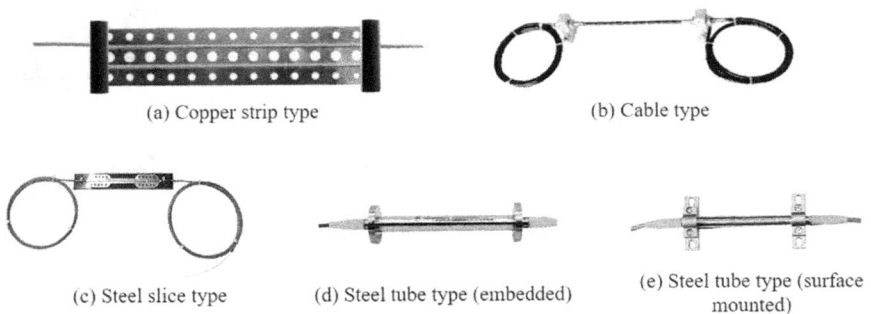

(a) Copper strip type   (b) Cable type

(c) Steel slice type   (d) Steel tube type (embedded)   (e) Steel tube type (surface mounted)

Figure 6.2 Typical FBG strain sensors.

and titanium alloy; (iii) embedded packaging, generally with the help of polymer materials, FRP reinforcement, etc.; and (iv) metal tube packaging, such as a stainless steel tube. In addition, clamping components or flange devices are usually installed at both ends of the sensor to ensure that the strain of the object to be measured is sufficiently transmitted to the grating. Sometimes, clamping sensors are also supplemented by laser welding and rivet connections.

6.3.2 FBG temperature sensor

In engineering practice, there is often a need for temperature measurements, such as the frost heave of frozen soils, the state of energy piles, and the temperature evolution characteristics of sliding zones in landslides. As the readings of most sensors are affected by temperature, temperature measurements are needed to perform temperature compensation. When the temperature changes, the shift of the central wavelength of an FBG occurs due to the thermo-optical and thermal expansion effects. The former causes a change in the effective refractive index of the optical fiber, and the latter causes a change in the grating period. The temperature sensitivity coefficient of a bare FBG is generally 0.011.7 pm/°C. If the measurement accuracy of the central wavelength is 1 pm, the temperature measurement accuracy is approximately 0.1°C. Similar to strain sensors, FBG temperature sensors also need a detailed packaging protection. These packaging processes can prevent the FBG temperature sensor from being damaged during installation and monitoring and sometimes play a role in temperature sensitization. The packaging methods include substrate, metal tube, and embedded packaging. Common FBG temperature sensors are shown in Figure 6.3.

6.3.3 FBG displacement transducer

FBG displacement transducers are usually divided into the bar type and bending beam type (see Figure 6.4). The main principle is to convert strains into displacements. The bar-type displacement transducer generally consists of an FBG strain sensor and a spring. This sensor can be connected in series to make a multi-point displacement transducer to realize

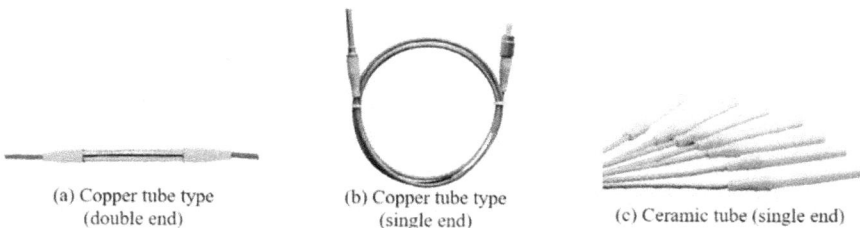

(a) Copper tube type (double end)   (b) Copper tube type (single end)   (c) Ceramic tube (single end)

Figure 6.3 Typical FBG temperature sensors.

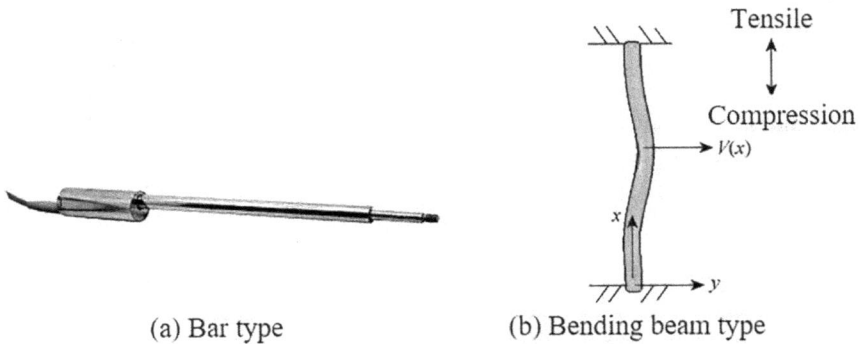

(a) Bar type (b) Bending beam type

Figure 6.4 Typical FBG displacement transducers.

quasi-distributed displacement monitoring. The bending beam-type sensor generally comprises an elastic beam body and FBG sensing array adhered to its surface along the length (Metje et al., 2008; Xu et al., 2013). The bending beam is designed to be surface-mounted or directly embedded. Based on Euler-Bernoulli bending beam theory and the assumption of boundary conditions, the conversion from tensile and compressive strains to beam deflection can be realized. In this way, an FBG in-situ inclinometer can be made by gluing FBGs on the surface of the inclinometer casing (Ho et al., 2006; Zhu et al., 2012; Wang et al., 2015). The disadvantage of this sensor is that the measurement error will be introduced when performing secondary integration of strain data. This problem can be solved by measuring deflections and angles section by section.

6.3.4 FBG pressure sensor

It is often necessary to measure the interface contact pressure for subgrades, retaining walls, embankments, pile foundations, and other geotechnical structures. Sometimes, the earth pressure cell is used to measure the stress of the soil. The traditional earth pressure cell is divided into hydraulic and electric types. The former gauges the earth pressure by measuring the liquid pressure in the sensor box, and the latter quantifies the earth pressure by measuring the deflection of the elastic diaphragm. The deficiency of these sensors lies in their poor performance of anti-electromagnetic interference and anti-corrosion, which is very important for long-term monitoring.

Figure 6.5a–c shows the typical FBG earth pressure cells. Two problems need to be solved in this kind of sensor design: (i) establishing the linear relationship between reading and pressure value and (ii) avoiding excessive disturbance of the cell to the in-situ stress field to ensure the measurement accuracy.

The pore water pressure gauge is mainly used to measure the groundwater level and pore water pressure of the soil. This kind of sensor mainly

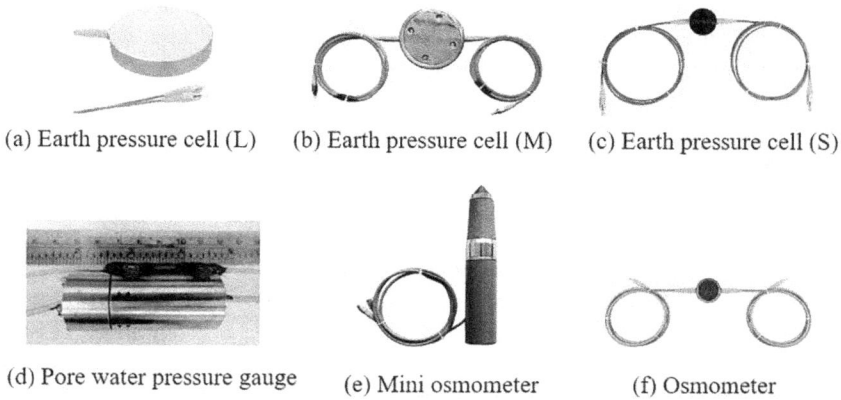

(a) Earth pressure cell (L) (b) Earth pressure cell (M) (c) Earth pressure cell (S)

(d) Pore water pressure gauge (e) Mini osmometer (f) Osmometer

Figure 6.5 Typical FBG pressure sensors.

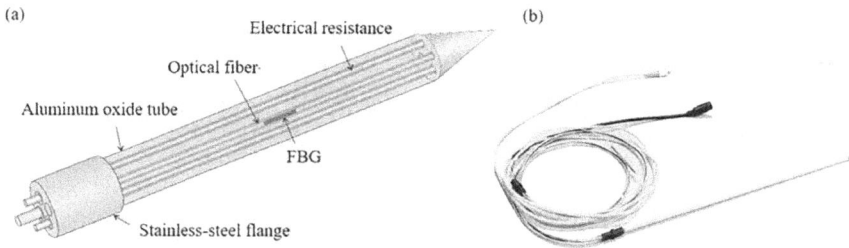

Figure 6.6 FBG soil moisture sensor. (a) Conceptual diagram. (b) Photograph.

adopts a diaphragm structure. To separate water from soil particles, the sensor is generally built in a permeable plate. Recently, Huang et al. developed a temperature-compensated FBG osmometer by using FBG to monitor the flexural deformation of an elastic circular plate and successfully applied it to a landslide site for long-term monitoring of pore water pressure along the depth (Huang et al., 2012; Qin et al., 2021).

6.3.5 FBG moisture sensor

As shown in Figure 6.6, the FBG soil moisture sensor is usually composed of an encapsulated tube, a loose optical fiber, and an electrical resistance wire penetrating through the tube. The quasi-distributed FBGs along the fiber are insensitive to strain variations and only used as temperature-measuring elements. When the sensor is embedded in the soil, the resistance wire is actively heated for a short period and the temperature changes are recorded. Due to the different thermophysical parameters of water, soil particles, and air, the temperature response of soil caused by pulse heating differs in soils with different moisture contents (Cao et al., 2018, 2021). Therefore, the soil moisture content can be determined by calibrating the temperature characteristic value in the heating process.

6.4 CASE STUDIES

6.4.1 Luk Keng landslide

A trial was successfully conducted on a roadside slope in Luk Keng, Hong Kong, where FBG strain sensors, temperature sensors, and in-place inclinometers were used to perform field monitoring. The slope is located at Luk Keng Road, Sheung Shui, New Territories, Hong Kong, which has a height of 10 m and a slope angle of 35°. From top to bottom, the slope consists of layers of colluvium, completely decomposed tuff, extremely weak to moderately weak siltstone, and the underlying rock layer. The groundwater regime in this area is greatly affected by the tidal effects of Sha Tau Kok Hoi. Previous monitoring data reveal that there were significant groundwater flows after heavy rainfall and apparent deformations were observed at the slope toe. To conduct slope stabilization measures, soil nails, soldier piles, and drainage facilities were constructed by local contractors. At the same time, an FBG monitoring system was established in the field (Zhu et al., 2012). A 14-m-long soil nail was instrumented with ten surface-glued FBG strain sensors and ten tube-packaged FBG temperature sensors. One soldier pile was instrumented with 18 FBG strain sensors and 9 temperature sensors. An FBG in-place inclinometer was installed in a 120-mm-diameter and 15-m-deep drillhole to monitor the slope movements. On the in-place inclinometer, four lines of quasi-distributed FBG strain sensors at 1.5-m intervals were adhered to the inclinometer casing. The monitoring results in Figure 6.7 show that the soil nail strains were generally consistent with monthly rainfall, but such a relationship was not found for the soldier pile.

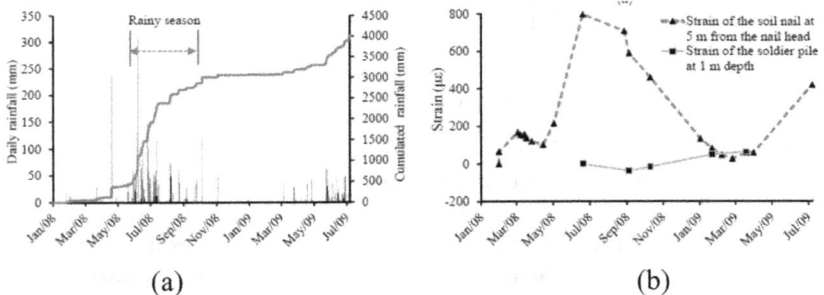

Figure 6.7 Long-term monitoring results of the slope site in Hong Kong. (a) Rainfall. (b) Strains of the soil nail and the soldier pile.

6.4.2 Xinpu landslide

The Xinpu landslide is located in Anping Town, Fengjie County, Chongqing stretch of the TGR area, which is on the right bank of the Yangtze River (see Figure 6.8). The overall slope is approximately 2 km long, with an average slope angle of 15°~20°. The volume of this massive landslide is $3792 \times 10^4 m^3$. The front edge of the landslide extends below the water level of the Yangtze River, which is a large wading landslide. Due to the multi-phase landslide activities of the slope, the entire slope forms a multi-level platform landform. According to ground surveys and boreholes, the bed-rock is the Upper Triassic Xujiahe Formation (T3xj) and the Lower Jurassic Zhenzhuchong Formation (J1z), with an occurrence of 345°∠21°. The slip surface is mainly divided into three layers from top to bottom: (i) the local weak structural surface in the slope, (ii) the interface between the crushed stone soil and the cataclastic rock, and (iii) the interface between the bed-rock and the soils. The rear part of the landslide is currently in a relatively

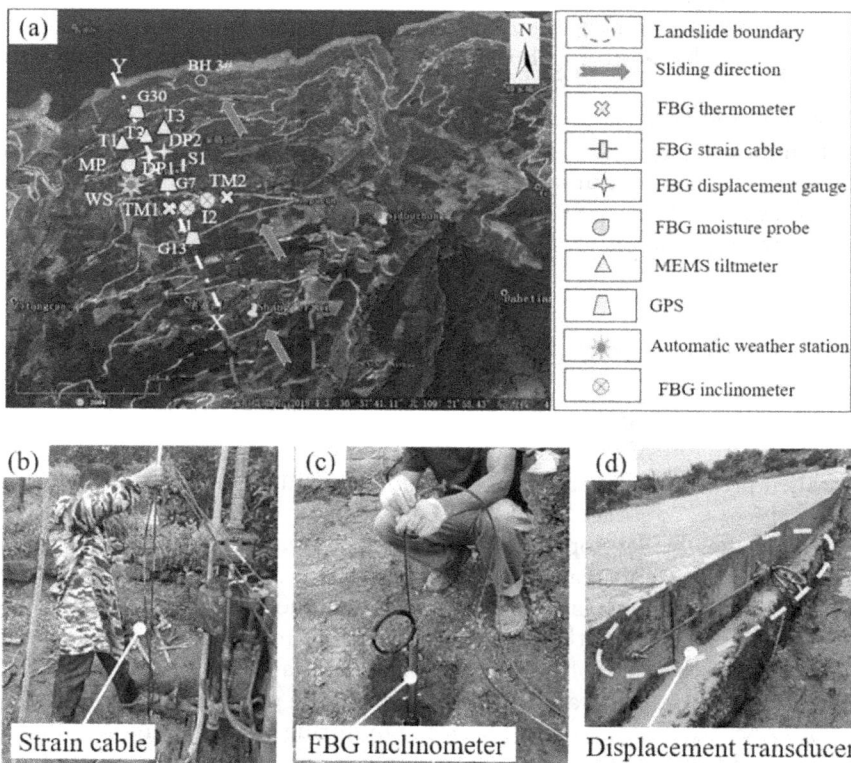

Figure 6.8 Monitoring arrangement of the Xinpu landslide. (a) Geomorphology and field monitoring deployment. (b) Installation of a strain-sensing cable. (c) Installation of the FBG inclinometer. (d) Installation of a displacement transducer.

stable state, whereas the deformation of the front part still develops and even deteriorates. Therefore, the dynamic changes in slope stability have received significant attention (Sun et al., 2020).

To explore the deformation mechanism and assess the failure phase of the landslide, a multi-source and multi-physical monitoring system dominated by FBG technology was recently established. Pipawan was selected as the first monitoring station location, where a maximum deformation of almost 400 mm occurred in June 2020. A quasi-distributed real-time monitoring system involving the surface and subsurface is composed of an FBG strain-sensing cable, thermometers, moisture probes, in-place inclinometers, and displacement gauges, and an advanced MEMS-based tiltmeter was also installed at a depth of 0.5 m. It should be noted that the FBG strain-sensing cable was installed in a 20-m-deep borehole to submerge the sliding surface. For other sensors, shallow information within a depth of 4.0 m is the focus. An automated weather station was also installed to obtain sufficient meteorological data for investigating the landslide triggering mechanism.

To obtain a more detailed subsurface information, a UWFBG-based multi-physical monitoring borehole was also implemented. The spacing of the sensing elements of temperature and strain was 1 m, which provided significantly enhanced spatial resolution and can demonstrate spatiotemporal evolution of subsurface multi-physics. The monitoring data can provide meaningful insights into the landslide mechanism. As stated by Ye et al. (2022), to improve the level of early detection, monitoring, and warning of landslides, it is crucial to obtain real-time multi-physical monitoring data.

6.5 CONCLUSIONS

In this chapter, the design and development of FBG sensors are introduced. Two case studies of surface and subsurface monitoring using FBG sensors are then presented. The following conclusions can be drawn:

1. The newly developed FBG sensors are feasible to perform quasi-distributed and real-time monitoring of landslides regarding strain, temperature, displacement, pressure, and moisture content.
2. UWFBG enables the improvement of the multiplexing number and sensing distance of grating sensors, especially for the real-time and accurate multi-parameter measurement of a giant landslide.
3. Long time-series multi-physics monitoring data obtained by FBG sensors can produce a crucial database to relate them to the changing boundary conditions for reaching a detailed interpretation of the overall stability condition. In this process, artificial intelligence algorithms will be important tools.

ACKNOWLEDGMENTS

This research was funded by the National Natural Science Foundation of China (Grant No. 42077235) and the National Key Research and Development Program of China (Grant No. 2018YFC1505104).

REFERENCES

Cao, D.F., Shi, B., Zhu, H.H., Inyang, H.I., Wei, G.Q., Duan, C.Z.: A soil moisture estimation method using actively heated fiber Bragg grating sensors. *Engineering Geology* 242, 142–149 (2018).

Cao, D.F., Zhu, H.H., Wu, B., Wang, J.C., Shukla, S.K.: Investigating temperature and moisture profiles of seasonally frozen soil under different land covers using actively heated fiber Bragg grating sensors. *Engineering Geology* 290, 106197 (2021).

Chae, B.G., Park, H.J., Catani, F., Simoni, A., Matteo, B.: Landslide prediction, monitoring and early warning: a concise review of state-of-the-art. *Geoscience Journal* 21, 1033–70 (2017).

Hill, K. O., Fujii, Y., Johnson, D.C., Kawasaki, B.S.: Photosensitivity in optical fiber waveguides: Application to reflection filter fabrication. *Applied Physics Letters* 32(10), 647–649 (1978).

Ho, Y.T., Huang, A.B., Lee, J.T.: Development of a fibre Bragg grating sensored ground movement monitoring system. *Measurement Science & Technology* 17, 1733–1740 (2006).

Huang, A.B., Lee, J.T., Ho, Y.T., Chiu, Y.F., Cheng, S.Y.: Stability monitoring of rainfall-induced deep landslides through pore pressure profile measurements. *Soils and Foundations* 52, 737–747 (2012).

Jiang, J.P., Gan, W.B., Hu, Y., Li, S., Deng, J., Yue, L.N., Yang, Y., Nan, Q.M., Pan, J.J., Liu, F., Wang, H.H.: Real-time monitoring method for unauthorized working activities above the subway tunnel based on ultra-weak fiber Bragg grating vibration sensing array. *Measurement* 182, 109744 (2021).

Li, H.J., Zhu, H.H., Li, Y.H., et al.: Fiber Bragg grating–based flume test to study the initiation of landslide-debris flows induced by concentrated runoff. *Geotechnical Testing Journal* 44(4), 986–999 (2021).

Li, W., Zhang, J.: Distributed weak fiber Bragg grating vibration sensing system based on 3×3 fiber coupler. *Photonic Sensors* 8, 146–56 (2018).

Meltz, G., Morey, W. W., Glenn, W. H.: Formation of Bragg gratings in optical fibers by a transverse holographic method. *Optics Letters* 14(15), 823–825 (1989).

Metje, N., Chapman, D.N., Rogers, C.D.F., Henderson, P., Beth, M.: An optical fiber sensor system for remote displacement monitoring of structures-prototype tests in the laboratory. *Structural Health Monitoring* 7(1), 51–63 (2008).

Othonos, A., Kalli, K.: *Fiber Bragg Gratings: Fundamentals and Applications in Telecommunications and Sensing*. Artech House, London, UK (1999).

Othonos, A., Kalli, K., Pureur, D., Mugnier, A.: Fibre Bragg Gratings. In: Venghaus H. (eds) *Wavelength Filters in Fibre Optics*. Springer Series in Optical Sciences, 189–202. Springer, Berlin, Heidelberg (2006).

Peng, B.J., Zhao, Y., Zhao, Y., Yang, J.: Tilt sensor with FBG technology and matched FBG demodulating method. *IEEE Sensors Journal* 6, 63–66 (2006).

Qin, Y., Wang, Q., Xu, D., Yan, J., Zhang, S.: A fiber Bragg grating based earth and water pressures transducer with three-dimensional fused deposition modeling for soil mass. *Journal of Rock Mechanics and Geotechnical Engineering* 14(2), 663–669 (2021).

Rao, Y.J.: Recent progress in applications of in-fibre Bragg grating sensors. *Optical Laser Engineering* 31, 297–324 (1999).

Song, K., Wang, F.W., Yi, Q.L., Lu, S.Q.: Landslide deformation behavior influenced by water level fluctuations of the Three Gorges Reservoir (China). *Engineering Geology* 247, 58–68 (2018).

Sun, Y., Cao, S., Xu H, Zhou X.: Application of distributed fiber optic sensing technique to monitor stability of a geogrid-reinforced model slope. *International Journal of Geosynthetics and Ground Engineering* 6(2), 1–11 (2020).

Tang, H.M., Wasowski, J. Juang C.H.: Geohazards in the three Gorges Reservoir Area, China-Lessons learned from decades of research. *Engineering Geology* 261, 105267 (2019).

Wang, Y.L., Shi, B., Zhang, T.L., Zhu, H.H., Jie, Q., Sun, Q.: Introduction to an FBG-based inclinometer and its application to landslide monitoring. *Journal of Civil Structural Health Monitoring* 5 645–653 (2015).

Wu, P., Tan, D., Lin, S., Chen, W., Yin, J., Malik, N., Li, A.: Development of a monitoring and warning system based on optical fiber sensing technology for masonry retaining walls and trees. *Journal of Rock Mechanics and Geotechnical Engineering*, in press (2022).

Xu, D.S., Yin, J.H., Cao, Z.Z., Wang, Y.L., Zhu, H.H., Pei, H.F.: A new flexible FBG sensing beam for measuring dynamic lateral displacements of soil in a shaking table test. *Measurement* 46(1), 200–209 (2013).

Xu, H.B., Zheng, X.Y., Zhao, W.G., Xu, S., Li, F., Du, Y.L., Liu, B., Gao, Y.: High precision, small size and flexible FBG strain sensor for slope model monitoring. *Sensors* 19(12), 2716 (2019).

Ye, X., Zhu, H.H., Wang, J., Zhang, Q., Shi, B., Schenato, L., Pasuto, A.: Subsurface multi-physical monitoring of a reservoir landslide with the fiber-optic nerve system, *Geophysical Research Letters*, 49, e2022GL098211 (2022).

Zhang, L., Shi, B., Zhu, H., Yu, X.B., Han, H.M., Fan, X.D.: PSO-SVM-based deep displacement prediction of Majiagou landslide considering the deformation hysteresis effect. *Landslides* 18, 179–193 (2021).

Zhu, H.H., Ho, A.N.L., Yin, J.H., Sun, H.W., Pei, H.F., Hong, C.Y.: An optical fibre monitoring system for evaluating the performance of a soil nailed slope. *Smart Structures and Systems* 9, 393–410 (2012).

Zhu, H.H., Shi, B., Yan, J.F., Zhang, J., Wang, J.: Investigation of the evolutionary process of a reinforced model slope using a fiber-optic monitoring network. *Engineering Geology* 186, 34–43 (2015).

Zhu, H.H., Shi, B., Yan, J.F., Zhang, J., Zhang, C.C., Wang, B.J.: Fiber Bragg grating-based performance monitoring of a slope model subjected to seepage. *Smart Materials and Structures* 23, 095027 (2014).

Zhu, H.H., Shi, B., Zhang, C.C.: FBG-based monitoring of geohazards: current status and trends. *Sensors* 17, 452 (2017).

Chapter 7

Small and medium enterprises and sustainable development

Ayon Chakraborty
Federation University

CONTENTS

7.1 INTRODUCTION

Small and medium enterprises (SMEs) have become an important driver of growth for not only developed but also for emerging economies. In OECD countries, SMEs account for 60%–70% of employment generation (OECD, 2000). Thus, as a major employment generator, SMEs also stimulate savings, income as well as investment in the economy. Further development happens in terms of promoting markets, developing infrastructure facilities and regional economy, and growth in export potential (Vashishth et al., 2021). SMEs follow different ownership structures such as cooperative societies, partnerships or limited company, and sole proprietorship along with different forms such as factories, workshops, trading, or service companies (Kuzmisin & Kuzmisinova, 2017; Watson, 2010).

The growing importance of SMEs has also brought several challenges in the form of global competition, generating awareness related to environmental issues, and struggle due to businesses opening up globally. This global supply chain is creating vulnerable situation for SMEs due to resource constraints. SMEs in this scenario have to rethink and redesign their business models in order to respond and overcome these challenges (Jaeger & Upadhyay, 2020; Stewart, 2004; Watson, 2010). In this scenario, circular

DOI: 10.1201/9781003368335-7

economy (CE) concept will be really useful for SMEs in facing these challenges and ensuring economic growth.

Generally, economy has followed a linear pattern of production, consumption, and disposal. This linear pattern starts with industry acquiring raw materials for production in the *take phase*, which are non-renewable. Then, in the *make phase*, product is manufactured and provided to the customer in *use phase*. After the end-of-life for the product, in *throw phase*, the customer disposes the product, which is then taken to the landfills. This *take-make-use-throw* model has resulted in increased unsustainability (Thorley et al., 2019). The consumption pattern of raw materials and use of non-renewable energy sources have aggravated depletion of Earth's resources. Further, the waste management policy through landfills has adversely affected our ecology (Ghisellini et al., 2016).

In order to break away from this linear model, CE came as a new paradigm to optimize environmental, economic, and social factors of the business and transforming the entire society in becoming more sustainable by involving all the concerned stakeholders (Dey et al., 2020). Circular economy is defined as "an economic system that represents a change of paradigm in the way that human society is interrelated with nature and aims to prevent the depletion of resources, close energy and materials loops, and facilitates sustainable development through its implementation at the micro (enterprises and consumers), meso (economic agents integrated in symbiosis) and macro (city, regions and governments) levels. Attaining this circular model requires cyclical and regenerative environmental innovations in the way society legislates, produces and consumes" (Dey et al., 2020, p. 2; Ormazabal et al., 2018, p. 12). CE has been widely applied and studied in large corporations in China and European Union, but its study in SMEs is still limited (D'Amato et al., 2017). Thus, there is a need to have further research on understanding and developing frameworks to facilitate CE adoption in SMEs. Recent studies have started focusing on this aspect but more from supply chain perspective such as Meherishi et al. (2019), Kalverkamp and Young (2019), and Nasir et al. (2017). There are several other studies that have focused on supply chains with focus on drivers, enablers, barriers, challenges, practices, business models, strategies, policies, benefits, etc. (Dey et al., 2019; Govindan & Hasanagic, 2018; Govindan et al., 2016; Gupta & Palsule-Desai, 2011; Lewandowski, 2016; Saidani et al., 2019). There is still lack of research on CE adoption framework for SMEs. Thus, this study wants to overcome this gap by proposing a conceptual framework based on drivers, barriers, actions, and practices to facilitate CE adoption for SMEs.

7.2 THEORETICAL BACKGROUND

The Ellen McArthur foundation propose adoption of six actions by businesses for transitions to CE. They term it as ReSOLVE (Regenerate, Share, Optimize, Loop, Virtualize and Exchange) framework. This framework promotes CE by reducing the overall demand for products from a physical ownership to sharing, virtualizing, or exchanging products (MacArthur, 2013; The Circular Economy In Detail, n.d.). Similarly, they also focus on extending the life of a product and devising ways in which a product at the end of its life can be re-utilized as raw material in creating new products. Circular economy is a circular system where waste, whether physical or in energy form, is re-utilized in a new creation cycle, rather than finding its way to a landfill. This reduced waste helps prevent environmental degradation in terms of pollution as well as unsustainable drain on environment resources. Creating a safer environment would result in a healthier society – and shifting consumption models from ownership to shared use would reduce the divide between the haves and have nots – thus improving overall societal health (Cheshire, 2019; MacArthur, 2013).

While both CE and sustainability focus on the balance between the triple bottom lines, CE places the economic contributions at its core while trying to meet environmental and societal benefits. This places private businesses and policy makers to steer policy and implementation (Merli et al., 2018). For these stakeholders, sustainable resource consumption and environmental protection are thrust areas as they help them achieve superior economic or financial advantage. On the other hand, sustainability, as a concept, places all three, i.e., economic, environment, and society at its core (Geissdoerfer et al., 2017; Rizos et al., 2017; Türkeli et al., 2018). This results in a dilution of responsibilities as it becomes difficult to achieve an interest alignment between the potential stakeholders. Thus, achieving sustainable development through CE modes holds greater promise (Geissdoerfer et al., 2017).

Despite available knowledge about CE, its adoption is still limited in SMEs. This is surprising given SMEs being backbone of both developed and emerging economies and plays a significant role in employment generation and economic development (Cantú et al., 2021a; Ormazabal et al., 2018). SMEs mostly represent small enterprises. They have been defined in different ways based on number of employees, annual revenue generated, and value of fixed assets in financial settings. According to the World Bank, SMEs are identified based on annual sales, asset size, and employee numbers (refer Table 7.1).

In most of the countries, SMEs are defined based on their employee numbers, which signifies that SMEs are smaller replica of larger corporations. So, in practice, researchers have to contextualize the definition of SMEs based on the focus of their study (Jones et al., 2007). SMEs are mostly

Table 7.1 SME criteria in general

Size of organization	Average employee numbers	Total assets ($)	Annual turnover ($)
Micro	0–10	Less than 10,000	Less than 10,000
Small	11–50	Less than 3,000,000	Less than 3,000,000
Medium	51–299	Less than 15,000,000	Less than 15,000,000

categorized as small and medium enterprises. In some contexts, microenterprises are also incorporated. All these different categories are significantly heterogeneous as they depend on the type of industry, ownership structure, stage of development, and area of operation (Canevari-Luzardo, 2019; Kato & Charoenrat, 2018). Type of ownership structure is an important consideration in the categorization of SMEs as this decides on the type of funding or capital in the form of equity or debt the enterprises can avail from external sources. For example, firms with proprietorship or partnership cannot receive external equity, and this restricts growth prospects both at start-up and expansion stage (Storey, 2016). One of the most widely used ownership structures is proprietorship due to its limited legal overhead requirements. This is followed by structures such as partnership, cooperative, and public or private limited companies (Jones, 2007). The research in SMEs is considered multidisciplinary, and several disciplines are reinforcing their claim toward conceptual underpinning and developing research approaches in this field (Landström & Johannisson, 2001). As the research on SMEs is approaching toward maturity, more theory is getting generated with research focusing primarily on the aspects of entrepreneurship, which, on the other hand, is limiting its practical application. Given the applied nature of CE, it is important to develop a framework which will be applicable and assist SMEs in adoption of CE.

Welsh and White (1991) suggested that SMEs are not *little big business* and so there are lots of considerations to be made before prescribing implementation of any new initiatives such as CE. These differences are clearly visible in the decisions related to the way of making policies, organizational structure, and resource availability (Jones, 2007; Welsh & White, 1991). Further, SMEs also differ from large organizations in terms of competition, uncertainties in demand, and also issue due to cash flow, skill shortage, and high employee turnover (Dey et al., 2020). So, SMEs mostly focus on economic performance rather than on environment or social performance. Thus, while going for CE adoption, SMEs face numerous challenges and understanding the differences with large enterprises helps in developing a framework aligned with the requirements of SMEs.

The literature on CE adoption is primarily focused on challenges or barriers, followed by drivers and enablers, and practices. Research by Ormazabal et al. (2018) highlighted several challenges faced by SMEs such as lack of financial support, lack of technical and financial resources, inadequate technology support, and lack of structure for proper information management. There are also lack of support internally (top management) and externally (government policy, public institutions) as well as absence of knowledgeable environment professional within the organization.

Along with the challenges, there are also several benefits and opportunities for SMEs through CE adoption. Some benefits and opportunities highlighted in the literature are in the form of better image, cost savings, growth

Table 7.2 Actions and strategies

Actions	Strategies
Take	Raw material and supplier selection with green image; avoiding use of toxic materials; transparency in process and product; take-up use of sustainable and fully recoverable materials
Make	Employee training; reducing environmental impact; applying sustainable energy sources; avoiding environmental damages, eco-design; and implementing zero waste production processes
Distribute	Optimization of stocks, routes, and space for both forward and reverse logistics; develop collaborative initiatives from all the concerned stakeholders
Use	Communicating with customers about sustainability initiatives such as zero waste certification and eco-labeling; green marketing strategy; market segmentation; product system service implementation
Recover	Effective adoption of CE through appropriate synergy among different stakeholders such as supply chain partners and policy makers

Source: Author's Conception: Adapted from Dey et al. (2020); Ormazabal et al. (2018)

in business, better productivity, becoming environment friendly through better waste management, and thus achieving greater sustainability (Rizos et al., 2017; Thorley et al., 2019). There are also several strategies that can be considered along with the actions for CE adoption in SMEs. Table 7.2 summarizes the strategies and actions as suggested in the literature.

7.2.1 Research gaps

The review highlighted the following research gaps:

a. There is considerable interest among researchers about CE adoption in supply chain in general but specific focus on SMEs is missing.
b. There are several government policies toward CE adoption, but they are not contextualized toward SMEs. So, there is a need for more focused regulations and policies to facilitate CE adoption in SMEs.
c. The frameworks around enablers, barriers, practices, strategies, and business models are more aligned with CE adoption in supply chain but concentrated efforts in developing a conceptual framework for CE adoption in SMEs is still the missing element.

In summary, CE adoption in SMEs is mainly dependent on the drivers and/or enablers, overcoming barriers and/or challenges, incorporating various strategies and practices to achieve sustainable performance from environmental, economic, and social aspects. Thus, this chapter – through literature review of drivers/enablers, barriers/challenges, strategies, actions, and practices – proposes and develops a framework to facilitate CE adoption in SMEs.

7.3 METHODOLOGY

A systematic literature review approach has been employed. The review included articles for the last two decades. The articles were searched in two databases: Scopus and Web of Science. It was observed that there were more articles in Web of Science than in Scopus. Share of the articles from both the databases is provided in Figure 7.1. The review also showed that there is growing interest in the studies on CE adoption in supply chains in general, with specific focus on SMEs. There is a growing trend seen in the number of articles from 2000 to 2019 with a slight drop in 2020 (refer Figure 7.2).

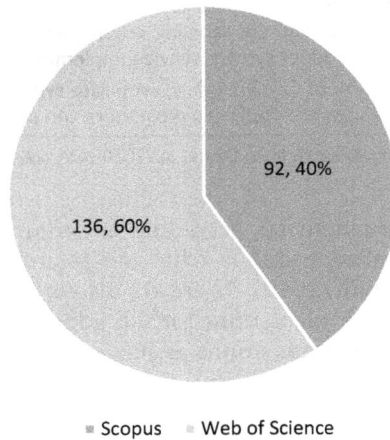

92, 40%

136, 60%

▪ Scopus   ▪ Web of Science

Figure 7.1 Article share in two databases.

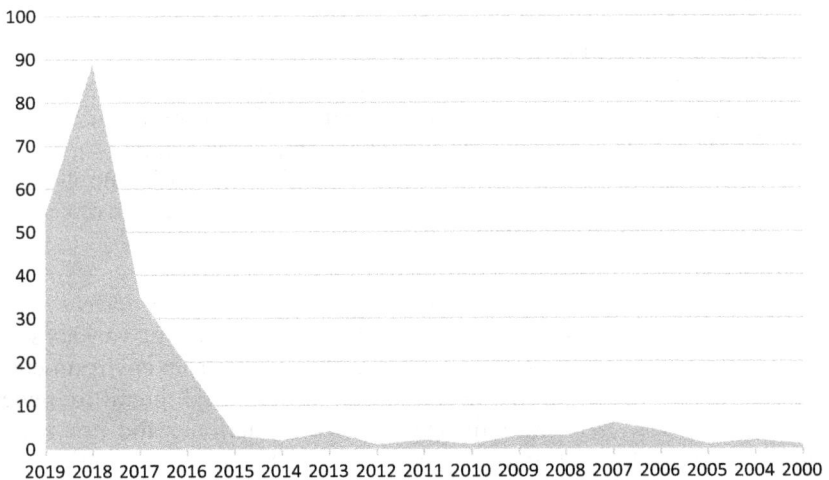

Figure 7.2 Circular economy research article trend.

7.4 CONCEPTUAL FRAMEWORK

A conceptual framework helps in providing a direction to the research problem by identifying the research variables and also clarifying the relationship between the variables. Conceptual framework mostly *"sets the stage"* for further investigation of the research problem (McGaghie et al., 2001). In case of CE adoption for SMEs, the variables that can be considered are drivers or enablers, challenges or barriers, strategies, opportunities, and practices.

7.4.1 Life-cycle stages and Implementation approaches

From SME perspective, organizational growth and development have been raising interest among organizational scholars for decades (Hanks et al., 1994; Quinn & Cameron, 1983). The research by Hanks et al. (1994) has shown that SMEs success progressively as they mature and develop through the life-cycle stages. So, life-cycle stage of a SME is an important consideration while focusing on initiatives such as CE. Further, as has been observed in quality management literature, adoption or implementation of any initiative goes through different phases. It has also been observed in case of Quality Management Standards that the benefits derived by two SMEs are not identical (Vashishth et al., 2021). If results derived due to implementation of standards are not identical for two firms, then it can be safely assumed that the benefits derived through CE adoption will vary between SMEs significantly. So, borrowing from quality management literature, the implementation approaches are discussed, which will be beneficial to understand while developing a framework for CE adoption in SMEs.

7.4.1.1 Reactive approach

In this approach, the SMEs follow a mechanistic approach toward adoption of new initiatives. They feel it as an unnecessary burden and just follow the norms just to maintain customers. In case of ISO14001, it has been observed that SMEs following this approach have a scant consideration toward environmental impact (Vashishth, 2019). The SMEs in this category can be termed as *minimalist* as they make enough changes to just comply with the existing environmental standards (Sammalisto, 2001; Tilley, 1999).

7.4.1.2 Coactive approach

The approach is again driven primarily by customer requirements. The certification or initiatives are mostly at a level where compliance is followed with some training and education. The training to the employees is mostly informative rather than equipping them with tools and techniques (Sammalisto, 2001). This shows that the firms do not take this as a learning process and thus failing to utilize the full potential of the certifications or initiatives.

7.4.1.3 Process-oriented approach

The SMEs following this approach look at certifications or initiatives as a tool and also focus on overall employee training to change their orientation toward initiatives (Sammalisto, 2001). This approach is more management motivated, and SME itself initiates the effort by responding to the expected customer demands and trends. The staff involvement is more than just fulfilling day-to-day work. Management involvement and continuous improvement are two main features of this approach (Dolorems Oreno-Luzon, 1993). The chances of high commitment toward CE will be possible if environmental vision is inculcated within the firm. The employees of such firms will be engaged in better environmental practices as they feel belonging to the company as a family (Vashishth, 2019).

The life-cycle study of SMEs along with the implementation approaches provides a holistic understanding of how SMEs at different stages of life cycle and with particular approaches toward implementing an initiative might behave. This will help both researchers and practitioners to develop directions for SMEs while focusing on CE adoption. Thus, life-cycle stages and implementation approaches (refer) form two important building blocks of the framework toward CE adoption for SMEs.

7.4.2 Enablers/drivers, challenges/barriers, practices, and strategies

The research on CE adoption has seen several articles in last few years focusing on enablers/drivers, challenges/barriers, practices, and strategies. As larger organizations were mainly providing the impetus toward CE adoption for SMEs, so most of the articles focus on whole of supply chain rather than SMEs (Sohal et al., 2022). Further, the articles that focused on

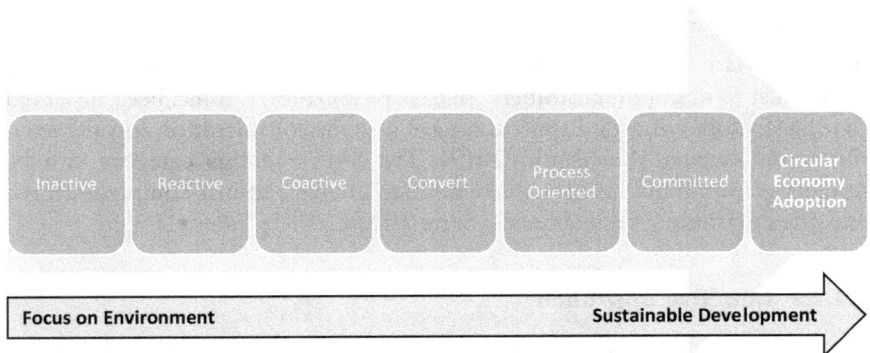

| Inactive | Reactive | Coactive | Convert | Process Oriented | Committed | Circular Economy Adoption |

Focus on Environment   **Sustainable Development**

Figure 7.3 Implementation approaches toward circular economy.

Source: Author's Conception: Adapted from Brown & van der Wiele (1996); Sammalisto (2001) and Vashishth (2019).

SMEs are mostly on a particular sector such as textile, paper, food, construction, etc. There is still very limited focus on SME-specific literature on CE adoption.

In the last five years, there are around 22 articles that have discussed about drivers, enablers, success factors, or motivators. In this chapter and the framework, the term "driver" is used instead of other terms for maintaining uniformity in the discussion. These articles mostly focused on supply chains as a whole; but many of the drivers discussed will be equally applicable to SMEs. The drivers are also classified into different categories or clusters by several researchers such as artificial intelligence, policy framework, manufacturing ecosystem, network agility, and self-automation (Rajput & Singh, 2019); economic, social, and environmental (Salim et al., 2019); policy and economy, health, environmental protection, society, and product development (Govindan & Hasanagic, 2018). These classifications are done on the basis of either technology, sustainability, or supply chain perspective.

Similar to drivers, the term "barriers" will be used to signify obstacles, challenges, or issues that are highlighted in the literature. There are classification schemes also suggested for barriers such as external and internal (Dey et al., 2019); economic and financial viability, market and competition, product characteristics, standards and regulation, supply chain management, technology, and user's behavior (Bressanelli et al., 2019; Sohal & De Vass, 2022); interface designing, technology upgradation, and synergy model (Rajput & Singh, 2019); collection of used products, manufacturing level, and enterprise level (Shi et al., 2019); policy and economy, social, market, environment, and infrastructure (Salim et al., 2019); and government, economic, technological, knowledge and skill, market, and CE framework (Govindan & Hasanagic, 2018).

There are not many articles on practices, and one of the most prominent reviews is by Govindan and Hasanagic (2018). They did a review about CE adoption from supply chain perspective and categorized the practices into eight clusters. The clusters suggested are government initiatives, economic initiatives, cleaner production, product development, management support, infrastructure, knowledge, and social and culture. The CE actions are take, make, distribute, use, and recover (Dey et al., 2020; MacArthur, 2013; Ormazabal et al., 2018). Table 7.3 summarizes the drivers, barriers, and practices from the perspective of CE adoption in SMEs. These drivers, barriers, and practices are synthesized from the existing literature and are vital part of our conceptual framework.

Based on the discussion so far, it can be observed that the literature is still limited about CE adoption in SMEs. Thus, through this work, there is an attempt to develop a conceptual framework that can facilitate CE implementation in SMEs. The life-cycle stages and implementation approach by SMEs need to be considered along with the drivers, barriers, practices, and actions for CE adoption leading to sustainable development. The proposed framework is presented in Figure 7.4.

Table 7.3 Drivers, barriers, and practices for CE adoption in SMEs

Drivers	Using organization's resources efficiently, increasing the products' value by increasing the quality, collaborating with stakeholders (NGOs, governments) and within the supply chain, support for employee training and personal development, using environment-friendly products, better information sharing, leadership, and commitment from top management
Barriers	Difficulty in measuring CE benefits due to lack of tools and methods, lack of clear guidelines to define sustainability, lack of regulations for intellectual property protection, insufficient implementation of circular economy laws, lack of technology and technical skills, limited innovation capacity, lack of top management commitment, lack of environmental education within the enterprise, lack of business models and frameworks for CE implementation in SMEs, sustainability is perceived as cost and not as an investment, high production cost, lack of market mechanism for recovery
Practices adapted from (Govindan & Hasanagic, 2018)	Pilot projects for CE, performance indicators on recycling, reuse and remanufacture, increase environmental accounting, proper costing of the product, increase eco-efficiency in production, sustainable purchasing of raw materials, develop metrics to measure environmental performance, appropriate support from top management for CE implantation, sustainable infrastructure development, developing efficient information management system, providing training and education to employees regarding CE

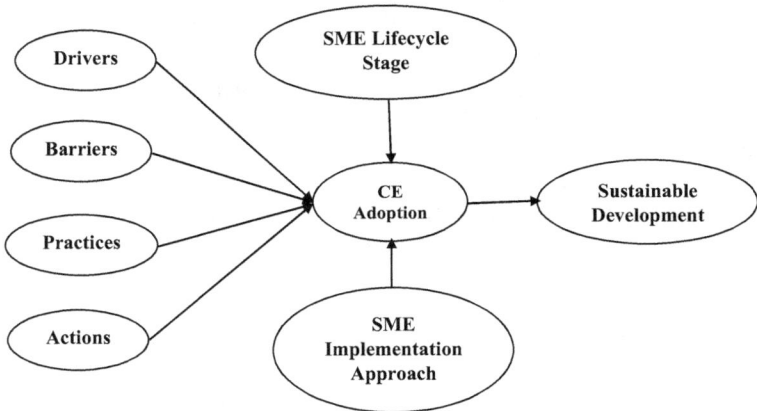

Figure 7.4 Conceptual framework for CE adoption in SMEs.

7.5 CONCLUSION

The purpose of this study was to explore the existing body of work and under-stand the necessary ingredients that is required for successful CE adoption by the SMEs. Literature suggests various drivers, barriers, practices, strategies, and actions that need to be understood in the context of CE adoption in organizations. There is still limited literature about CE adoption in SMEs. In

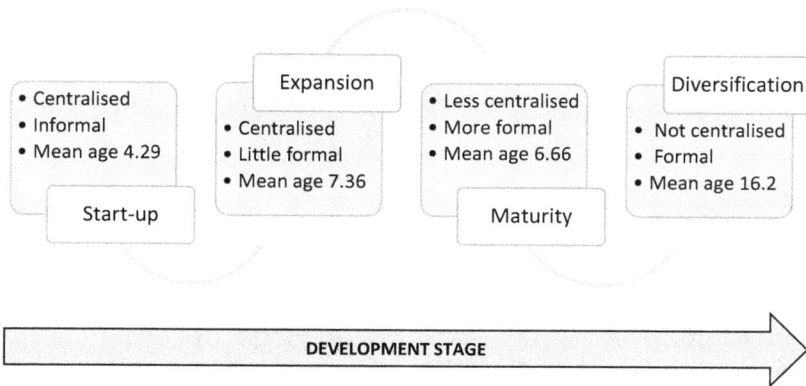

Figure 7.5 Life Cycle Stages of SMEs.

Source: Author's Conception: Adapted from enterprise life-cycle model (Hanks et al., 1994).

other words, CE adoption in SMEs is not well explored. Thus, through the exploration of literature, the study comes out with a conceptual framework, which will facilitate CE adoption in SMEs. Consideration of life-cycle stages of SMEs as well as implementation approaches borrowed from quality management literature brings in a perspective, which is not yet explored while developing frameworks. While working with SMEs, it is important to understand at what stage of life cycle (refer Figure 7.5) they are in as that will help in deciding the proper CE adoption approach to be applied.

Along with life-cycle stages and implementation approaches, focus on appropriate drivers such as top management commitment, and proper government policies and procedures will be important considerations for SMEs while adopting CE (Govindan et al., 2016; Mura et al., 2020). There is also a need to understand the challenges that will arise due to lack of technology know-how, and also lack of training related to sustainability. There is also an absence of culture, related to environment management and certifications, making it further difficult for SMEs to go forward with the adoption of CE (Cantú et al., 2021b; Vashishth et al., 2021). Finally, aligning practices with strategies and actions will help SMEs to focus on each stage of take, make, distribute, use, and recover and align them with environmental, economic, and social aspect of sustainable development. Study by Dey et al. (2020) suggests that SMEs are generally focused on economic contribution across all CE fields of action but the same is not applicable for environment and social contributions. They conclude that there is a huge scope of aligning environment and social contributions with CE fields of action.

The proposed framework needs to be further tested and investigated empirically. Further analysis and exploration can be done on the constructs of drivers, barriers, practices, actions, life-cycle stages, and implementation approaches. Country- or industry sector-specific further research can be conducted using this framework. There are possibilities of combining other theoretical frameworks from another field such as the one used in this study from quality management literature. Finally, further exploration through qualitative case studies can be done to see the validity and applicability of the framework in the real-world scenario.

REFERENCES

Bressanelli, G., Perona, M., & Saccani, N. (2019). Challenges in supply chain redesign for the Circular Economy: a literature review and a multiple case study. *International Journal of Production Research*, 57(23), 7395–7422. https://doi.org/10.1080/00207543.2018.1542176

Brown, A., & van der Wiele, T. (1996). A typology of approaches to ISO certification and TQM. *Australian Journal of Management*, 21(1), 57–72. https://doi.org/10.1177/031289629602100107

Canevari-Luzardo, L. M. (2019). Value chain climate resilience and adaptive capacity in micro, small and medium agribusiness in Jamaica: a network approach. *Regional Environmental Change*, 19(8), 2535–2550. https://doi.org/10.1007/s10113-019-01561-0

Cantú, A., Aguiñaga, E., & Scheel, C. (2021a). Learning from failure and success: the challenges for circular economy implementation in SMEs in an emerging economy. *Sustainability (Switzerland)*, 13(3), 1–34. https://doi.org/10.3390/su13031529

Cantú, A., Aguiñaga, E., & Scheel, C. (2021b). Learning from failure and success: the challenges for circular economy implementation in SMEs in an emerging economy. *Sustainability (Switzerland)*, 13(3), 1529–1564. https://doi.org/10.3390/su13031529

Cheshire, D. (2019). What is a circular economy? In *Building Revolutions* (1st ed.). Routledge. https://doi.org/10.4324/9780429346712-2

D'Amato, D., Droste, N., Allen, B., Kettunen, M., Lähtinen, K., Korhonen, J., Leskinen, P., Matthies, B. D., & Toppinen, A. (2017). Green, circular, bio economy: a comparative analysis of sustainability avenues. *Journal of Cleaner Production, 168*, 716–734. https://doi.org/10.1016/j.jclepro.2017.09.053

Dey, P. K., Malesios, C., De, D., Budhwar, P., Chowdhury, S., & Cheffi, W. (2020). Circular economy to enhance sustainability of small and medium-sized enterprises. *Business Strategy and the Environment*, 29(6), 2145–2169. https://doi.org/10.1002/bse.2492

Dey, P. K., Malesios, C., De, D., Chowdhury, S., & Abdelaziz, F. Ben. (2019). Could lean practices and process innovation enhance supply chain sustainability of small and medium-sized enterprises? *Business Strategy and the Environment*, 28(4), 582–598. https://doi.org/10.1002/bse.2266

Dolorems Oreno-Luzon, M. (1993). Can total quality management make small firms competitive? *Total Quality Management*, 4(2), 165–182. https://doi.org/10.1080/09544129300000027

Geissdoerfer, M., Savaget, P., Bocken, N. M. P., & Hultink, E. J. (2017). The circular economy – A new sustainability paradigm? In *Journal of Cleaner Production* (pp. 757–768). https://doi.org/10.1016/j.jclepro.2016.12.048

Ghisellini, P., Cialani, C., & Ulgiati, S. (2016). A review on circular economy: the expected transition to a balanced interplay of environmental and economic systems. *Journal of Cleaner Production, 114*, 11–32. https://doi.org/10.1016/j.jclepro.2015.09.007

Govindan, K., & Hasanagic, M. (2018). A systematic review on drivers, barriers, and practices towards circular economy: a supply chain perspective. *International Journal of Production Research, 56*(1–2), 278–311. https://doi.org/10.1080/00207543.2017.1402141

Govindan, K., Jha, P. C., & Garg, K. (2016). Product recovery optimization in closed-loop supply chain to improve sustainability in manufacturing. *International Journal of Production Research, 54*(5), 1463–1486. https://doi.org/10.1080/0207543.2015.1083625

Gupta, S., & Palsule-Desai, O. D. (2011). Sustainable supply chain management: review and research opportunities. *IIMB Management Review, 23*(4), 234–245. https://doi.org/10.1016/j.iimb.2011.09.002

Hanks, S. H., Watson, C. J., Jansen, E., & Chandler, G. N. (1994). Tightening the life-cycle construct: a taxonomic study of growth stage configurations in high-technology organizations. *Entrepreneurship Theory and Practice, 18*(2), 5–29. https://doi.org/10.1177/104225879401800201

Jaeger, B., & Upadhyay, A. (2020). Understanding barriers to circular economy: cases from the manufacturing industry. *Journal of Enterprise Information Management, 33*(4), 729–745. https://doi.org/10.1108/JEIM-02-2019–0047

Jones, O. (2007). Understanding the small business sector. *International Journal of Entrepreneurial Behaviour & Research, 13*(6), 36–38. https://doi.org/10.1108/ijebr.2007.16013faa.001

Jones, O., Macpherson, A., Thorpe, R., & Ghecham, A. (2007). The evolution of business knowledge in SMEs: conceptualizing strategic space. *Strategic Change, 16*(6), 281–294. https://doi.org/10.1002/jsc.803

Kalverkamp, M., & Young, S. B. (2019). In support of open-loop supply chains: expanding the scope of environmental sustainability in reverse supply chains. *Journal of Cleaner Production, 214*, 573–582. https://doi.org/10.1016/j.jclepro.2019.01.006

Kato, M., & Charoenrat, T. (2018). Business continuity management of small and medium sized enterprises: evidence from Thailand. *International Journal of Disaster Risk Reduction, 27*, 577–587. https://doi.org/10.1016/j.ijdrr.2017.10.002

Kuzmisin, P., & Kuzmisinova, V. (2017). Small and medium-sized enterprises in global value chains. *Economic Annals-XXI, 162*, 22–27. https://doi.org/10.21003/ea.V162–05

Landström, H., & Johannisson, B. (2001). Theoretical foundations of Swedish entrepreneurship and small-business research. *Scandinavian Journal of Management, 17*(2), 225–248. https://doi.org/10.1016/S0956–5221(99)00030–5

Lewandowski, M. (2016). Designing the business models for circular economy-towards the conceptual framework. *Sustainability (Switzerland), 8*(1), 1–28. https://doi.org/10.3390/su8010043

MacArthur, E. (2013). Towards the circular economy: opportunities for the consumer goods sector. *Ellen MacArthur Foundation*, 1–112. https://ellenmacarthurfoundation.org/towards-the-circular-economy-vol-2-opportunities-for-the-consumer-goods

McGaghie, W. C., Bordage, G., & Shea, J. A. (2001). Problem statement, conceptual framework, and research question. *Academic Medicine, 76*(9), 923–924. https://doi.org/10.1097/00001888–200109000–00021

Meherishi, L., Narayana, S. A., & Ranjani, K. S. (2019). Sustainable packaging for supply chain management in the circular economy: a review. *Journal of Cleaner Production, 237,* 1175–1182. https://doi.org/10.1016/j.jclepro.2019.07.057

Merli, R., Preziosi, M., & Acampora, A. (2018). How do scholars approach the circular economy? A systematic literature review. *Journal of Cleaner Production, 178,* 703–722. https://doi.org/10.1016/j.jclepro.2017.12.112

Mura, M., Longo, M., & Zanni, S. (2020). Circular economy in Italian SMEs: a multi-method study. *Journal of Cleaner Production, 245,* 118821. https://doi.org/10.1016/j.jclepro.2019.118821

Nasir, M. H. A., Genovese, A., Acquaye, A. A., Koh, S. C. L., & Yamoah, F. (2017). Comparing linear and circular supply chains: a case study from the construction industry. *International Journal of Production Economics, 183,* 443–457. https://doi.org/10.1016/j.ijpe.2016.06.008

OECD. (2000). *OECD Small and Medium Enterprise Outlook.* OECD, Head of Publications Service, Paris.

Ormazabal, M., Prieto-Sandoval, V., Puga-Leal, R., & Jaca, C. (2018). Circular Economy in Spanish SMEs: challenges and opportunities. *Journal of Cleaner Production, 185,* 157–167. https://doi.org/10.1016/j.jclepro.2018.03.031

Quinn, R. E., & Cameron, K. (1983). Organizational life cycles and shifting criteria of effectiveess: some preliminary evidence. *Management Science, 29*(1), 33–51. https://doi.org/10.1287/mnsc.29.1.33

Rajput, S., & Singh, S. P. (2019). Connecting circular economy and industry 4.0. *International Journal of Information Management, 49,* 98–113. https://doi.org/10.1016/j.ijinfomgt.2019.03.002

Rizos, V., Tuokko, K., & Behrens, A. (2017). The circular economy, a review of definitions, processes and impacts. *Centre for European Policy Studies* (Brussels, Belgium).

Saidani, M., Yannou, B., Leroy, Y., Cluzel, F., & Kendall, A. (2019). A taxonomy of circular economy indicators. *Journal of Cleaner Production, 207,* 542–559. https://doi.org/10.1016/j.jclepro.2018.10.014

Salim, H. K., Stewart, R. A., Sahin, O., & Dudley, M. (2019). Drivers, barriers and enablers to end-of-life management of solar photovoltaic and battery energy storage systems: a systematic literature review. *Journal of Cleaner Production, 211,* 537–554. https://doi.org/10.1016/j.jclepro.2018.11.229

Sammalisto, K. (2001). *Developing TQEM in SMEs - Management Systems Approach.* Lund University.

Shi, J., Zhou, J., & Zhu, Q. (2019). Barriers of a closed-loop cartridge remanufacturing supply chain for urban waste recovery governance in China. *Journal of Cleaner Production, 212,* 1544–1553. https://doi.org/10.1016/j.jclepro.2018.12.114

Sohal, A., & De Vass, T. (2022). Australian SME's experience in transitioning to circular economy. *Journal of Business Research, 142,* 594–604.

Sohal, A., Nand, A. A., Goyal, P., & Bhattacharya, A. (2022). Developing a circular economy: an examination of SME's role in India. *Journal of Business Research, 142,* 435–447.

Stewart, K. S. (2004). The relationship between strategic planning and growth in small businesses. *ProQuest Dissertations and Theses.*

Storey, D. J. (2016). Understanding the small business sector. *Understanding The Small Business Sector*. https://doi.org/10.4324/9781315544335

The Circular Economy In Detail. (n.d.). Retrieved November 27, 2020, from https://www.ellenmacarthurfoundation.org/explore/the-circular-economy-in-detail

Thorley, J., Garza-Reyes, J. A., & Anosike, A. (2019). The circular economy impact on small to medium enterprises. *WIT Transactions on Ecology and the Environment*, *231*, 257–267. https://doi.org/10.2495/WM180241

Tilley, F. (1999). The gap between the environmental attitudes and the environmental behaviour of small firms. *Business Strategy and the Environment*, *8*(4), 238–248. https://doi.org/10.1002/(SICI)1099-0836(199907/08)8:4<238::AID-BSE197>3.0.CO;2–M

Türkeli, S., Kemp, R., Huang, B., Bleischwitz, R., & McDowall, W. (2018). Circular economy scientific knowledge in the European Union and China: a bibliometric, network and survey analysis (2006–2016). *Journal of Cleaner Production*, *197*, 1244–1261. https://doi.org/10.1016/j.jclepro.2018.06.118

Vashishth, A. (2019). *Integrated Management System Implementation in Indian SMEs*. IIM Tiruchirappalli.

Vashishth, A., Chakraborty, A., Gouda, S. K., & Gajanand, M. S. (2021). Integrated management systems maturity: drivers and benefits in Indian SMEs. *Journal of Cleaner Production*, *293*, 126243. https://doi.org/10.1016/j.jclepro.2021.126243

Watson, R. (2010). Small and medium size enterprises and the knowledge economy: assessing the relevance of intangible asset valuation, reporting and management initiatives. *Journal of Financial Regulation and Compliance, 18*(2), 131–143. https://doi.org/10.1108/13581981011033998

Welsh, J., & White, J. (1991). A small business is not a little big business. *Harvard Business Review*, *59*(4), 18–32.

Chapter 8

Application of GIS techniques and environmental flow norms for sustainable mitigation measure for hydropower impact on ecosystem

C Prakasam
Assam University Diphu Campus, Assam

Manvi Kanwar
Chitkara University

Saravanan R
Ecofirst Services Limited, Tata Enterprise, Bangalore, Karnataka

Aravinth R
Bharathi Vidyapeeth University

Varinder Singh Kanwar
Chitkara University

M K Sharma
National Institute of Hydrology

CONTENTS

DOI: 10.1201/9781003368335-8

123

8.1 INTRODUCTION

Hydropower is one of the important widely recognized sustainable sources of energy like solar energy, wind energy, etc. Furthermore, the hydropower plant is considered a flawless, inexhaustible, and non-polluting source of energy. The understanding of the importance of the environmental flow is still scarce in many nations. The definition of the environmental flow is not remarkable yet. The definition changes based on their usage. The source has higher effectiveness in terms of management and financial prevalence for a longer period. In Himachal Pradesh, India, the accessibility of hydropower is massive and is highly dependent upon the perineal nature of the riverine system. The Beas River basin has an increasing and enormous potential to produce hydropower up to 20,000 MW as per the reports (State of India's Rivers for India Rivers Week, 2016). Intending to produce hydropower, the Himachal Government has installed several projects in the Beas River basin. However, the Government often overlooks the impact created by the hydropower project construction and operation upon the surrounding environment. To highlight and reduce the impact, the law passed by the National Green Tribunal (NGT) recommends that "15% of the average lean season flow should be maintained as the environmental flow". Environmental flow is termed as the minimum flow that has to be sustained or released downstream of the hydropower project to maintain the health of the ecosystem. Therefore, the law/or norms for maintaining environmental flow have become a big concern because of the hydropower impact. In the study area, the Larji Hydropower Project at the Pandoh Dam is located downstream. To the downstream of the Pandoh Dam, the Patikari hydropower dam site has been constructed. Therefore, the point of maintaining the ecological flow is questionable keeping this construction in view. In addition, various other factors contribute significantly to the potential hydropower project site. Another question mark is the amount of environmental flow to be maintained. The value changes with respect to various factors.

In this research work, the delineation of the potential sites involves providing weights to the various parameters determined from the DEM. For that purpose, various works of literature have been studied related to groundwater, aquifer, hydropower, and rainwater harvesting potential zone mapping. There are a number of studies on identifying a suitable location for hydropower generation. For example, Tarife et al. (2017) modeled suitable hydrological sites using GIS applications using meteorological and topographical data. The soil, DEM, weather data, land use/land cover (LULC), and watershed boundary were used as inputs. They showed about 62% as potential site micro-hydroelectric projects (5–100 kW) and 38% less than 5 kW for constructing the hydropower project. Larentis et al. (2010) delineated the potential hydropower sites using the regional streamflow data and using the GIS-based computational program called Hydrospot. This Hydrospot helps in identifying more probable sites along with the river than the traditional survey method for spotting different hydropower sites. Similarly, Sahu and Prasad (2018) analyzed the Digital Elevation Model for Hasdeo Bango Hydroelectric Project in Korba district, Chhattisgarh, to create watershed (fill, direction, accumulation map), stream network, and hydrological data (runoff). Lakshmi et al. (2018) have stated that the significant parameters for suitable site selections are soil type, LULC, topography, contour, rainfall data, and availability of water in the river. The satellite and other data sources can also be used to select small hydropower plants. Saha et al. (2018) attempted to identify a suitable location for water harvesting structures. They spotted a suitable site in Mandri watershed, Chhattisgarh, for water resource development such as gabion, gully plug, check dams, boulder check, bori bandhan, and stop dams using GIS and remote sensing techniques. The procedure followed the multi-criteria decision analysis (MCDA) and an analytical hierarchy process (AHP). Iftikhar et al. (2016) incorporated geological maps, discharge data, rainfall, Landsat 5, and ASTER Global DEM of Swat valley to evaluate the feasibility of selected locations to overcome the energy crisis, water shortage, and natural misfortunes that reduced the agricultural efficiency in Pakistan. Jasrotia et al. (2019) used the aquifer and terrain characteristics to delineate the artificial recharge zones for the Northern Western Himalayas. The suitable sites were delineated using groundwater modeling and geospatial technologies. Haque et al. (2020) and Khan et al. (2020) used different thematic inventory of hydrology, geology, and geomorphological layers using geospatial technologies to delineate the groundwater resources zone. The resultant zone was validated using the topography data. The use of remote sensing techniques paves the way toward a cost-effective survey for groundwater zonation mapping. Gnanachandrasamy et al. (2018) for the groundwater potential zone identification used thematic layers such as drainage, geology, drainage density, slope, and rainfall with the inverse distance weightage (IDW) methods. Goyal et al. (2015) reported that with the increasing energy demand, the demand for constructing small hydropower projects also increases. Overall, from the literature review above, it is observed that with the advent of the GIS and remote sensing technique, the identification of hydropower project sites has become a time and cost-effective. Wang et al. (2014) reported that the use of RS and GIS enhanced our capability to explore the water resources based on satellite data. Singh et al. (2017) and Sarkar et al. (2020) reported that the zonation map may help in decision-making in the planning of regional renewable energy by incorporating data on regional potentials and restrictions to various stallholders. Samanta and Aiau (2015) reported that the advantage of weighted overlay analysis is that the user can specify the relative weights as percentages, decimals, or relative weightings. Sammartano et al. (2019) carried out the detection of the run of river plant potential location by coupling the hydrological model and GIS techniques. This could effectively improve the process of delineating the hydropower sites (Jena et al. 2020). The thematic layer's lineament density and drainage density are least sensitive to the potential zone identification. The layers such as soil and LULC show higher sensitivity to the groundwater potential zonation. Nithya et al. (2019) selected eight influencing factors, which were analyzed using the AHP technique. The GIS-based AHP method is effectively performed for the identification of groundwater potential zone in the hard rock terrain, and the information sources can be used for planning sustainable groundwater management. Al-Ruzouq et al. (2019) stated that the process involves three stages. The first stage is the collection of raw data. Processing the raw data into various thematic layers would be the second stage, and modeling the thematic maps be the third stage. Kumar et al. (2008) used remote sensing and GIS techniques with different decision-making strategies, like fuzzy logic, AHP, weighted overlay analysis, and AI, for the flexibility of incorporating spatial information (Chen et al. 2013). The thematic layers were reclassified into five or ten classes to standardize the units of each thematic layer using the Jenks technique. Hassan et al. (2020) integrated the AHP with the GIS technique and weighted overlay analysis for the land suitability analysis in the Jammu and Kashmir region. The results show that the AHP and GIS integration is efficient.

Present research reviews the various minimal flow norms and regulations followed by various countries and in India. The GIS application has been coupled with this research work to find the potential location for constructing the hydropower project. The results from these two objectives will be straight down to an optimal environmental framework pertaining to the environmental conditions and other factors for the study area.

8.2 STUDY AREA

The Beas Basin starts from the Indus River with a hydropower potential of 5,995 MW with a catchment area of 20,303 km². It is 460 km in total length, originates from Rohtang Pass, flows from Larji to Talwara through a valley, and mixes into the Sutlej at Harike, Punjab. About 9,125 MW of major hydropower projects (>25 MW) and 61.8 MW of small hydropower projects (<25 MW) are under operation in this basin. Under the Himachal Pradesh State Electricity Board Ltd, Bhakra Beas Management Board, and Punjab State Power Corporation Limited, nine major hydropower projects are up and running. About ten small hydropower projects are under operation with less than 25 MW capacity. There are 18 hydropower projects under construction and still, more to go.

The Beas River basin was delineated from the topo sheets, and the location of the hydropower projects was obtained from Google Earth (Figure 8.1). The Larji Hydropower Project of 126 MW capacity and the Binwa Hydropower Project of 6 MW capacity were chosen as a study area to represent the Beas River basin. Larji catchment area is spread over an area of 4,921 sq. km. The large dam was constructed in 1984 and completed

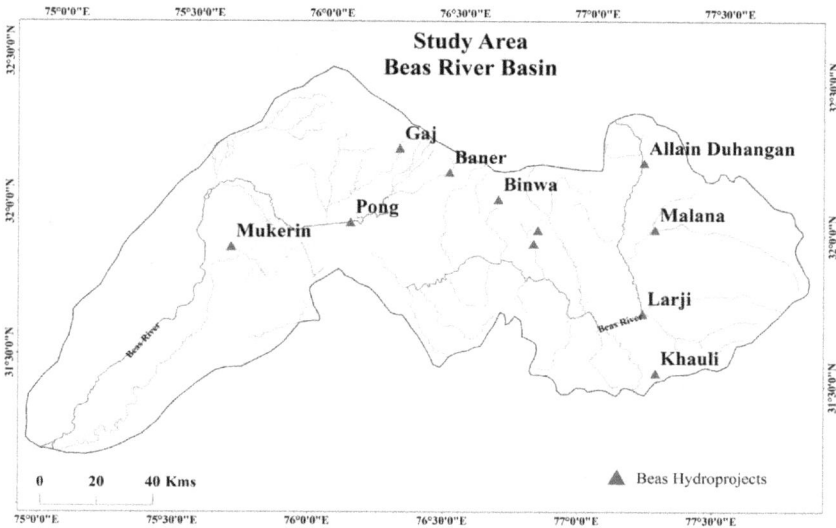

Figure 8.1 Location map of the various hydropower projects in Beas River basin. (Toposheet.)

in September 2007. It is located at an altitude of 2,299 m mean sea level (MSL). The project has a design head of 56.84 m, a discharging capacity of 250 m³/s, and a live storage capacity of a 230-hectare meter. The Binwa watershed is located in Kangra district of Himachal Pradesh. It is between 31° 53′ 15″ to 32° 11′ 58″ N latitude and 76° 34′ 08″ to 76° 45′ 53″ E longitude between Lesser Himalayas and Shiwalik hilly slopes. The watershed is an Agro-Eco Region of 340.1 km².

8.3 MATERIALS AND METHODS

There are two objectives: first one reviews the existing norms for the environmental flow, and the second one identifies the potential location for the construction of the hydropower project.

8.3.1 Environmental flows law and guidelines in foreign countries

The term for environmental flow is different in different nations, such as minimal flow, instream, and so on. The regulation must provide a practical, legitimate, and sensible action of water for ecological flows. As of today, at the international level, no understanding is especially stressed over the importance of ecological flows (Table 8.1).

Table 8.1 Environmental flows law and guidelines in foreign countries

Country	Reference	Environmental flow law and guidelines
European Union (EU)	Acreman et al. (2010) & ("EU: DIRECTIVE 2000/60/EC, Official Journal of the European Communities (2000), 22.12.2000").	WFD (Water System Directive) states that the discharge of 7.5%–35% of available flow regime.
France	Souchon et al. (2001).	Legal minimum flow (1/40) ADF (average daily flow) below dams built before the law.
The United Kingdom	Souchon et al. (2001).	The authority can decide the minimal flow requirements based upon the demand and constraints.
Germany	Water (Federal Water Act-WHG), March 2010	Various factors are involved in legalizing a definite minimal flow.
Switzerland	Swiss Water Protection Act. (http://www.bafu.admin.ch).	The law revolves around preserving the aquatic ecosystem's health and provides sufficient water for their livelihood
The United States	Water Resources Act 1963. (http://www.swfwmd.state.fl.us)	Based on their requirements, the environmental flow will be decided by the state water administration.
Australia	Halliday et al. (2001) & Environmental flow Guidelines (2006)	The ecological flow should be maintained to overcome extreme flow events in rare cases.
South Africa	Van Wyk et al. (2006) & South African National Water Act (NWA) (1998)	The water essential to guarantee the river ecosystems. It is called an "ecological standby".
China	De-sheng (2010)	The suggestion is that water will be given to maintain imperative ecosystem resources and limits during ensuing water management operations.

8.3.2 Minimal flow norms in India

To set up the environmental flow norm, many considerations and procedures like an ecosystem, water needs, future, etc., should be involved. There are different laws in India that have stayed fruitless and difficult. In India, the importance of environmental flow was understood in the 1980s, and Center Water Commission first coined the term "minimum flow" in 1992. The progress of minimal flow stipulations in India is represented in Table 8.2.

Recently, NGT has passed a law for the environmental flow that 15% of the average lean season flow should be maintained as the environmental flow. The environmental flow recommendations for various projects are given in Table 8.3.

Table 8.2 Minimum flow stipulations

Year	Minimum flow stipulations
1999	10 m³/s as the environmental flow (Yamuna River)
2007–2008	10% of the base flow
2010	20% of ecological flow
2012	The authorities decide on the environmental flow
2017–2018	15%–20% of the average lean season flow

Source: Jain et al. (2014).

Table 8.3 E-flow recommendations for various projects

Recommendation	Project	Reference
Environmental flow of 7 cumecs	Nathpa Jhakri Hydroelectric Project (1,500 MW)	Kumar (2009)
Environmental flow of 4–10 cumecs	RHEP (Rampur Hydroelectric Project) (412 MW)	DHI (2006)
70% all the time and 30% during the lean period	The Upper Ganges basin (2007)	WWF (Worldwide Fund for Nature) (2012)
50%–60% of the total flow	Yamuna River (Delhi)	Soni et.al. (2013)
Andhiyakore 0.388 cumecs, Sundergarh 0.96 cumecs	Mahanadi River – A case study	Sundaray (2013)
225 cumecs	Triveni Sangam, Allahabad	WWF (2013)
4.70 cumecs	Kalagarh, Ram Ganga River	Sundaray (2013)
Belur (0.07–9.05) m³/s, Kollegal (18.24–364.88) m³/s.	Cauvery River (2014)	Durbude et al. (2014)
542 m³/s during lean period	The Ganges River (2010)	Akter (2010).

8.3.3 Identification of potential location for hydropower project

The process of delineating the potential zones involves the collection of raw data, processing of the raw data, and analyzing the processed data in the model. The ASTERDEM data has been downloaded from the www.earthdata.nasa.gov/ with 30 m resolution and the Landsat 8 data from the www.earthexplorer.com/ with 30 m resolution. Using ASTERDEM, the terrain features such as slope and stream order for the study area were derived. The forthcoming steps are done for the suitable site analysis: (i) spatial data preparation, (ii) processing of the data, (iii) selection of criteria for providing the weight, (iv) analysis of overlay, (v) relative weights for the total weight calculation, and (vi) locating the selected sites by mapping. The weightage was given based on various pieces of literature, expert advice, and the property of each parameter to serve the purpose.

The criteria map definition and their calculation for an allocation of weights is the most challenging part. The threshold values define the streams whereby following it decides the stream network density (Imran et al. 2019).

The glacial, fluvial processes and periglacial, which are the climatic influences that are sensitive to the process domains, show interrelationships with topographic parameters such as aspect, altitude, or slope angle, in particular. Additionally, the position of slope and its angle represent the snow cover on the permafrost. The Beas River flowing in the catchment area was demarcated using the topo sheets (Figure 8.2). Land use map is one of the important influencing factors in deciding the location of the hydropower projects (Figure 8.3). The Landsat 8 data has been used to perform the supervised classification with five classes. The water bodies were given higher weightage. Slope (Figure 8.4) as principal descriptors of form, altitude, slope, and contributing drainage area as influential factors of process activity (in the case of fluvial and gravitational processes), and surface roughness as an indicator of surface material characteristics (Otto et al. 2018). The slope is analyzed in the degree units and reclassified into five categories. The topography with less slope degree would be more suitable for the hydropower plant. The Strahler method has been used to derive the stream order from the DEM data (Figure 8.5). The process involves fill, flow direction, flow accumulation, and drainage calculation using map algebra. The stream order has been derived up to the seventh order, which represents the river. The higher-order steam will be given more importance. To define the catchment area, the stream segmentation process is important against an individual stream as well as total discharge. This process is considered in different forms of rate, magnitude, and very complex, a variety of dynamics in its scale (Paudel et al. 2016).

IDW as a method of interpolation is defined as deterministic interpolation to allocate the interpolated values for the creation of a surface from the point data by the obtained results of mathematical formulas to surrounding measured values (Earls et al. 2007). The common approach of remote sensing calculation is the normalized difference vegetation index (NDVI) (Figure 8.6), which is used to assess the green biomass and in addition to water stress and nutrients (Glenn et al. 2019). The normalized difference water index (NDWI) (Figure 8.7) has been modified by the replacement of NDWI used near the infra-red band concerning a middle infra-red band such as LANDSAT TM band 5 (Xu 2006). This index is structured to (i) by green wavelength maximum reflectance of water, (ii) low water reflectance minimized in NIR, and (iii) benefit from the NIR of high vegetation and soil reflectance. As a result, all the water characteristics will have positive values and are enhanced. Therefore, soil and vegetation typically have zero or negative values and are suppressed. The processed thematic layers have been analyzed using the weighted overlay analysis by providing suitable weightage for each parameter for delineating the potential location of the hydropower projects.

Major Streams in the Basin
Larji, Mandi district & Binwa, Kangra district, HP

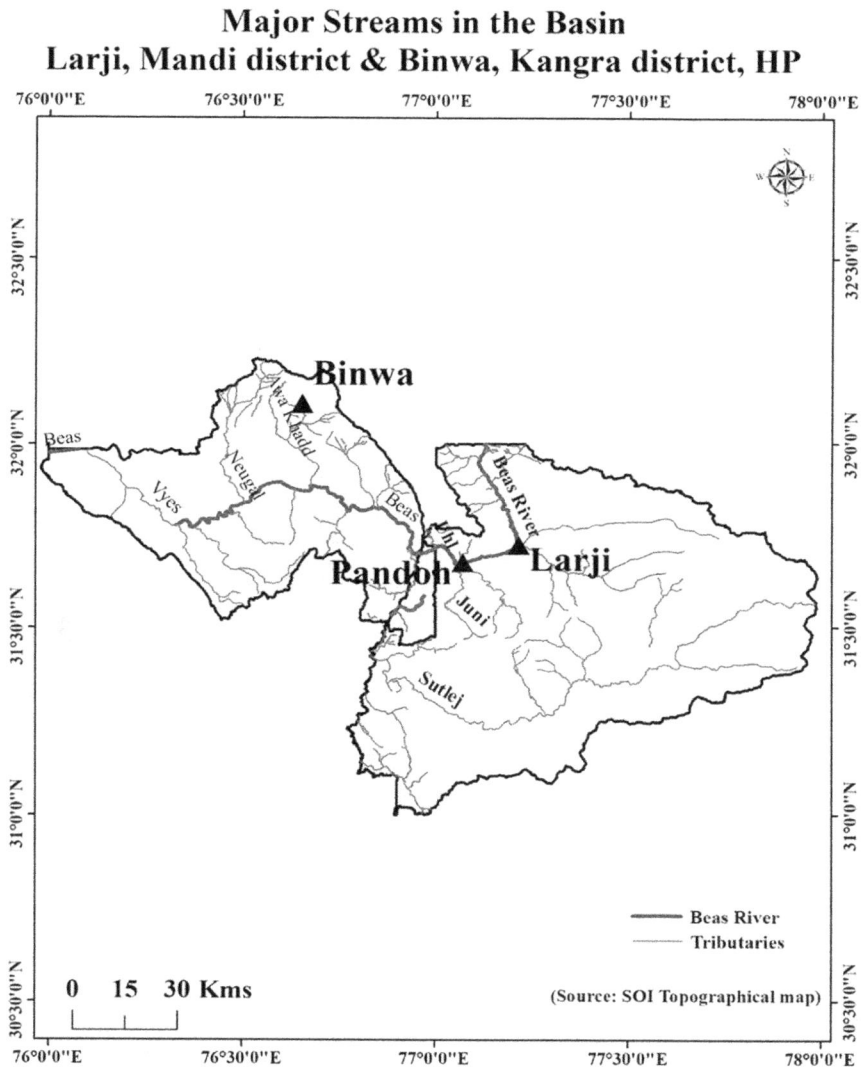

Figure 8.2 Major streams in the study area.

8.4 RESULTS AND DISCUSSION

It is a challenging process to identify the likely location of the hydropower project zones using remote sensing and GIS. This research work is mainly for a suitable site for the installation of hydropower plants. In the past times, by using the map, the hydropower plant was carried out with

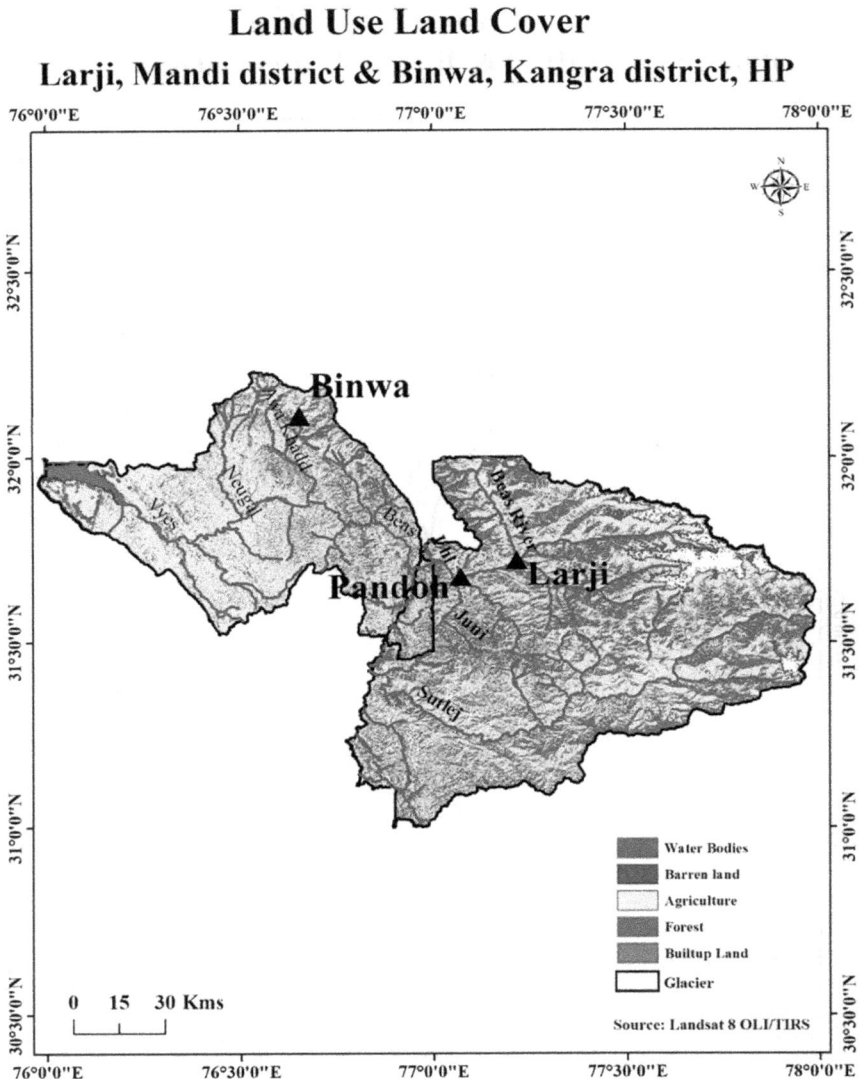

Figure 8.3 LULC map of the study area.

the requirement of more time and cost. This paper explains the selection of suitable site ability by different methods approach with the help of remote sensing and GIS. GIS-based geospatial technology is involved in the identification of suitable sites for hydropower plant development with the usage of the weighted overlay method. The weighted overlay method can be economical and simulated within a shorter period. This method

Slope
Larji,Mandi District & Binwa, Kangra District , HP

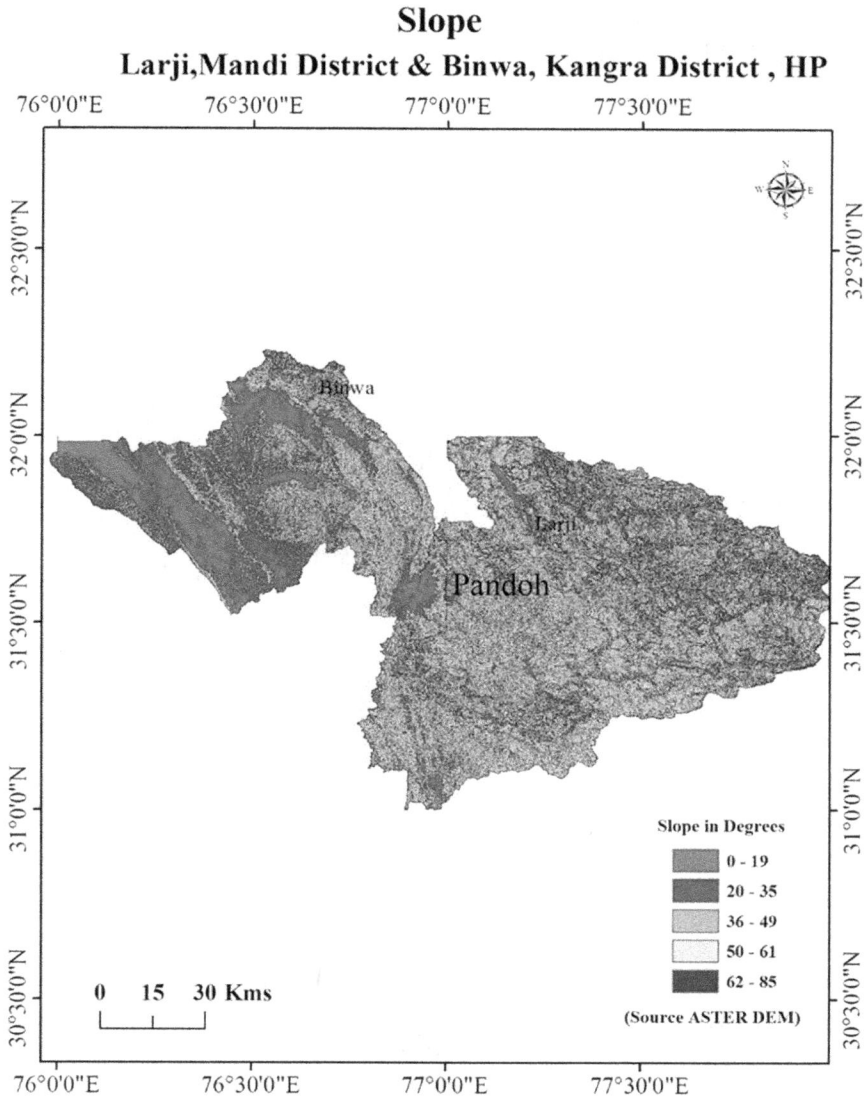

Figure 8.4 Slope in degrees.

applies to hilly areas or mountainous regions. The weighted overlay with the data layers is DEM classification based on elevation data, LULC, and stream order and number.

The study area is a moderate slope of 34% of the area coverage. The zero degrees' slopes are along the river where agriculture takes place in large. The glacier prevails in both the basins of about 31% along the hilly regions.

Major Streams in the Basin
Larji, Mandi district & Binwa, Kangra district, HP

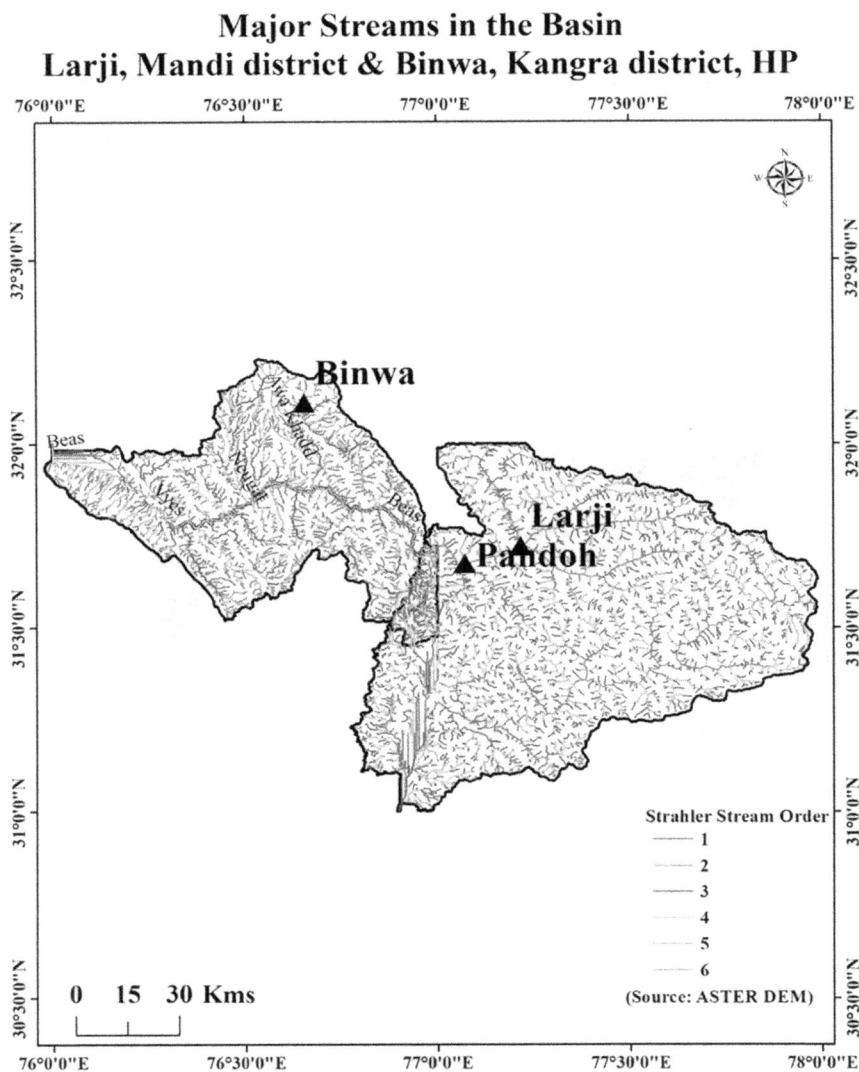

Figure 8.5 Strahler stream order map.

The stream order ranges up to the seventh order, indicating the amount of water flowing to the river, i.e., the seventh-order stream will be given more weightage and suitable for hydropower projects. The annual average rainfall ranges from 920 mm to 2,500 mm. The rainfall intensity decreases as we move from Dharamshala to Kullu. The Binwa Basin is mostly shrouded in forest and shrubland. The agricultural land area has impacted along the

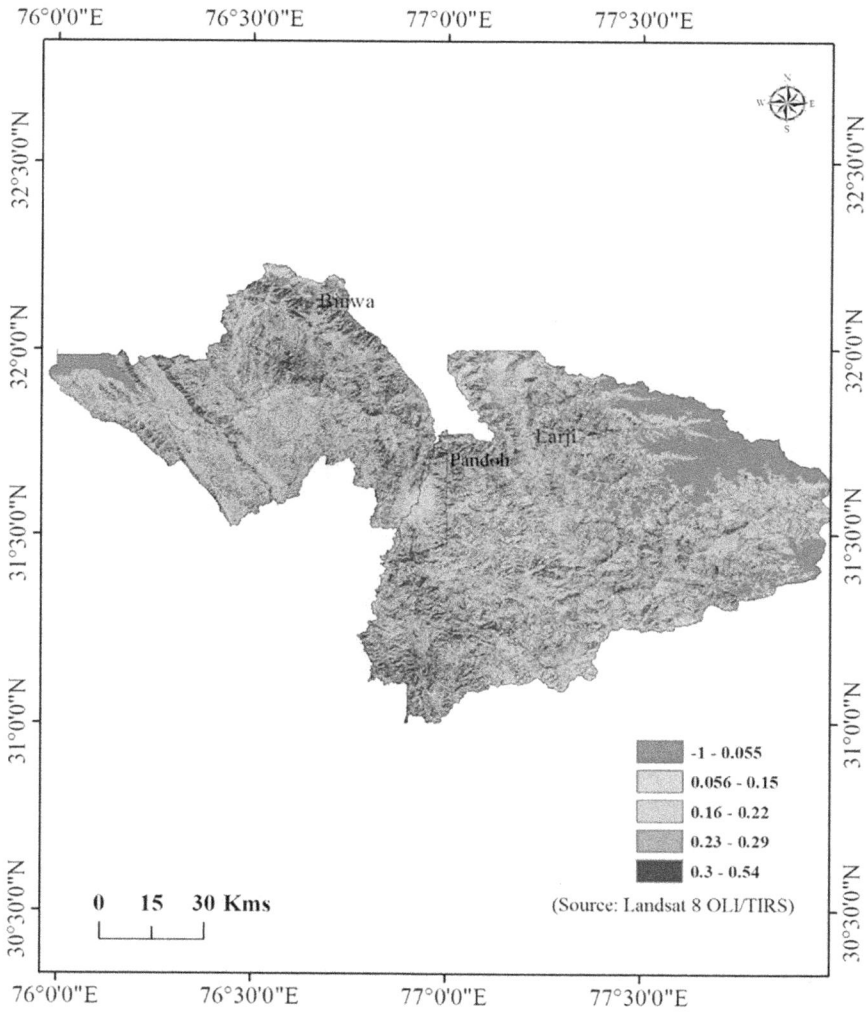

Normalized Difference Vegetation Index
Larji,Mandi District & Binwa, Kangra District , HP

Figure 8.6 NDVI map representing vegetation.

plain region, i.e., riverside up to 5.83%. The vegetation index shows that the greenness of the vegetation in the region is average in most of the area and high in the agricultural fields. The assigned weights are from 1 to 9 per thematic layers' importance and are normalized between 0 and 100 (Table 8.4). This figure represents the areas of potential zones to set up

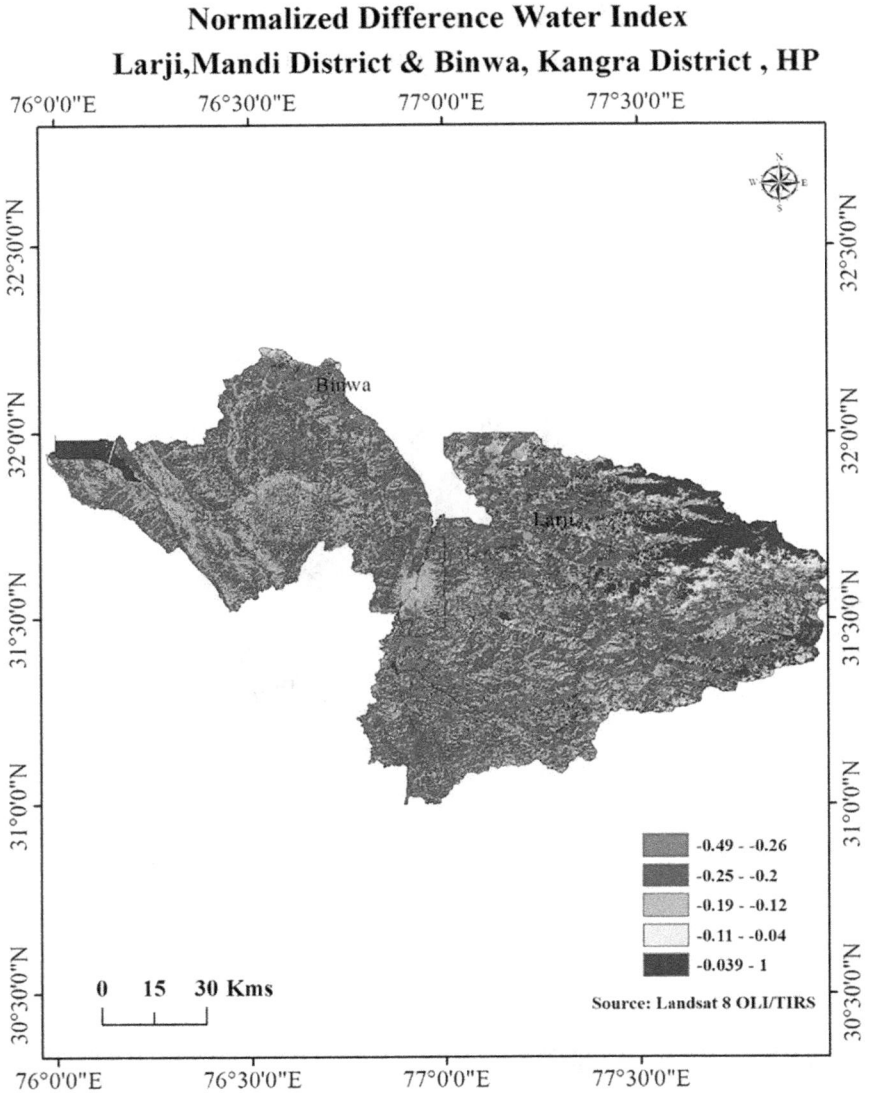

Figure 8.7 NDWI map representing water index of the study.

hydropower projects. Based on these, the sites were identified and catego-rized as a low, medium, high, and very high (Figure 8.8). Very high zones indicate potential location for the construction of the hydropower project. This location is rich in resources that serve the purpose in a sustainable way. To support the result, existing hydropower projects are marked in the study area where the potential zone classification is high.

Table 8.4 Weights for various parameters

Parameter	Class	Weightage	Parameter	Class	Weightage
River	Beas	5	Land use land cover	Forest	3
	Beas River	5		Vegetation	3
	Beas Sutlej	4		Barren land	2
	Pandoh	4		Water	5
	Parvathy	2		Glacier	1
	Sutlej	1	NDVI	(−1)- 0.055	5
	Uhl	4		0.056–0.15	4
Slope (Degree)	0–19	1		0.16–0.22	2
	20–35	2		0.23–0.29	2
	36–49	3		0.3–0.54	1
	50–61	4	NDWI	(−0.49)- (−0.26)	1
	62–85	5		(−0.25)- (−0.2)	2
Stream order	7th order	5		(−0.19)- (−0.12)	2
	6th order	5		(−0.11)- (−0.04)	4
	5th order	4		(−0.039)- 1	5
	4th order	3			
	3rd order	2			
	2nd order	1			
	1st order	1			

8.5 DISCUSSION

Most of the hydropower projects are built upon the mountain region; mountain communities depend upon honey cultivation, tree bark trade, and other herbal shrubs cultivation and trading; the construction phase affects the environment by deforestation, exploitation of land, and other resources upon which the mountain community regions rely upon. Likewise, the operation affects the mountain communities who primarily depend upon fish, agribusiness, and cattle by overexploiting the water for the production of hydropower. The concentration was more on that a wide number of projects are completed; anyway, what influences neighborhood living of remotely located individuals and effect upon ecosystems in different means are overlooked by government.

Irrespective of the minimal flow law passed, government, private, and individual people must go hand in hand to preserve the environment. Recently, the NGT has passed the law of 15%–20% of average lean season flow as the environmental flow to be maintained in the downstream side of the river. The next thing is the construction of the hydropower project in the identified locations. The role of local institutions, government schemes, and governance has their respective goals.

Potential Locations for the Hydropower Project
Larji,Mandi District & Binwa, Kangra District , HP

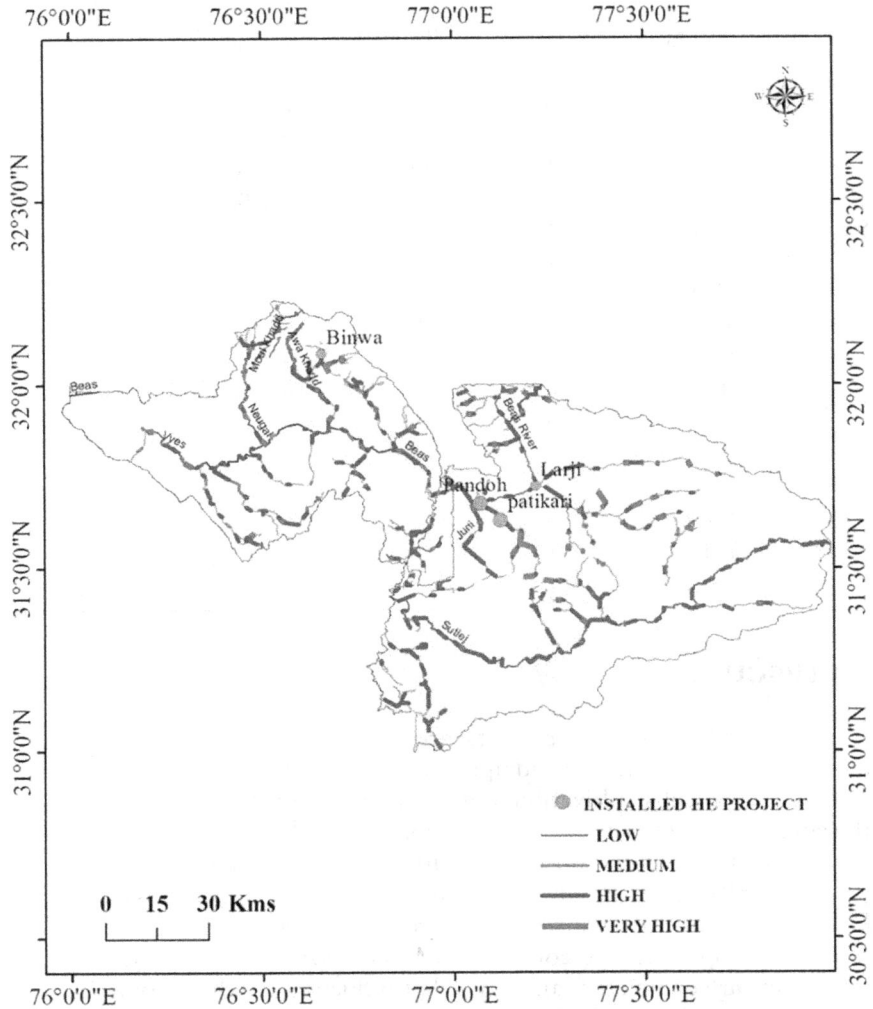

Figure 8.8 Hydropower project potential zones.

8.6 CONCLUSION

The construction and operation of the hydropower projects will have a potential impact on the environment. The framework for a sustainable environment should be practiced by the authorities concerned. The hydropower projects require a large amount of water for power production; hence, the amount of water released in the river might be insufficient. As a mitigation measure, the NGT has passed a law that 15% of the average lean season flow should be maintained/released into the river by the hydropower projects. The small hydropower projects were given an exception from this rule. Considering a hypothetical situation where three to four small hydropower projects are being constructed in the place of a large hydropower project, then the environmental flow maintenance might be neglected. The location for the construction of the hydropower projects wherein the impact will be less should also be delineated by the authorities. Traditional survey methods, suitable precautions, and assessments might have been employed to select the sites for the construction of the existing hydropower projects. The incorporation of the GIS technique to delineate the potential zones using the weighted overlay method might save cost and time. The weighted overlay methodology has been employed to identify the potential zone for the hydropower project. The criterion for using various layers is primarily established on the watershed characteristics that primarily create the runoff like slope, drainage, etc. NDWI, NDVI, land use, slope, and IMSD guidelines have been used to define the stream order priority. The GIS tools have been used to prepare the thematic layers to identify the suitable locations in the Larji and Binwa Basin in the Kullu and Kangra district of Himachal Pradesh, India. The resolution of the DEM and Landsat data can affect the results for the slope, stream order, LULC, and NDVI. For this reason, the field study has to be conducted to validate the results from the GIS technique. The GIS technique reduces the time and cost for squaring out the potential zone. The delineated zone should be surveyed again before finalizing the zonation. In a notion to produce a large hydropower potential, the government is aiming at constructing hydropower projects wherever possible. These sites were selected based on various other factors. This research work has been designed to overcome the environmental impact, i.e., avoid lack of water in the river for farming, livelihood, etc. The topographical, hydrological, and vegetation parameters are considered in mapping these regions. Hence, the suggestion can be proposed to construct the upcoming hydropower project in the delineated location for a sustainable and healthy environment.

ACKNOWLEDGMENT

The authors would like to express their gratitude toward DOES&T, Himachal Pradesh, for funding this research work.

REFERENCES

Acreman, M. C., & Ferguson, A. J. D. (2010). Environmental flows and the European water system directive. *Freshwater Biology*, 55(1), 32–48.

Akter, J. (2010). Environmental flow assessment for the Ganges River.

Al-Ruzouq, R., Shanableh, A., Yilmaz, A. G., Idris, A., Mukherjee, S., & Khalil, M. A., Gibril, M. B. A. (2019). Dam site suitability mapping and analysis using an integrated GIS and machine learning approach. *Water*, 11(9), 1880

Chen, J., Yang, S., Li, H., Zhang, B., & Lv J (2013) Research on geographical environment unit division based on the method of natural breaks (Jenks). *The International Archives of the Photogrammetry, Remote Sensing Spatial Information Sciences*, 3, 47–50

De-sheng, H. U. (2010). Water for the eco-environment: a perspective from International Law. *Journal of Xi'an Jiaotong University (Social Sciences)*, 2, 017.

DHI, (2006). *Managed River Flows for RHEP*. New Delhi: DHI Water and Environment.

Durbude, D. G., Jain, C. K., & Singh, O. (2014). Assessment of E-flows for a river in Southern India using hydrological index methods. *Journal of Indian Water Resources Society*, 34(3), 82–90.

Earls, J., & Dixon, B. (2007). Spatial interpolation of rainfall data using ArcGIS: A comparative study. *Proceedings of the 27th Annual ESRI International User Conference* Vol 31.

EU: DIRECTIVE 2000/60/EC, Official Journal of the European Communities (2000), 22.12.2000.

Glenn, D. M. & Tabb, A. (2019) Evaluation of five methods to measure normalized difference vegetation index (NDVI) in apple and citrus. *International Journal of Fruit Science*, 19(2), 191–210.

Gnanachandrasamy, G., Zhou, Y., Bagyaraj, M., Venkatramanan, S., Ramkumar, T., & Wang, S. (2018). Remote sensing and GIS based groundwater potential zone mapping in Ariyalur District, Tamil Nadu. *Journal of the Geological Society of India*, 92(4), 484–490

Goyal, M. K., Singh, V., & Meena, A. H. (2015) Geospatial and hydrological modeling to assess hydropower potential zones and site location over rainfall dependent Inland catchment. *Water Resource Management*, 29(8), 2875–2894.

Halliday, I., & Robins, J. (2001). Environmental flows for sub-tropical Estuaries. *Final Report FRDC Project*, (2001/022).

Haque, S., Kannaujiya, S., Taloor, A. K., Keshri, D., Bhunia, R. K., Ray, P. K. C., & Chauhan, P. (2020). Identification of groundwater resource zone in the active tectonic region of Himalaya through earth observatory techniques. *Groundwater for Sustainable Development*, 10, P100337

Hassan, I., Javed, M. A., Asif, M., Luqman, M., Ahmad, S. R., Ahmad, A., & Hussain, B. (2020). Weighted overlay based land suitability analysis of agriculture land in Azad Jammu and Kashmir using GIS and AHP. *Pakistan Journal of Agricultural Sciences*, 57(6), 1509–1519.

HMSO (1963). Water resources act 1963.

Iftikhar, S., Hassan, Z., & Shabbir, R. (2016). Site suitability analysis for small multipurpose dams using geospatial technologies. *Journal of Remote Sensing & GIS*, 5(2), 1–13.

Imran, R. M., Rehman, A., Khan, M. M., Rahat, M., Jamil, U. A., Mahmood, R. S., & Ehsan, R. M. (2019). Delineation of drainage network and estimation of total discharge using digital elevation model (DEM). *Science Technology*, 1(2), 50–61.

Jain, S. K., & Kumar, P. (2014). Environmental flows in India: towards sustainable water management. *Hydrological Sciences Journal*, 59(3–4), 751–769.

Jasrotia, A. S. Kumar, R., Taloor, A. K., & Saraf, A. K. (2019). Artificial recharge to groundwater using geospatial and groundwater modelling techniques in North Western Himalaya, India. *Arabian Journal of Geosciences*, 12(24), 774.

Jena, S., Panda, R. K., Ramadas, M., Mohanty, B. P., & Pattanaik, S. K. (2020). Delineation of groundwater storage and recharge potential zones using RS-GIS-AHP: Application in arable land expansion. *Remote Sensing Applications: Society and Environment*, 19, 100354.

Khan, A., Govil, H., Taloor, A. K., & Kumar, G. (2020). Identification of artificial groundwater recharge sites in parts of Yamuna river basin India based on remote sensing and geographical information system. *Groundwater Sustainable Development*, 100415.

Kumar, M. G., Agarwal, A. K., & Bali, R. (2008). Delineation of potential sites for water harvesting structures using remote sensing and GIS. *Journal of the Indian Society of Remote Sensing*, 36(4), 323–334.

Kumar, P. (2009). Environmental flow assessment for a hydropower project on a Himalayan river. *Thesis (PhD)*, Indian Institute of Technology Roorkee, Roorkee.

Lakshmi, S. V., & Sarvani, G. R. (2018). Selection of suitable sites for small hydro-power plants using geo-spatial technology. *International Journal of Pure and Applied Mathematics*, 119(17), 217–240.

Larentis, D. G., Collischonn, W., Olivera, F., & Tucci, C. E. (2010). Gis-based procedures for hydropower potential spotting. *Energy*, 35(10), 4237–4243

Nithya, C. N., Srinivas, Y., Magesh, N. S., & Kaliraj, S. (2019). Assessment of groundwater potential zones in Chittar basin, Southern India using GIS based AHP technique. *Remote Sensing Applications: Society and Environment*, 15, 100248.

Otto, J. C., Prasicek, G., Blöthe, J., & Schrott, L. (2018). GIS Applications in geomorphology. *Comprehensive Geographic Information Systems*, pp. 81–111. Elsevier.

Paudel, B., Zhang, Y. L., Li, S. C., Liu, L. S., Wu, X., & Khanal, N. R. (2016). Review of studies on land use and land cover change in Nepal. *Journal of Mountain Science*, 13(4), 643–660.

Saha, A., Patil, M., Karwariya, S., Pingale, S. M., Azmi, S., Goyal, V. C., & Rathore, D. S. (2018). Identification of potential sites for water harvesting structures using geospatial techniques and multi-criteria decision analysis. *The International Archives of the Photogrammetry, Remote Sensing and Spatial Information Sciences*, XLII-5, 329–334.

Sahu, I, & Prasad, A. D. (2018). Assessment of hydro potential using integrated tool in QGIS ISPRS. *Annals of the Photogrammetry, Remote Sensing and Spatial Information Sciences*, 45, 115–119.

Samanta, S., & Aiau, S. S. (2015). Spatial analysis of renewable energy in Papua New Guinea through remote sensing and GIS. *International Journal of Geosciences*, 6(08), 853.

Sammartano, V., Liuzzo, L., & Freni, G. (2019). Identification of potential locations for run-of-river hydropower plants using a GIS-based procedure. *Energies*, 12(18), 3446.

Sarkar, T., Kannaujiya, S., Taloor, A. K., Ray, P. K., & Chauhan, P. (2020). Integrated study of GRACE data derived interannual groundwater storage variability over water stressed Indian regions. *Groundwater Sustainable Development*, P100376. https://doi.org/10.1016/j.gsd.2020.100376

Singh, A. K., Jasrotia, A. S., Taloor, A. K., Kotlia, B. S,, Kumar, V., Roy, S., Ray, P. K., Singh, K. K., Singh, A. K., & Sharma, A. K. (2017). Estimation of quantitative measures of total water storage variation from GRACE and GLDAS-NOAH satellites using geospatial technology. *QuatInt*, 444, 191–200.

Soni, V., Shekhar, S., & Singh, D. 2013. Environmental flow for monsoon rivers in India: the Yamuna river as a case study [online]. Submitted on 12 June 2013 to the Cornell University. Available from: http://arxiv.org/ftp/arxiv/papers/1306/1306.2709.pdf [Accessed 4 November 2013].

Souchon, Y., & Keith, P. (2001). Freshwater fish habitat: science, management, and conservation in France. *Aquatic Ecosystem Health & Management*, 4(4), 401–412.

Sundaray, P. (2013). Environmental flow. *Doctoral dissertation.*

Tarife, R. P., Tahud, A. P., Gulben, E. J. G., Macalisang, H. A. R. C. P., Ignacio, M. T. T. (2017). Application of Geographic Information System (GIS) in hydropower resource assessment: a case study in Misamis Occidental, Philippines. *International Journal of Environment and Sustainable Development*, 8(7), 507.

Van Wyk, E., Breen, C. M., Roux, D. J., Rogers, K. H., Sherwill, T., & van Wilgen, B. W. (2006). The ecological reserve: towards a common understanding for river management in South Africa. *Water SA*, 32(3), 403–409.

Wang, Q., MIkiugu, M. M., & Kinoshita, I. (2014). A GIS-based approach in support of spatial planning for renewable energy: a case study of Fukushima, Japan. *Sustainability*, 6(4), 2087–2117.

WWF (Worldwide Fund for Nature) (2012). Summary report. Assessment of environment flows for Upper Ganga basin. New Delhi: WWF-India.

WWF (Worldwide Fund for Nature) (2013). Environmental flows for Kumbha 2013. Triveni Sangham. Allahabad: WWF-India.

Xu, H. (2006). Modification of normalised difference water index (NDWI) to enhance open water features in remotely sensed imagery. *International Journal of Remote Sensing*, 27(14), 3025–3033.

Chapter 9

Waste lime sludge
A potential additive in asphalt

Abhishek Kanoungo, Varinder Singh Kanwar,
and Ankush Tanta
Chitkara University

Shristi Kanoungo
Punjab Engineering College (Deemed to be University)

CONTENTS

9.1 INTRODUCTION

Using bituminous mixes in pavement construction is universal. About 98% of the roads have the top layer of bitumen in India due to its less time to construct and low initial cost [1]. Initially, asphalt roads tend to deteriorate long after construction, but then it deteriorate rapidly. The deterioration of flexible pavement usually occurs due to functional or structural factors or a combination of both [2]. When the pavement is unable to absorb and transmit wheel loads through the structure, structural failure occurs. Functional failure encompasses poor pavement performance in terms of slip resistance, structural ability, and ride quality or driver comfort [3]. However, pavement damage due to poor construction and design is caused by the inevitable wear and tear that occurs over time, climate change, tank loads, and

DOI: 10.1201/9781003368335-9

heavy traffic [4]. The presence of cracks in the surface significantly reduces the service life of pavements or soft roads [5]. Indeed, upper cracks are an important factor contributing to the occurrence of different types of road cracks in subsequent layers, leading to premature pavement failure. It has been found that the use of hydrated lime with hot mix asphalt (HMA) has multiple benefits. It is proven to be an effective anti-stripping agent to reduce damage caused by the ingress of water. It also provides resistance to fracture growth by enhancing the stiffness of the binder and HMA. The plastic properties also altered clay fines with lime to improve durability and moisture stability [6]. Thus, the use of industrial waste sludge containing lime as a prime component is highly favoured to improve the performance of the highway pavement. This will also resolve the disposal of waste sludge generated in a massive volume by several industries. These wastes have a high potential and can be used profitably as raw mixes or mixed ingredients. The search for an alternative road material for construction and industrial waste is holding a huge opportunity. The depletion of natural resources and the generation of waste are similar to global growth. Therefore, it is required to identify the potential areas to utilise these waste materials. In this study, the use of sludge from the paper and toothpaste industry as a modifier in binder or bitumen or asphalt was tested. The motives for this research were to enhance the physical and mechanical properties of modified asphalt using lime and resolve its waste disposal. Broadly, this work aimed to evaluate the performance of improved asphalt pavement, focusing on the residual lime mortar that can be used effectively with the appropriate particle size and optimal binder to improve properties and damage resistance.

9.2 EXPERIMENTAL STUDY

9.2.1 Methodology adopted

This study pivoted on the addition of lime sludge that is produced in large quantity from paper and toothpaste industry to bitumen, thereby upgrading the properties of bitumen. Figure 9.1 depicts the flowchart for the study.

9.2.2 Material selection

- Waste Sludge
    Sludge is a general term for the build-up from some processes in industries. The same is precisely applicable in the paper and toothpaste industry. It is important to audit how it is framed to comprehend its properties. Figure 9.2 depicts the strategy embraced by the paper and toothpaste industry to secure waste sludge. Lime sludge blended with bitumen is shown in Figure 9.3.

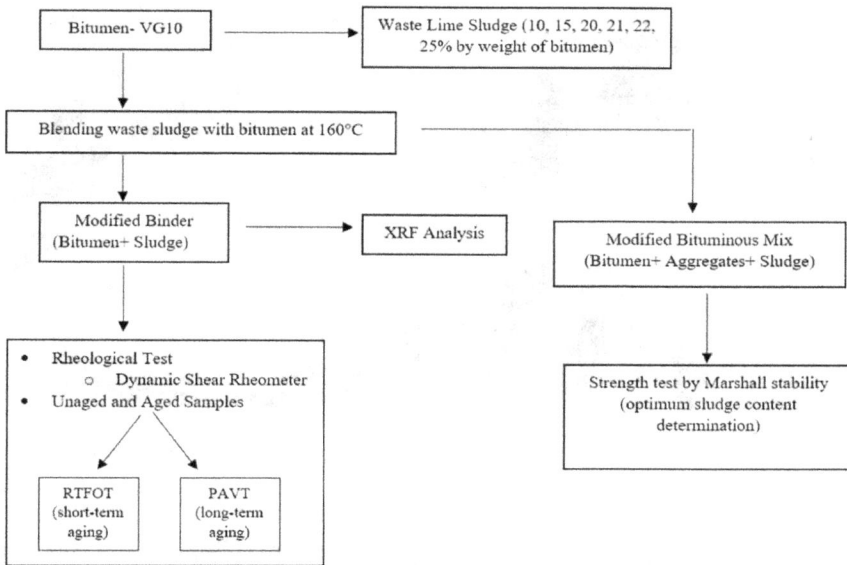

Figure 9.1 The flowchart of the experimental work carried out is depicted in this figure.

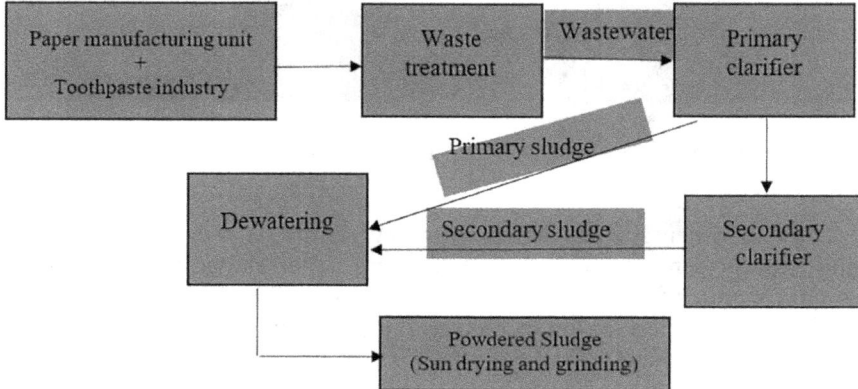

Figure 9.2 The method opted for procurement of sludge is depicted in this figure.

X-ray fluorescence (XRF), a non-destructive, scientific procedure to analyse the essential synthesis of materials, was used to trace the elemental analysis of industrial waste sludge collected from the paper and toothpaste industry. The XRF studies were carried out at Raja Ramanna Centre for Advanced Technology (RRCAT), Indore, India, on the industrial waste of toothpaste sludge using the experimental set-up of XRF BL-16, Indus-2, and synchrotron source. Fine pellets of size 0.2003 g/cm^2 were prepared from paper and toothpaste industry-generated sludge powder, as shown in Figure 9.4.

Figure 9.3 Waste lime sludge collected from the toothpaste and paper industry.

Figure 9.4 Pellet samples of toothpaste and paper sludge.

The experiments produced encouraging results. It was found that calcium was present in toothpaste sludge and paper sludge in a reasonable amount. Figures 9.5 and 9.6 depict the concentration of different elements present in toothpaste and paper sludge, respectively. No toxic element was detected in the sludge.

9.2.3 Preparation of hot mix asphalt (HMA)

The bitumen mix was cooled down to room temperature. Thirty-five Marshall samples of different mixes were prepared. Both sides were compacted with 75 blows. Binder content of 6% by weight of aggregates was used for the unmodified bitumen to achieve a stability value of 1,601.4 kg. The Marshall test tested the samples to assess the effect of lime on bitumen and evaluate the optimum bitumen and sludge content.

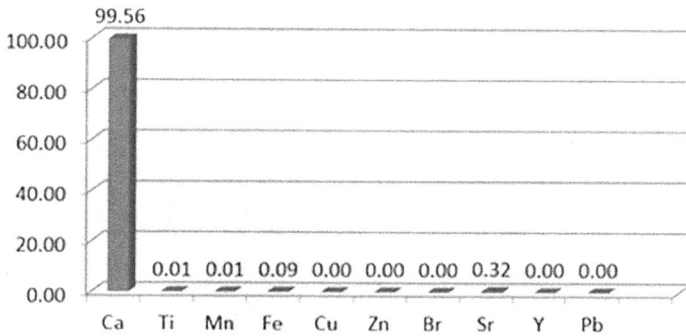

Figure 9.5 Percentage concentration of elements present in the toothpaste sludge.

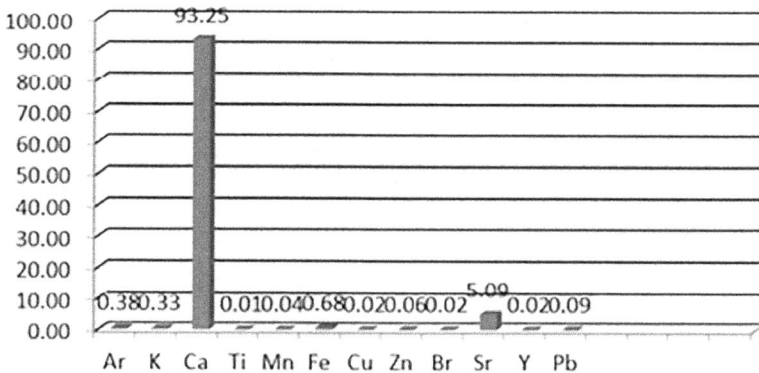

Figure 9.6 Percentage concentration of elements present in the paper sludge.

9.2.4 Tests conducted

- *Rolling thin-film oven test (RTFOT)*

  The modified and unmodified bitumen samples were subjected to short-term ageing as per RTFOT standards, at 163°C for 85 min following ASTM D2872 [7], as shown in Figure 9.7.

- *Pressure ageing vessel test (PAVT)*

  The accelerated long-term ageing test of binders in asphalt was conducted in the pressure ageing vessel test (PAVT) according to ASTM 6521-18 [8] standards, as shown in Figure 9.8.

- *Dynamic shear rheometer test (DSR)*

  The dynamic shear rheometer was used to determine the binder's rheological properties in accordance with ASTM D7175 [9], shown in Figure 9.9.

Figure 9.7 RTFOT set-up.

Figure 9.8 PAVT test apparatus.

Figure 9.9 DSR test apparatus.

9.3 EXAMINATION OF RESULTS AND DISCUSSION

9.3.1 Assessment of mechanical properties of sludge-modified bituminous mix

The bituminous mix specimens of different compositions were tested for their resistance against moisture damage, rutting, fatigue, penetration, ductility, softening, viscosity, and Marshall stability.

9.3.1.1 Marshall mix design

The Marshall stability test was conducted according to ASTM D6927-15 on 35 samples with VG 10 as the binder to determine the optimum bitumen and sludge content. The percentage content of binder utilised for different mix groups is considered according to the Marshall procedure of mix design. 75 blows were applied on both sides of the Marshall specimen. The mould used for preparing the specimen was 101.6 mm in diameter and 63.5 mm in height. All the required contents like aggregates, binder, and waste sludge were heated to a temperature of 160°C while mixing. The percentage content of the binder varied from 4% to 7% for different mix combinations. The value of bitumen content analogous to air void content as 4% was finally

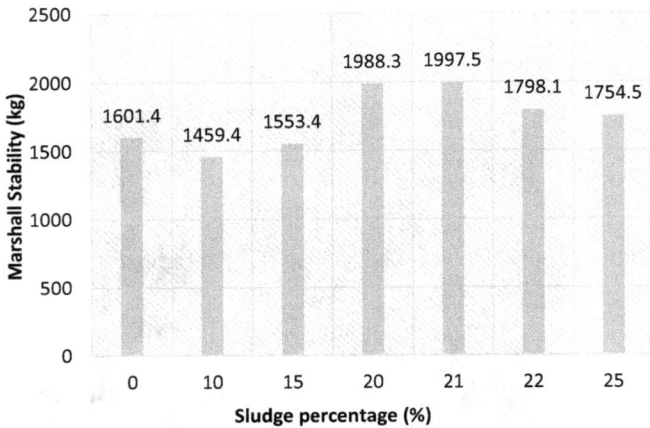

Figure 9.10 Comparative analysis of Marshall stability with sludge.

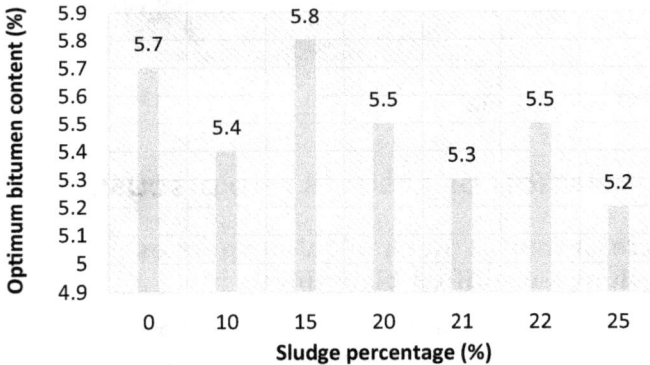

Figure 9.11 Comparative analysis of OBC with sludge.

approved for design. Figure 9.10 graphically represents Marshall results and the variation of stability with an increase in sludge. The stability of bitumen mixes with 20% and 21% sludge is comparable; hence, further tests need to be conducted to establish the optimum value of sludge.

Figure 9.11 graphically depicts the optimum binder content (OBC) determined by the Marshall test varying from 5% to 6% with different percentages of sludge. The value of OBC corresponding to 21% is 5.3%.

Figure 9.12 presents the variation of bulk specific gravity with respect to increasing percentages of sludge. The values corresponding to 21% and 20% sludge are 2.6 and 2.7, respectively. Further tests are needed to derive the optimum proportion of sludge.

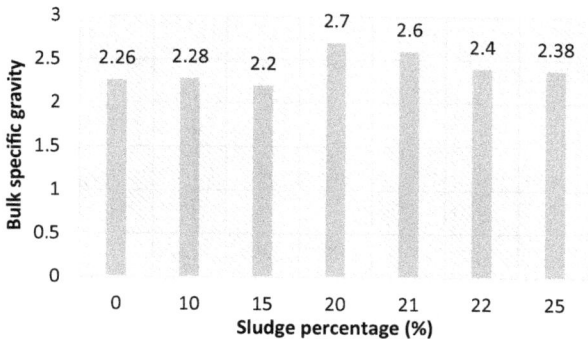

Figure 9.12 Comparative analysis of bulk specific gravity with sludge.

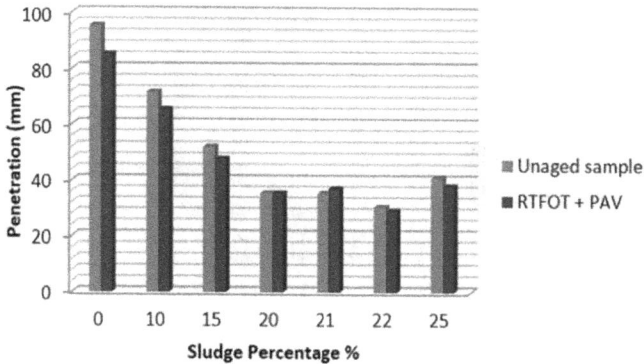

Figure 9.13 Comparative analysis of penetration for aged and unaged specimens.

9.3.2 Evaluation of rheological characteristics of sludge-modified bituminous binder

The tests were performed as per the relevant Indian standards. The properties influence its viscoelastic behaviour and bitumen flow. Before the rheological tests on the binder were conducted, the conventional tests on unaged modified samples and the results showed improved results with 21% sludge content. The results are elaborated in Figures 9.13–9.15.

The consistency of the mix decreased as the quantity of waste sludge increased. It is evident that after ageing, with the increase in the percentage of waste sludge up to 21%, the consistency decreased, but then consistency increased beyond 21%. As the waste sludge hampered the ageing process, therefore not much variation was observed in its consistency. The analysis showed that after ageing, the proportion of waste sludge was increased up to

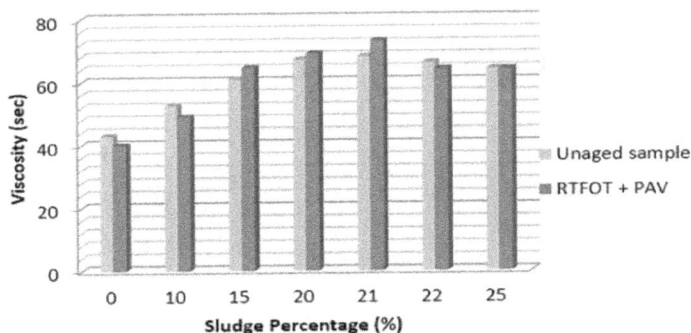

Figure 9.14 Comparative analysis of viscosity for aged and unaged specimens.

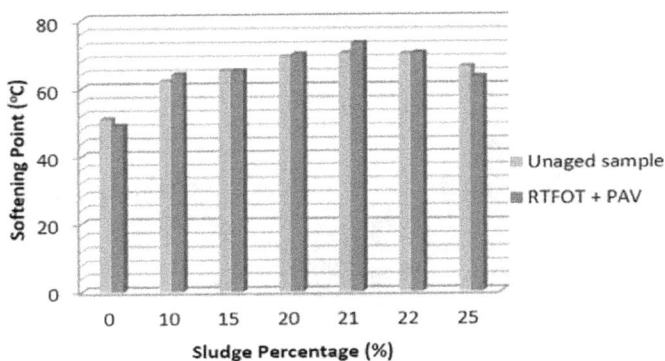

Figure 9.15 Comparative analysis of softening point for aged and unaged specimens.

21% as the viscosity and softening point of the modified binder increased, and no significant change was seen after that.

9.3.2.1 Fatigue and rutting resistance of bitumen

A 25-mm DSR plate was used to look into the rheological properties of aged samples with high-temperature values, whereas, for the PAV sample, an 8-mm plate with a lower temperature value was considered. The graphical representation of G^* and δ is presented in Figures 9.16 and 9.17, separately with different values of waste sludge for both unaged and modified bitumen.

An increase of waste sludge up to 21% to the binder increased the complex modulus. The phase angle of the modified binder followed a reverse trend. This was due to bitumen's hardening, which further upgraded elastic response for modified bitumen compared to the neat bitumen at high-value temperature ranges.

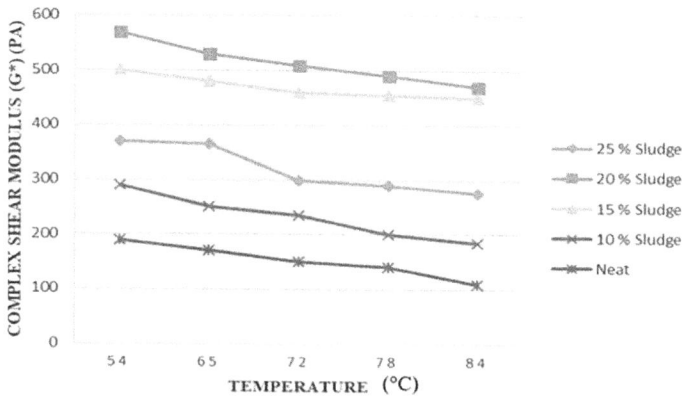

Figure 9.16 Comparative analysis between complex modulus (Pa) and temperature (°C) for unaged and modified bitumen.

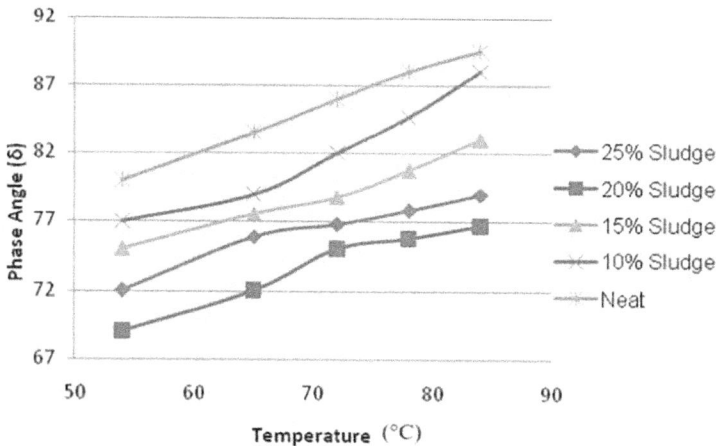

Figure 9.17 Comparative analysis between phase angle (δ) and temperature (OC) for unaged and modified bitumen.

Ratio $G*/\sin\delta$ could be a restraint used as a pointer for rutting resistance. It can be often understood from Figure 9.18 that the extreme value of this parameter is at 21% waste sludge, which induces greater resistance to permanent deformation. $G*/\sin\delta$ increases when ageing indicates that the changed bitumen has higher resistance than the standard quality bitumen.

$G*.\sin\delta$ (Pa) of fatigue resistance is a limit used as an indicator. As depicted in Figure 9.19, with an increase in waste sludge, the fatigue resistance increases. The aged samples show better resistance than the standard bitumen.

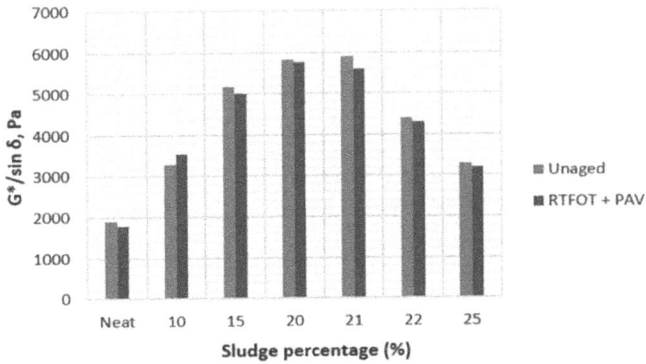

Figure 9.18 Comparative analysis between $G*/\sin \delta$ and sludge content for unaged and RTFOT specimen.

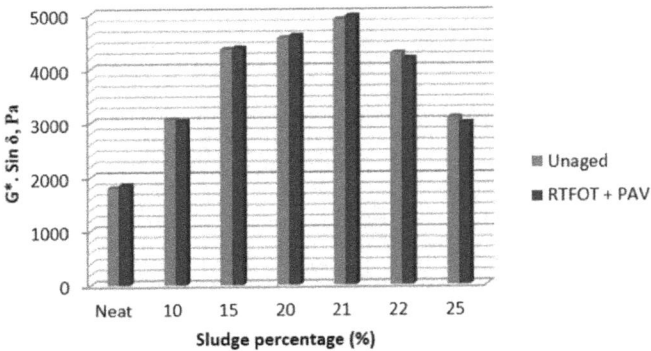

Figure 9.19 Comparative analysis between $G*.\sin \delta$ and sludge content for unaged and RTFOT specimen.

9.4 CONCLUSION AND FURTHER RECOMMENDATIONS

- The perfect percent of waste sludge that may be introduced to the bituminous blend is 21% of the weight of the binder.
- The Marshall stability of unmodified bitumen changed into 1,601.4 kg for strength and 3.21 mm for float value with optimum bitumen content material as 5.7%. The strength will increase 25% with 21% of waste sludge and OBC as 5.3%, and flow value decreases about 22%.
- The value of complex modulus with 21% sludge was highest due to the increased rigidity of binder due to waste sludge. Thus, it modified into obtrusive that the optimum percentage of the sludge content with in the binder should be restricted to 21%.

- The RTFOT+PAVT aged bitumen resulted in an improved overall performance after ageing. The rutting parameter gave the highest value with 21% sludge content.
- The complex modulus with 21% sludge becomes maximum because of the increased rigidity of binder. Thus, it becomes evident that the optimum percentage of the sludge content material should be limited to 21%.
- The RTFOT+PAVT aged bitumen led to improve overall performance after ageing. The rutting parameter gave the highest value with 21% sludge content.

REFERENCES

[1] Ghosh, D., Marasteanu, M., Falchetto, A.C. and Turos, M. (2017) Testing protocol to obtain failure properties of asphalt binders at low-temperature using creep compliance and stress-controlled strength test. *Road Materials and Pavement Design*, 18(2): 352–367.

[2] Lu, X. and Isacsson, U. (2002) Effect of ageing on bitumen chemistry and rheology. *Article in Construction and Building Materials*, 16(1): 15–22.

[3] Serkan, T., Cevik, A., Uşar, Ü. and Gülşan, E. (2013) Rutting prediction of asphalt mixtures modified by polypropylene fibres via repeated creep testing by utilizing genetic programming. *Materials Research*, 16(2): 277–292.

[4] Johnson, O.A, Napiah, M. and Kamaruddin, I. (2014) Potential uses of waste sludge in construction industry. *Research Journal of Applied Sciences, Engineering and Technology*, 8(4): 565–570. DOI:10.19026/rjaset.8.1006. ISSN: 2040-7459; e-ISSN: 2040-7467, July 21.

[5] Ziari, H., Babagoli, R. and Akbari, A. (2014) Investigation of fatigue and rutting performance of hot mix asphalt mixtures prepared by bentonite-modified bitumen. *Road Materials and Pavement Design, School of Civil Engineering, Iran University of Science and Technology*, Tehran, Iran, December 2: 667–674.

[6] Kanoungo, A., Kanwar, V. S., and Shukla, S. K. (2019) Effect of aging on characteristics of Bitumen modified with waste lime sludge. *Advances in Civil Engineering Materials*, 8(1): 298–307. Emerging Sources Citation Indexed (Web of Sciences). ISSN: 2379-1357; IF 1.61, https://doi.org/10..1520/ACEM20180166.

[7] Standard Test Method for Effect of Heat and Air on a Moving Film of Asphalt (Rolling Thin-Film Oven Test), ASTM D2872, (West Conshohocken, PA: ASTM International, 2012). https://doi.org/10.1520/D2872-12E01.

[8] Standard Practice for Accelerated Aging of Asphalt Binder Using a Pressurized Aging Vessel (PAV), ASTM D6521 (West Conshohocken, PA: ASTM International, 2018). https://doi.org/10.1520/D6521-18.

[9] Standard Test Method for Determining the Rheological Properties of Asphalt Binder Using a Dynamic Shear Rheometer, ASTM D7175 (West Conshohocken, PA: ASTM International, 2015). https://doi.org/10.1520/D7175-15.

Chapter 10

Sustainable solid waste management and factors influencing it

An assessment from Indian context

Pramod Kumar and Siby John
Punjab Engineering College

CONTENTS

10.1 INTRODUCTION

Solid waste management (SWM) is one of the issues that requires global attention with specific technical and socio-economic inputs. Also, this raises concerns about protecting the environment and natural resources. Some of the facts that were published in the World Bank Report (Kaza et al., 2018) "What a waste 2.0" are:

- The amount of waste generated in low-income nations is believed to reach three times the current waste generation by 2050. The present scenario states that the Pacific region and East Asia contribute about 23% of global waste, and North Africa and the Middle East generate the least, around 6%. However, the growth is observed fastest in low-income regions like South Asia and North Africa where total waste generation will reach two to three times of the present waste generation in 2050. Other areas also show similar trends in that direction.
- The percentage of organic waste crosses the 50% mark in low- and center-salary nations. However, in high-pay nations, the quantum of waste is almost identical in any case. On account of bigger bundling waste and other inorganic waste measures, the organics count is around 32%.

DOI: 10.1201/9781003368335-10

- Recyclables vary from 16% to 50% from low-income nations to high-income nations. It includes paper, plastic, cardboard, metal, and glass. The percentage of recyclables increases if we go down in the income level of the countries with the paper contribution to be the most in it.
- Recycling and composting methods employed recover 33% of waste in high-pay nations.
- Waste collection percentages vary by pay level. Upper-center to high-income nations have mostly 100% waste collection. Low-income countries' collection counts gather around 48% of waste in urban areas, yet collection counts go down to 26% outside urban regions. The situation of the rural regions fluctuates between 33% and 45% in middle-income countries.
- Globally, only 19% of the waste undergoes material recovery by reusing and composting process, and 11% goes to incineration process and rest 33% is dumped, and 37% is scraped out in landfills.
- Space allocation for offices, waste treatment process, and controlled landfilling of waste is provided in the high- and upper-middle-income countries, and the percent of open waste dumping is only 2%. However, in the case of lower-income countries, by and large, open dumping is the most suited option, and around 93% of waste is dumped in the open.

The managers and policymakers are trying to deal with the issue in the best possible option by applying a sustainability approach to waste management. The integration of strategies that will be sustainable and most effective in handling the waste management issue should be our aim and the first step toward sustainable development. Sustainable development is an approach that defines the development of a system, process, or country's growth to fulfill the requirements of the future generation in the same way as the present generation. It tests the development in three categories: social, economic, and environmental. The sustainability of any system can be defined by analyzing the system in these three criteria.

This chapter reviews the studies on SWM carried out around the world and describes the different factors affecting SWM. It tries to point out the factors that affect the SWM. The review points out the significance of the part played by the informal sector in SWM activities and concern over the negligence of this sector in the activity. The review also describes the factors that affect the strategies of the SWM system. The second part of the review describes the waste management scenario in the Indian subcontinent. It analyzes the current practices and describes the areas that need improvement and efforts for better management.

10.2 STUDIES ON WASTE MANAGEMENT

SWM is a concern that spreads worldwide and affects every living body on earth. With the increase in population and income level of the countries, their waste management situation also changes and magnifies. Growth in living standards and movement toward cities or areas are linked to substantial increase in per-capita waste production or generation. Moreover, rapid development that leads to population growth creates larger and compacted population hubs, making waste management activities and the allocation of land for treatment and disposal a challenging task. The annual municipal waste generation globally reaches 2.01 billion tonnes, out of which 33% is not managed in an environment-friendly or safe manner. The per-capita waste generation worldwide typically ranges from 0.11 to 4.54 and averages out to 0.74 kg (Kaza et al., 2018).

10.3 SUSTAINABILITY AND SOLID WASTE MANAGEMENT

Sustainability is a well-heard word today. The people around the globe – from local to global goods makers or manufacturers to service providers and to international policymakers – are using the term in their own suitable context or another. In the last 35 years, it has become an important topic of discussion. Sustainability as a global concept has been used and discussed on various platforms and societal fields, such as business organizations or education, and to discuss a range of crucial global issues related to the environment and society, such as climate change locally and globally, loss of ecosystem and biodiversity, excess energy production, and use, distribution and use of non-renewable resources, economic issues, and global equity and justice. Although sustainability is widely recognized and discoursed as an eminent topic, there is a propensity to unclearness and vagueness in discussions around sustainability approaches. It seems like there is no one way to the precise meaning of the concept; instead, there is a large diversity of meanings and usages of the term. In a general way, it can be stated as current extraction rates of natural resources are a thousand times higher than their generation rates or imbalance between the current use of natural resources and their recharge. Therefore, one might conclude that sustainability is a modern concept of dealing with energy issues or any other issue that is mostly used in a different meaning in different contexts.

Sustainability terms in SWM refer to the systems that are focused to achieve the sustainability goals by incorporating the 3R principle (i.e., reduce, reuse, and recycle) technologies and circular economy. However, the steps taken toward the improvements vary and depend on the economic status or amount of money a nation or organization can spend (Annepu, 2012; John,

2010; Shekdar et al., 2009). Developed countries like the USA, France, Japan, Germany, and South Korea can afford to spend to integrate 3R technologies in different systems. In contrast, countries like India cannot spend huge sums of money on this issue. If it is allotted, it is forced to focus around a set of socially centered priorities and goals that suits the central vision and objectives of the mission (Mani and Singh, 2016).

In developing nations like India, the sustainable urban solid waste management (USWM) system consists of many factors that go around formal and informal systems. The formal system comprises a municipal body responsible for ground activities like collection, transportation, and disposal of solid wastes. Processing, i.e., conversion of waste into a valuable product like converting waste into compost or fuel pellets, also forms a part of the formal system. It may be operated either by the municipality or by private bodies. The informal system goes around the cycle, which consists of many points and starts with itinerant buyers, waste-pickers to small scrap dealers, and wholesalers. They manage the recycling part of the solid waste generated in the cities. Sudhir et al. (1997) studied the system dynamics of formal and informal systems. They suggested a model in which the informal sector should be included after the municipal activities, and funds should be allotted in proportion to the requirement of the activity. Sembiring and Nitivattananon (2012) recommended that the inclusion and clustering of environmental, economic, and social benefits and informal sector recycling should be analyzed holistically in a manner that their consequences can be rectified and we can have a sustainable SWM (SSWM).

In developing nations, planning for sustainable USWM needs to resolve several factors and conflicts like public well-being, ecosystem, present and future expenses, and the sustenance of the people in the informal recycling system. These all factors are interdependent to each other (Sudhir et al., 1997). Morrissey and Browne (2004) reviewed different models. The model's application in SSWM observed that focusing on technical issues alone for any decision support frameworks for waste management will not be beneficial unless we increase and improve the participation of relevant stakeholders. Costi et al. (2003) proposed a decision support system that proposed an integrated approach for recycling and waste disposal. The model tries to achieve that goal by optimizing the MSW management plant in the treatment plants. The study focused on the environmental sustainability of the system and started to consider the environmental impacts before the installation of any treatment facility. Shekdar (2009) stated that the SSWM system should focus on the compatibility of a present society's financial capacity and the capacity of its ecosystem.

Mani and Singh (2016) analyzed the policy issues in SSWM and observed that there was an absence of vision in many policies and programs. There is a need to integrate awareness, clarity, and future vision among the stakeholders and generators. Also, the poor enforcement of policies leads to failure to achieve the objectives planned. Furthermore, constraints like infrastructure, sufficient funds, and lack of visionary strategies for effective SWM diminish the SWM. The major points that need to be considered while making policies are waste segregation at source, door-to-door segregated waste collection, easy-to-implement technologies for waste management activities, adequate land resources, and scientific and environment-friendly disposal methods. Moreover, the policymakers should focus on bringing out behavioral changes among people or citizens, policymakers, elected representatives, and decision-makers to minimize wastage at source and reduce littering on streets and parks and maximize reuse and recycling. This can help to achieve SSWM. The fundamental concept of SSWM is to bring changes by integrating people, environment, and capital on one page. Providing only technological-based solutions to every SWM problem will not always provide adequate result and will only delay good results and waste resources.

A widely accepted definition of SSWM states that a system could be called "sustainable" if it incorporates feedback loops, focuses on processes, emphasizes adaptability, and reduces wastes from disposal (Seadon, 2010). Ngoc and Schnitzer (2010) stated that a SSWM is defined as "Integrated waste management and its emphasis on seeking reduction of waste at source before it enters to the main cycle to management". More specifically, SSWM aims to reduce the waste to enter the full life cycle of the product by making significant changes in the design, which function without producing waste. The waste management process should be able to process waste to be used as raw material for any other process like compost or RDF. An organized effort is needed to improve the fundamental of waste management policies and the factor affecting them. The factors are public participation and awareness, technology, institutional arrangements, policy and legal frameworks, financial provisions, human resource development, operations management, and SWM systems.

10.3.1 Factors affecting solid waste management

SWM is an activity that should be carried out by considering all factors affecting this activity. Studies carried out indicate that stakeholders or people and organizations play an important role in effective waste management. The stakeholders reported are national and local authorities, households, city corporations, municipalities (Shekdar, 2009); private contractors (Geng et al., 2009); recycling companies (Tai et al., 2011); non-governmental organizations (NGOs) (Sujauddin et al., 2008); and the informal sector. Factors influencing the process or activities of waste management systems are also reported in the studies. The generation of waste at source is influenced by education level of the people, family size, and monthly income and expenditure (Sujauddin et al., 2008). The quantity of waste generated also depends upon other factors such as standard of living of community, seasons in a year, food habits, and degree of commercial activities (Sharholy et al., 2007;

Zhuang et al., 2008). External factors influencing the waste composition and characteristics are geographical location, energy source and weather population, and standard of living (Ngoc and Schnitzer, 2009). Collection of waste is affected by the service provided, method of collection, equipment used, and user fee. The waste collection services should be organized in a manner that will take into account the role of informal sector and build a network of microenterprises (Sharholy et al., 2008). Transportation of waste depends upon the type of waste, the vehicle number and capacity, and the number of trips they can make in a day. One of the major factors affecting the effective treatment of waste is the lack of knowledge and communication of processing or treatment facilities among the authorities (Chung and Lo, 2008). Improper bin collection systems that influence the waste collection, its transfer, and transport practices are reported as factors affecting the transportation of waste. Other factors affecting the collection system are lack of information about collection schedules, poor route planning (González-Torre and Adenso-Díaz, 2005; Hazra and Goel, 2009), insufficient infrastructure, poor roads and route optimization, and the number of vehicles for waste collection and its transportation (Henry et al., 2006). All of the above factors play a crucial role in managing waste in any area or locality. A comprehensive analysis of these factors can resolve the critical source segregation of waste. It will bring steps to be taken to improve the source segregation issue.

SWM is also affected by other crucial factors in deciding strategies and facilitating the system's performance. The factor comprises technical, environmental, financial, socio-cultural, institutional, and legal. If considered, these factors in making strategies or decisions regarding SWM systems can lead us to a better and more effective management plan.

Technical factors that influence the SWM are the inadequate technical knowledge and skills among municipal and government personnel. Without having technical knowledge of any treatment facility, the system will ultimately fail and burden the municipalities. As one of the important aspects of the improper and unhealthy management of solid waste, the environment leads to many causes resulting in climate change, such as global warming (Nissim et al., 2005). In developing countries, the factors influencing the environmental aspect of SWM are the absence of environmental control systems and evaluation of the real impacts and also sometimes negligence to their impact (Matete and Trois, 2008, Asase et al., 2009). Quantification of impacts of each management strategy is vital in deciding its implementation on the ground. Ekere et al. (2009) proposed incorporating people's involvement; their positive influence in all environmental organizations is necessary to have better systems. It will provide us an insight into the loopholes in the management and ill effects of improper solid waste handling on the environment. One of the other reasons for the failure of SWM is inadequate capital, the capital need is a fundamental requirement for any system. A huge sum of capital is necessary to provide the service to the community, with the absence of capital support or inadequate resources due to the stingy behavior, or unwillingness of the users to pay for the user fee (Sujauddin et al., 2008), and lack of proper use of budgets and funds allotted to authorities have distorted the waste management system and services. Sharholy et al. (2007) observed that the involvement of the private partners in SWM activities could enhance the efficiency of the system and can resolve the capital issue. The tendency among the people that is generally observed is that waste management is the sole duty of local authorities and the government, and the public is not expected to participate or contribute to the activity (Vidanaarachchi et al., 2006). A SWM system's operational and overall working efficiency depends on the municipal authority, private partner, and citizen. Therefore, social aspects like public participation and community awareness are mentioned in some studies (Sharholy et al., 2007) as important points in deciding SWM strategies.

The municipalities often observe a lack of management support or dependencies on management for making small decisions (Moghadam et al., 2009). In some studies, it stated that the institutional factors that affect the system are lack of organizational capacity, i.e., professional knowledge and leadership. Another factor affecting the SWM is data regarding the waste generation and collection and its availability freely in the public domain (Chung and Lo, 2008). The negligible free data available is outdated and not reliable or lacks continuity, or is scattered around different agencies concerned. All the above reasons make it extremely difficult to gain an idea or insight into the issues in municipal SWM (Seng et al., 2011). Another factor of concern is the condition of the informal sector. People in the informal sector are generally treated with less credibility. People working in that sector are associated with a low social status that gives and generates distress and low motivation among the solid waste employees (Vidanaarachchi et al., 2006). The politician's view toward solid waste activities is comparatively low concerning other municipal activities (Pokhrel and Viraraghavan, 2005; Moghadam et al., 2009), leading to limited trained and skilled personnel in the municipalities. Integration of technology with solid waste is implemented in monitoring vehicles routes and timely disposal of waste. Positive factors that could enhance the solid waste activities and system are open support from local municipal officials and authorities and people cooperation (Zurbrügg et al., 2005) and holistic plans for waste management that allow monitoring and evaluating the system (Asase et al., 2009) annually. Systematic and regular monitoring of solid waste can infer a treatment system best effective for any particular area. Researchers have reported that an adequate legal framework contributes to steps forward toward the positive development of integrated waste management systems (Asase et al., 2009), while the absence of such policies and weak regulations (Seng et al., 2011) is disastrous.

An approach in developing a SSWM instead of an SWM for a sustainable society would be the best way to tackle the issue. The SWM system should take care of the financial capacity of a given society by managing the social and environmental factors in check. It should maintain the harmony of social, financial, and environmental factors that will lead to the sustainable management of solid waste.

10.4 WASTE MANAGEMENT IN INDIA

In India, waste is managed primarily by the municipal corporation authorized for a predefined area or locality. They have the authority to deal with the waste within their respective area in a way suited by them. However, the authorities have taken all the processes dealing with waste management by themselves or hired private partners to do the job. Municipalities are doing the jobs by using a conventional way of managing waste, combining three major activities, i.e., collection of waste, transportation, and disposal. The collection systems in cities typically involve the placement of the container in different locations as suited by the authorities and citizens or door or door collection of waste and collecting waste at primary collection points. Initiation of the door-to-door collection method, in return for a user fee, has made the work of citizens a little bit easier and tries to minimize the issue of local dumping. The collection process is carried out by the contractor or private companies in exchange for the user fee and charges from authorities who collect and dump the waste at a common point or any point directed by the authorities. Transportation usually refers to the transportation of waste using large vehicles, trucks, dumpers, tractors, and compactors to collect waste from the transfer stations or secondary collection points and carry it to the final disposal site. Finally, waste disposal comprises activities of dumping waste on open areas on city outskirts or vacant lands unhygienic and without any water, soil, and air pollution measures. Until waste management rules 2000, not many changes were observed in the above scenario, and not much emphasis was put on the three activities. The introduction of waste management rules 2000 brought a framework of responsibility for municipalities to carry out the managing activities of municipal solid waste collection, its transportation, treatment, and disposal. The rules were ineffective in implementing effective treatment of municipal solid waste and its disposal. Also, it does not emphasize different components of waste like plastic waste, hazardous waste, and sanitary waste.

The major modifications in the SWM rule were introduced in the year 2016 (SWM, 2016). The rules emphasize the duties of the waste generators to the ministry and the monitoring agencies. Also, the solid waste is classified into various categories, such as municipal solid waste, construction and demolition waste, E-waste, biomedical waste, and hazardous waste and rules for the management of each category of waste specified (SWM, 2016) . The definition of solid waste states that "Solid waste comprises solid or semi-solid domestic waste, commercial waste, institutional waste, sanitary waste, catering, and market waste, and other non-residential wastes. The street sweepings, silt removed or collected from the surface drains, horticulture waste, dairy waste, agricultural and treated biomedical waste excluding biomedical waste, industrial waste, E-waste, battery waste, radioactive waste generated in the area under the local authorities" (SWM rules, 2016) (Table 10.1).

After introducing SWM rules 2016, major changes spiked in waste processing areas. Data on SWM in India show that in December 2018, 145687 MT/day of waste was generated in India, out of which 50.19% is processed. Many states like Andhra Pradesh, Uttar Pradesh, Tripura, Madhya Pradesh, Gujarat, Rajasthan, Uttarakhand, Tamil Nadu, and Punjab have shown a significant improvement in waste processing. However, states/ UTs like Delhi, Haryana, and Karnataka have shown a negligible improvement in waste processing. Figure 10.1 shows the condition of state/UT on waste processing. It shows that in 17 states/UTs of India, less than 50% of waste is processed, and there are only three states/UTs with more than 70% waste processing. The reason behind such data shows that in India, the infrastructure required for SWM is not adequate, or the SWM activity is not taking place at utmost priority. According to Pujara et al., (2019), 85% of municipal solid waste is collected by the municipal corporation. The rest, 15%, goes directly to the dumping site, 30% of collected waste goes for treatment, and 70% of waste goes for landfilling options. The waste management in India is contributing 11% of GHGs emissions. India has the capacity of generating 1,460 MW of electricity from municipal solid waste, but less than 1.5% of total potential has been achieved (Seadon, 2010).

The estimation by Central Pollution Control Board in 2013 (CPCB, 2013) states that a significant portion of about 8%–9% of India's total municipal solid waste comprises plastic waste. About 60% undergoes recycling, most of it carried out by the informal sector. The PET recycling rate in India is 90%, as estimated by National Chemical Laboratory, Pune (2017), which is significantly higher than 31% in the USA, 48% in Europe, and 72% in Japan. However, the recycling rate of plastic is much better in India as compared to global average of only 15%. But still there remains a weighty amount of plastic waste left un-recycled largely due to intermixing of different components of waste, which will end up clogging the sewers or drains or polluting the groundwater source and will be dumped in landfill.

Paper and paper-based products and their end-of-cycle waste in India are recycled at a rate of 27%, a much lower value as compared to industrialized countries such as the USA (49%), Japan (60%), Sweden (69%), and Germany (73%) (CPPRI 2013), where recyclers are mostly exported. According to an assessment, 50% of India's business organizations' paper requirements are fulfilled by import of waste paper, one-third of which is imported.

Table 10.1 Waste data from 2016 to 2018 (OGD platform India)

States/UTs	Total waste generation (MT/Day) – December 2016	Total waste generation (MT/Day) – December 2017	Total waste generation (MT/Day) – December 2018	Total waste processing (%) – December 2016	Total waste processing (%) – December 2017	Total waste processing (%) – December 2018
Andhra Pradesh	6,440	6,525	6,384	8	7	31
Andaman & Nicobar Islands	100	115	100	35	23	54
Arunachal Pradesh	181	181	181	15	0	20
Assam	650	1,134	1,134	10	7	41
Bihar	14,820	1,318	2,389	40	15	43
Chandigarh	340	462	446	100	30	85
Chhattisgarh	1,896	1,680	1,649	0	60	84
Daman & Diu	85	23	32	0	0	65
Dadra & Nagar Haveli	35	58	35	0	0	0
NCT of Delhi	8,400	10,500	10,500	52	55	55
Goa	183	260	260	52	62	65
Gujarat	9,277	10,145	10,721	28	23	69
Haryana	3,490	4,514	4,514	25	6	26
Himachal Pradesh	300	342	342	25	20	45
Jammu & Kashmir	1,792	1,374	1,415	2	1	8
Jharkhand	2,350	2,327	2,126	15	2	52
Karnataka	8,784	10,000	10,000	40	22	32
Kerala	1,576	1,463	624	50	45	61
Madhya Pradesh	5,079	6,424	6,424	14	18	68
Maharashtra	26,820	22,570	22,570	10	35	57

(Continued)

Table 10.1 (Continued) Waste data from 2016 to 2018 (OGD platform India)

States/UTs	Total waste generation (MT/Day) – December 2016	Total waste generation (MT/Day) – December 2017	Total waste generation (MT/Day) – December 2018	Total waste processing (%) – December 2016	Total waste processing (%) – December 2017	Total waste processing (%) – December 2018
Manipur	176	176	174	50	50	50
Meghalaya	268	268	268	58	58	58
Mizoram	253	201	201	4	4	4
Nagaland	270	342	342	0	15	52
Odisha	2,460	2,650	2,720	2	1	12
Puducherry	495	350	350	20	3	10
Punjab	4,100	4,100	4,100	10	10	38
Rajasthan	5,247	6,500	6,500	16	10	56
Sikkim	49	89	89	0	66	66
Tamil Nadu	15,272	15,437	15,437	16	8	57
Telangana	6,628	8,071	8,634	49	67	73
Tripura	407	420	420	0	57	45
Uttar Pradesh	19,180	15,288	15,500	13	20	57
Uttarakhand	1,400	1,406	1,406	0.7	0	38
West Bengal	8,675	7,700	7,700	6	5	5
Total/Average	157,478	144,413	145,687	21.51	23.73	50.19

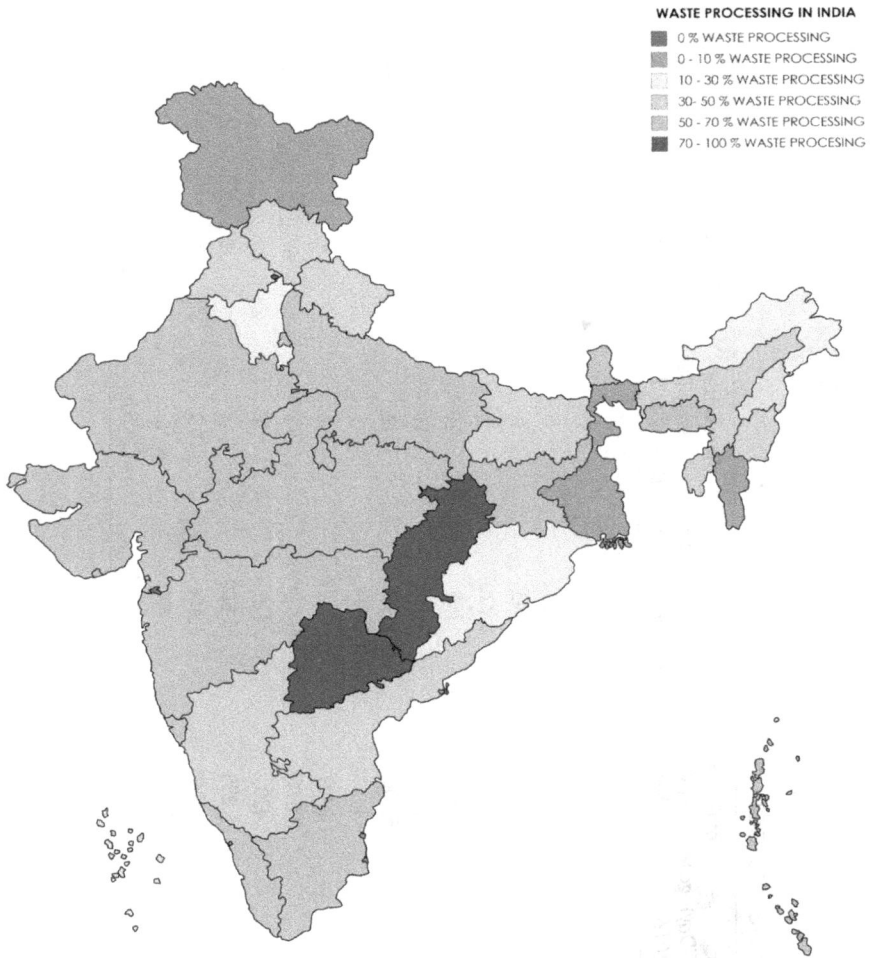

WASTE PROCESSING IN INDIA

- ▪ 0 % WASTE PROCESSING
- ▪ 0 - 10 % WASTE PROCESSING
- ▫ 10 - 30 % WASTE PROCESSING
- ▫ 30- 50 % WASTE PROCESSING
- ▪ 50 - 70 % WASTE PROCESSING
- ▪ 70 - 100 % WASTE PROCESING

Figure 10.1 Waste processing in India, December 2018 (OGD platform India).

Shifting to recycling waste paper also saves more than 70% of natural resources such as water and making paper from wood pulp. In the case of plastic, major issues are absence of segregation at collection point or sometimes at household level and inadequate or no primary collection facility for recyclables that leads to easy soiling of the paper. Systematic handling and allocating infrastructure to waste paper would reduce the toll on natural resources like water and trees and save a significant foreign exchange. It also avoids greenhouse gas emissions emitted in the production process plastic from virgin material (Ahluwalia and Patel, 2018). It was reported that on an annual basis, near about 20 kilotonnes of greenhouse gas emissions

Table 10.2 Installed capacity of recycling plants in India (OGD platform India)

S. No.	States/UTs	Capacity in maximum output (TPD)
1	Andhra Pradesh	388
2	Assam	100
3	Bihar	24
4	Chandigarh	50
5	Chhattisgarh	362
6	Delhi	30
7	Goa	28
8	Gujarat	1,116
9	Haryana	107
10	Himachal Pradesh	96
11	Jharkhand	16
12	Karnataka	652
13	Kerala	7
14	Madhya Pradesh	402
15	Maharashtra	990
16	Manipur	7
17	Nagaland	1
18	Odisha	20
19	Puducherry	22
20	Punjab	160
21	Rajasthan	53
22	Sikkim	6
23	Tamil Nadu	418
24	Telangana	4,494
25	Tripura	2
26	Uttar Pradesh	401
27	Uttarakhand	50
28	West Bengal	291
TOTAL		10,293

could be avoided directly for every 1% increase in waste paper recovery, as per a study conducted by the CPPRI 2013. The assessment shows that the informal sector in India plays an important role in the management of recyclables. Attention to them and formal inclusion in the waste management policy could bring promising results in SWM (Table 10.2).

According to the Planning Commission (2014) report, in India, municipal solid waste consists of 51% biodegradable, 10% plastic, 7% paper, and the rest 32% as textile, metal, glass, street sweepings., drain, silt, and inert. From Table 9.2, it can be seen that 28 states/UTs have installed plants of a total capacity of 10293 TPD for recycling, which is also not reported to be working at full capacity (Ahluwalia and Patel, 2018). The total capacity of recycling plants is insufficient as 17% of municipal solid waste comprises plastic and paper. It means out of 1.45 lakh tonne/day of waste, 0.24 lakh tonne (approximately 24000 TPD) is recyclables; excluding the material collected by the informal sector, we do not even have 50% infrastructure for the recycling of waste. This result emphasized working on speedy infrastructure development and enhancement empowerment of the existing informal sector. The major steps are taken in the SWM rules 2016, including the informal sector in the management policies, should be taken care of with the utmost care. Monitoring this aspect would also be a challenging task for the authorities (Table 10.3).

The maximum percentage of waste includes wet biodegradable waste. The emphasis on developing a treatment facility for this part of the waste is much needed. Under the waste to energy scheme, 597,120 m³/day biogas is produced in 74 plants in 19 state/UTs, and 46,628 kg/day purified biogas in 12 plants is produced. Apart from that, 49.57 lakhs of family-type plants are installed in 19 states/UTs of 1–6 m³/day biogas production. The data regarding the biogas plant are shown in Table 9.3. The initiative brings promising results in rural areas where family-type biogas plants can effectively solve energy and waste treatment issues. The point to look at in this area is the feasibility of such plants. The operational conditions of such plants are specific to the waste type and, if not met, will be ineffective after a certain duration of the operation. In India, we also have 95 compost plants having a capacity of 2,367,480 tonnes/year, and they have an operational capacity of only 14%. The reason for the underutilization of the plant could be the lack of feasibility studies conducted before the construction of plants. The composting plant requires certain operational conditions that could not be available in every part of the country. Another possible reason for the underutilization of the plant is the non-availability of segregated wet waste to the plant. The above data show a lack of infrastructure in India, and much effort has to be put on infrastructure for handling such a huge quantity of waste (Table 10.4).

10.5 KEY ISSUES IN SOLID WASTE MANAGEMENT IN INDIA

The studies on SWM point out certain issues in India. The key issues are:

- The present scenario of SWM in India is an issue of concern because the most suited techniques for waste collection, and its transport to disposal, are not being implemented or used. A region-wise management plan that suited the conditions could bring significant results.

Table 10.3 Biogas generation plants from biodegradable waste in India (OGD platform India)

S. No.	Name of state/UT	Biogas generation plant – m³/day (nos. of plants)	Purified biogas (BioCNG) generation plant – kg/day (nos. of plants)	Family-type biogas generation plant (1–6 m³/day/plant) – (nos. in lakh)
1	Andhra Pradesh	74,640 (6)	NA	5.49
2	Arunachal Pradesh	NA	NA	0.03
3	Assam	NA	NA	1.28
4	Bihar	12,000 (1)	NA	1.3
5	Chhattisgarh	NA	NA	0.54
6	Goa	NA	NA	0.04
7	Gujarat	24,840 (4)	12,538 (2)	4.33
8	Haryana	NA	2,050 (2)	0.62
9	Himachal Pradesh	12,000 (1)	NA	0.48
10	Jammu & Kashmir	NA	NA	0.03
11	Jharkhand	NA	NA	0.07
12	Karnataka	58,080 (3)	NA	4.9
13	Kerala	2,760 (1)	NA	1.49
14	Madhya Pradesh	5,640 (3)	1,200 (1)	3.64
15	Maharashtra	73,080 (8)	19,533 (3)	8.99
16	Manipur	NA	NA	0.02
17	Meghalaya	NA	NA	0.1
18	Mizoram	NA	NA	0.05
19	Nagaland	NA	NA	0.08
20	Orissa	NA	NA	2.7
21	Punjab	33,720 (5)	1,847 (1)	1.77
22	Rajasthan	NA	4,000 (2)	0.71
23	Sikkim	NA	NA	0.09
24	Tamil Nadu	142,920 (27)	NA	2.23
25	Telangana	30,000 (4)	NA	0
26	Tripura	NA	NA	0.04
27	Uttar Pradesh	46,200 (4)	NA	4.41
28	Uttarakhand	67,200 (5)	5,460 (1)	0.21
29	West Bengal	14,040 (2)	NA	3.67
30	Andaman & Nicobar Islands	NA	NA	0
31	Chandigarh	NA	NA	0
32	Dadar & Nagar Haveli	NA	NA	0
33	Daman & Diu	NA	NA	0
34	Delhi	NA	NA	0.01
35	Lakshadweep	NA	NA	0
36	Puducherry	NA	NA	0.01
37	Others	NA	NA	0.02
Total		597,120 (74)	46,628 (12)	49.57

Table 10.4 Installed and operational capacity of compost plants in India by State

State	Number of plants	Installed capacity (kilotonnes/year)	Operational capacity (%)
Andaman & Nicobar Islands	1	0.090	–
Andhra Pradesh	2	2.4	20
Assam	1	15.0	15
Chhattisgarh	1	1.2	20
Daman & Diu	1	4.05	–
Delhi	4	168.0	16.1
Goa	1	1.2	20
Gujarat	15	174.3	19.5
Haryana	4	18.6	15.3
Karnataka	18	473.4	10.1
Kerala	3	156.0	20
Madhya Pradesh	1	36.0	15
Maharashtra	13	488.4	12.5
Punjab	2	19.2	15
Rajasthan	1	180.0	15
Tamil Nadu	9	67.68	15.8
Telangana	5	192.0	15
Tripura	1	75.0	6
Uttar Pradesh	7	124.560	15.2
West Bengal	5	1.70.400	15
Total	95	2367.480	14

Source: Ahluwalia and Patel (2018).

- The lack of training and personal expert in this field is quite less. The lack of accountability for SWM throughout India makes the situation worse.
- The quantum of solid waste data available conflicts with the real quantum of total waste generation and collection in urban India because there is no systematic data collection on waste generation. Without quantifying the issue, no measures taken will be effective enough to resolve the problem.
- Lack of holistic MSW plans and lack of facilities or infrastructure for collection, processing, and treatment. .
- The general perception of "NOT in my Backyard" of the people and their perception that the municipal authority has to handle the waste makes things more severe.
- People's participation and inclusion in these activities are very less and inadequate. As the process is related to people's waste, their inclusion is vital for management activity.

- Implementation of policies on the ground level is still not monitored, and regulatory action on non-compliance of policies has not been stated in the rules. The policy framework in India was more focused on assigning duties to every authority rather than implementing policies on the ground level.
- The informal sector has been overlooked in the planning of solid waste activities. It can play a crucial role in recycling material as they have been doing this from a very old time.
- Meager environmental awareness and little motivation have inhibited innovative ideas and the adaptation of new reforms and advancements in technologies that could lead to the transformation of waste management in India.

10.6 CONCLUSION AND RECOMMENDATION

India is a country facing an increase in population and economic issues and the most prominent issue of development of metropolitan cities in a rapid manner. Lack of the basic infrastructure required for SWM and migration from rural to cities has made this issue severe and complex. The current scenario of India is that it completely relies on the wisdom of municipalities and infrastructure suggested by them. Policies have been made for the systematic way of handling and monitoring SWM activities. However, implementing those policies is far from what is defined to do. The slow decision-making process and implementation of strategies have made this issue more complex. Negligence due to lack of awareness of the informal sector is also one of the issues that affect improper SWM.

The following recommendations are drawn to overcome the current situation of SWM:

- An independent authority for regulating waste management to improve the situation is needed. Without strict regulation, monitoring, and enforcement, meliorations will not happen.
- The need of the present is to develop an inventory on current and future quantum and characterization of wastes in different streams. It is essential to have the quantum of different waste components to apply suitable treatment options.
- There is an urgent need to nurture societal awareness and know the people's perspective toward all solid waste activities and try to change the attitude of common people toward waste as they are one of the stakeholders of these activities.
- Inclusion of the informal sector in these activities, specifically in recycling activities, as they hold much experience in the field, and their integration will increase the system's performance. Also, their capacity building and improved awareness are essential.

- Specific emphasis in waste management should be given to fundamental issues like waste segregation at the source to allow much more efficient value extraction and recycling. Separating dry (inorganic) and wet (biodegradable) waste would have significant benefits and should be the waste generator's responsibility.
- More emphasis should be put on research studies that would bring technological and engineering advancement for solid waste activities.
- The legal framework should be revised periodically based on results and inputs from various stakeholders. Policies more focused on-ground work implementation should be made instead of having paper formalities.
- Training and awareness programs should be organized periodically to fulfill the requirement of skilled persons in solid waste activities.
- More effort should be made to reduce financial dependencies on the system by including community and a public or third party or independent societies, which would help manage these activities.

Until the fundamental reforms and requirements are achieved, India will continue to struggle with improper waste management and its impacts on the environment and public health. The discrepancy between ground reality and paperwork is large enough to suffer from this issue for a long time until we take down those discrepancies.

REFERENCES

Ahluwalia, I. J. and Patel, U. (2018) 'Working Paper No. 356 solid waste management in India an assessment of resource recovery and environmental impact Isher Judge Ahluwalia', *Indian Council for Research on International Economic Relations*, (356), pp. 1–48.

Annepu, R. K. (2012) 'Sustainable solid waste management in India', *MS Dissertation*. doi: 10.1007/978-981-4451-73-4.

Asase, M., Yanful, E.K., Mensah, M., Stanford, J., Amponsah, S. (2009) 'Comparison of municipal solid waste management systems in Canada and Ghana: a case study of the cities of London, Ontario, and Kumasi, Ghana', *Waste Management*, 29(10), pp. 2779–2786. doi: 10.1016/j.wasman.2009.06.019.

Chung, S. S. and Lo, C. W. H. (2008) 'Local waste management constraints and waste administrators in China', *Waste Management*, 28(2), pp. 272–281. doi: 10.1016/j.wasman.2006.11.013.

Ekere, W., Mugisha, J. and Drake, L. (2009) 'Factors influencing waste separation and utilization among households in the Lake Victoria Crescent, Uganda', *Waste Management*, 29(12), pp. 3047–3051. doi: 10.1016/j.wasman.2009.08.001.

Geng, Y., Zhu, Q., Doberstein, B., Fujita, T. (2009) 'Implementing China's circular economy concept at the regional level: A review of progress in Dalian, China', *Waste Management*, 29(2), pp. 996–1002. doi: 10.1016/j.wasman.2008.06.036.

González-Torre, P. L. and Adenso-Díaz, B. (2005) 'Influence of distance on the motivation and frequency of household recycling', *Waste Management*, 25(1), pp. 15–23. doi: 10.1016/j.wasman.2004.08.007.

Hazra, T. and Goel, S. (2009) 'Solid waste management in Kolkata, India: practices and challenges', *Waste Management*, 29(1), pp. 470–478. doi: 10.1016/j.wasman.2008.01.023.

Henry, R. K., Yongsheng, Z. and Jun, D. (2006) 'Municipal solid waste management challenges in developing countries - Kenyan case study', *Waste Management*, 26(1), pp. 92–100. doi: 10.1016/j.wasman.2005.03.007.

John, S. (2010) 'Sustainability-based decision-support system for solid waste management', *International Journal of Environment and Waste Management*, 6 (1/2), pp 51–70.

Kaza, S; Yao, L C.; Bhada-Tata, Perinaz; Van Woerden, F. (2018) '*What a Waste 2.0 : A Global Snapshot of Solid Waste Management to 2050*'. Washington, DC: World Bank, *https://openknowledge.worldbank.org/handle/10986/30317*

Matete, N. and Trois, C. (2008) 'Towards zero waste in emerging countries - a South African experience', *Waste Management*, 28(8), pp. 1480–1492. doi: 10.1016/j.wasman.2007.06.006.

Moghadam, M. R. A., Mokhtarani, N., and Mokhtarani, B. (2009) 'Municipal solid waste management in Rasht City, Iran', *Waste Management*, 29(1), pp. 485–489. doi: 10.1016/j.wasman.2008.02.029.

Ngoc, U. N. and Schnitzer, H. (2009) 'Sustainable solutions for solid waste management in Southeast Asian countries', *Waste Management*, 29(6), pp. 1982–1995. doi: 10.1016/j.wasman.2008.08.031.

Nissim, I., Shohat, T. and Inbar, Y. (2005) 'From dumping to sanitary landfills - solid waste management in Israel', *Waste Management*, 25(3), pp. 323–327. doi: 10.1016/j.wasman.2004.06.004.

OGD Platform of India. https://data.gov.in/resource/population-housing-and-basic-amenities-urban-areas

Planning Commission, Government of India (2014). Report of the Task Force on Waste to Energy. http://swachhbharaturban.gov.in/writereaddata/Task_force_report_on_WTE.pdf

Pokhrel, D. and Viraraghavan, T. (2005) 'Municipal solid waste management in Nepal: practices and challenges', *Waste Management*, 25(5), pp. 555–562. doi: 10.1016/j.wasman.2005.01.020.

Pujara, Y., Pathak, P., Sharma, A., and Govani, J. (2019) 'Review on Indian municipal solid waste management practices for reduction of environmental impacts to achieve sustainable development goals', *Journal of Environmental Management*, 248, p. 109238. doi: 10.1016/j.jenvman.2019.07.009.

Scheinberg, A., Spies, S., Simpson, M.H., and Mol, A.P.J. (2011) 'Assessing urban recycling in low- and middle-income countries: building on modernised mixtures', *Habitat International*. 35(2), pp. 188–198. doi: 10.1016/j.habitatint.2010.08.004.

Seadon, J. K. (2010) 'Sustainable waste management systems', *Journal of Cleaner Production*. 18(16–17), pp. 1639–1651. doi: 10.1016/j.jclepro.2010.07.009.

Sembiring, E. and Nitivattananon, V. (2010) 'Sustainable solid waste management toward an inclusive society: integration of the informal sector', *Resources, Conservation and Recycling*, 54(11), pp. 802–809. doi: 10.1016/j.resconrec.2009.12.010.

Seng, B., Kaneko, H., Hirayama, K., and Katayama-Hirayama, K (2011) 'Municipal solid waste management in Phnom Penh, capital city of Cambodia', *Waste Management and Research*, 29(5), pp. 491–500. doi: 10.1177/0734242X10380994.

Sharholy, M., Ahmed, K., Vaishya, R.C., and Gupta, R.D. (2007) 'Municipal solid waste characteristics and management in Allahabad, India', *Waste Management*, 27(4), pp. 490–496. doi: 10.1016/j.wasman.2006.03.001.

Shekdar, A. V. (2009) 'Sustainable solid waste management: an integrated approach for Asian countries', *Waste Management*. 29(4), pp. 1438–1448. doi: 10.1016/j.wasman.2008.08.025.

SWM (2016) Solid Waste Management (SWM) rules, 2016 Ministry of Environment, Forest and Climate Change. Government of India. New Delhi

Sudhir, V., Srinivasan, G. and Muraleedharan, V. R. (1997) 'Planning for sustainable solid waste management in urban India', *System Dynamics Review*, 13(3), pp. 223–246. doi: 10.1002/(sici)1099-1727(199723)13:3<223::aid-sdr127>3.3.co;2-h.

Sujauddin, M., Huda, S. M. S. and Hoque, A. T. M. R. (2008) 'Household solid waste characteristics and management in Chittagong, Bangladesh', *Waste Management*, 28(9), pp. 1688–1695. doi: 10.1016/j.wasman.2007.06.013.

Tai, J., Zhang, W., Che, Y., and Feng, D. (2011) 'Municipal solid waste source-separated collection in China: a comparative analysis', *Waste Management*. 31(8), pp. 1673–1682. doi: 10.1016/j.wasman.2011.03.014.

Vidanaarachchi, C. K., Yuen, S. T. S. and Pilapitiya, S. (2006) 'Municipal solid waste management in the Southern Province of Sri Lanka: problems, issues and challenges', *Waste Management*, 26(8), pp. 920–930. doi: 10.1016/j.wasman.2005.09.013.

Zhuang, Y., Song, W., Wang, Y., Wu, W., and Chen, Y. (2008) 'Source separation of household waste: a case study in China', *Waste Management*, 28(10), pp. 2022–2030. doi: 10.1016/j.wasman.2007.08.012.

Zurbrügg, C., Drescher, S, Rytz, I., Sinha, A. H. M., and Enayetullah, I (2005) 'Decentralised composting in Bangladesh, a win-win situation for all stakeholders', *Resources, Conservation and Recycling*, 43(3), pp. 281–292. doi: 10.1016/j.resconrec.2004.06.005.

Chapter 11

Carbon-neutral construction
Assessing the potential for carbon capture in an integrated pavement system

*Simon Beecham, Asif Iqbal, and
Md Mizanur Rahman*
University of South Australia

CONTENTS

11.1 INTRODUCTION

Paved areas constitute a major spatial proportion of a typical urban setting. The construction of paved areas is energy intensive because of the energy requirement in material extraction, cement production, and urban construction (Abey & Kolathayar, 2020; Hammond & Jones, 2008a, 2008b; Singh et al., 2020). Consequently, the carbon footprint of pavements is also very high. There is now an urgent need to reduce this carbon footprint in order to reduce future global warming and to attain the goals of the Paris Agreement (UNFCCC, 2021).

Replacing conventional impermeable pavements with permeable pavements in an urban environment is a widely accepted water-sensitive urban design (WSUD) strategy, which has the capacity for runoff flood control, water quality treatment, and stormwater harvesting and reuse (DPLG, 2010; Kumar et al., 2016; Kuruppu et al., 2019; Sanicola et al., 2018). Permeable pavements can also support the growth of trees grown in or around the paved areas (Beecham, 2020; Mullaney et al., 2016), which

DOI: 10.1201/9781003368335-11

can provide the additional benefit of carbon sequestration. Also, the trees watered by permeable pavements can reduce the urban heat island effect through both evapotranspiration and shading (Liu et al., 2020), ensuring more comfort to urban dwellers. There is also increased interest in using recycled aggregates in place of raw materials, particularly in the base course gravel layer as well as in the paver manufacturing process for both impermeable (CCAA, 2008; Kang et al., 2011) and permeable pavements (Cai et al., 2020; Lei et al., 2020; Mohammadinia et al., 2018; Monrose et al., 2021; Rahman et al., 2015). The use of recycled materials can also increase the sustainability of the pavement system by reducing the embodied carbon (Rahman et al., 2020a, b).

This chapter presents a numerical modelling analysis of a hypothetical pavement scenario to demonstrate the potential to reach a net-zero carbon pavement. Permeable pavements are designed using recycled aggregates as the base course material and with trees planted adjacent to the pavement. The aim of this integrated pavement design is to reduce the embodied carbon by using recycled materials and by promoting carbon capture through uptake of carbon dioxide (CO_2) by trees, leading to storage of carbon through tree growth. Both mechanistic and hydraulic modelling are employed for designing the pavement thickness, and this is combined with a carbon accounting methodology to examine the sustainability of the integrated pavement design and whether this approach can be used more broadly as an urban design tool.

11.2 METHODOLOGY

Mechanistic-empirical methodologies have been developed worldwide for the effective design of pavements, including flexible and concrete block pavements (CBP) (Shackel, 1980, 1986, 1990, 1992, 2000; Shackel & Arora, 1978). These two pavement types are different, as follows:

- In a flexible pavement, the deflection tolerance of bituminous layers is usually very low (to reduce fatigue cracking). However, CPBs can stand a movement of maximum 2 mm without distress (Shackel, 1980; Shackel & Arora, 1978).
- It is common for CPBs to interlock during operation immediately after construction, which enhances its performance (Houben et al., 1984; Mahapatra & Kalita, 2018; Shackel, 2000). A CBP considers changes in the vertical elastic modulus of the block paved layers during its initial stages as well. (Shackel & Lim, 2003).
- Unlike flexible pavements, actual traffic spectrum loads are considered in the CBP design instead of relying on the equivalent axle loads.

In this chapter, the carbon footprint for CBP is assessed, considering the life cycle assessment of the constituent materials, the quantity of which depends on the pavement design requirements. The latest Australian construction industry mechanistic design approach to calculate the thickness requirements of the various pavement layers is described in the following section.

11.2.1 Mechanistic design methodology

Figure 11.1 schematically presents the mechanistic-empirical design algorithm, which has four basic steps:

1. Estimating the traffic load;
2. Choosing the pavement layer system and material characteristics for the layers;

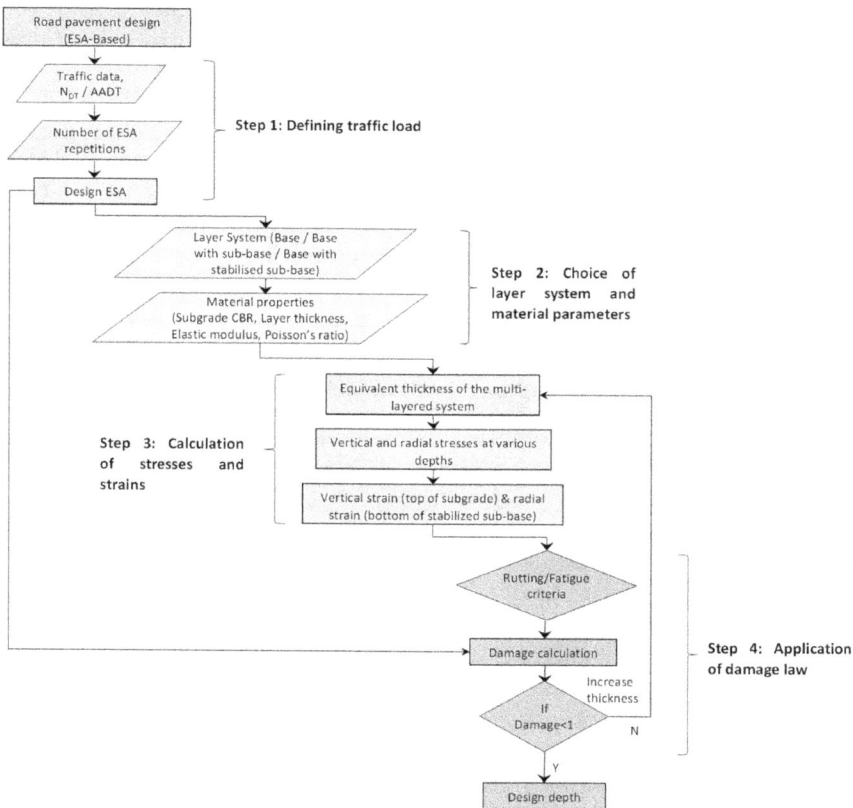

Figure 11.1 Design algorithm to estimate the base/sub-base thickness for varying pavement materials and traffic loadings. (After Rahman et al., 2020a.)

3. Calculating stresses and strains at the top of the semi-infinite sub-grade layer;
4. Implementing the damage law for the load from design traffic or traffic distribution (using Miner's rule).

11.2.1.1 Traffic load

Firstly, total axle group repetitions for the design life (N_{DT}) is estimated, using the equation proposed in Austroads (2017):

$$N_{DT} = 365 * AADT * DF * \left(\% \frac{HV}{100}\right) * LDF * CGF * N_{HVAG} \tag{11.1}$$

where AADT is the annual average daily traffic, DF is the direction factor (proportion of two-way traffic travelling in the direction of the design lane), %HV is the average percentage of heavy vehicles, LDF is the land distribution factor (proportion of heavy vehicles in the design lane), CGF is the cumulative growth factor, and N_{HVAG} is the average number of axle groups per heavy vehicle. According to Austroads (2017), $CGF = \dfrac{(1+0.01R)^P - 1}{0.01R}$ for $R>0$, where, R is the annual growth rate (%), P is the design period (years), and assuming $CGF = P$ when $R=0$. Traffic load spectrum or ESA (equivalent standard axle load) can represent the traffic load in practice; however, ESA was considered in this chapter. Design ESA (DESA) was calculated by multiplying N_{DT} with ESA/heavy vehicle axle groups (HVAG). This eventually adjusts the pavement damage that would result from the traffic load distribution (Rahman et al., 2018).

11.2.1.2 Pavement layer system and material properties

The CBP layers generally comprise concrete block pavers (60–120 mm) at the surface. A 20- to 30-mm bedding layer of sand (for impermeable pavements) or a 20- to 30-mm bedding layer of 2–6 mm gravel (for permeable pavements) lies under the paver layer. Underneath the bedding sand layer, a base course gravel layer is placed with a minimum of 100 mm thickness. A sub-base course layer can also be set (in between base and subgrade layer) subject to the traffic load above, the expected design life of the pavement, and the strength of the subgrade (indicated by the California Bearing Ratio (CBR)). A simple representation of the layering configuration is shown in Figure 11.2. Selecting appropriate layers is crucial in the mechanistic-empirical approach, as it regulates the stresses and strains at the subgrade layer.

In this chapter, the layer system with only a granular base course was considered (Figure 11.2a). The corresponding layer system parameters are shown in Table 11.1, which were used to calculate the base course thickness. For the analysis, a particular traffic load (10^6 ESA) and two different subgrade strengths (CBR of 2% or 4%) were considered, along with typical raw aggregates with an elastic modulus of 200 MPa.

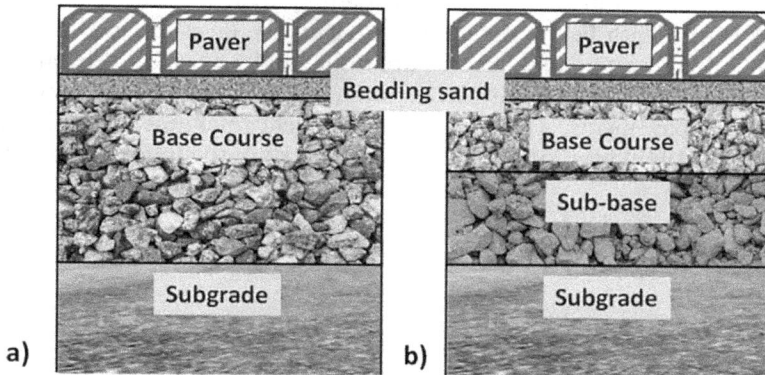

Figure 11.2 Examples of CBP layer systems. (a) Pavement with a granular base course only. (b) Pavement with granular base and sub-base.

Table 11.1 Layer system characteristics for a pavement containing a base course only

Layers	Typical thickness (mm)	Elastic modulus (MPa)	Poison's ratio
Paver	80 (for vehicle loading)	3,200	0.30
Bedding sand	20	200	0.35
Base course	Estimated[a]	200	0.35
Subgrade	-	40	0.40

[a] Calculated design thickness of the granular base course.

11.2.1.3 Estimation of stresses and strains

The circular wheel loads were estimated based on the stress calculation for a point load at the surface, which was initially provided by Boussinesq (1885). The equation below represents the stresses and strains for the wheel load:

$$\sigma_z = P \left[\frac{z^3}{\left(a^2 + z^2\right)^{3/2}} - 1 \right] \tag{11.2a}$$

$$\sigma_r = \sigma_t = \frac{P}{2} \left[\frac{2z(1+)}{\left(a^2 + z^2\right)^{1/2}} - \frac{z^3}{\left(a^2 + z^2\right)^{3/2}} - (1+2) \right] \tag{11.2b}$$

$$\varepsilon_z = \frac{1}{E} \left[\sigma_z - \left(\sigma_r + \sigma_t\right) \right] \tag{11.2c}$$

where a is the radius of the circular loading area, p is the contact pressure, z is the depth below the surface, and   is the Poisson ratio. The vertical, radial, and tangential stresses are σ_z, σ_r, and σ_t, respectively, and ε_z is the vertical strain. Equations (11.2a) and (11.2b) are not directly related to multi-layer pavements. Therefore, the equivalent thickness method (ETM) (Odemark, 1949) was considered, as represented in the following equation:

$$h_{eq} = nh_i \left(\frac{E_i}{E_m} \right)^{0.33} \sqrt[3]{\frac{1 - \frac{2}{m}}{1 - \frac{2}{i}}}$$
(11.3)

where $n = 0.90$, E_i and E_m are the Poisson ratios of the top layer and half space, respectively, h_i is the thickness of the top layer, and E_i and E_m are the vertical elastic moduli of the top layer and half space, respectively. Rahman et al. (2020a) showed the justifications of using Equations (11.2a, 11.2c) and (11.3) when compared with the estimated stress and strains by finite element method (FEM).

11.2.1.4 Estimating pavement life: damage laws

A failure might occur in the pavements as a result of cumulative rutting deformation in the subgrade layer. Various methods can assess the magnitude of the permanent strain against the number of elastic strain repetitions under wheel loads (Azam et al., 2015). The method suggested by Edwards and Valkering (1974) is commonly employed to estimate the design life of CBPs in Australia and is also used in this study (Equation 11.4):

$$\varepsilon = 2.8 \times 10^{-2} \times N^{-0.25}$$
(11.4)

where ε is the permissible subgrade compressive strain and N is the number of strain repetitions.

For ESA-based designs, strains are calculated under a standard axle load. Fatigue laws are then used to assess the pavement life pavement for the calculated strains. For a lower estimated service life than the expected design life, the pavement thickness is increased progressively until obtaining the desired service life.

11.2.2 Hydraulic design methodology

For the hydraulic design of the pavement, it is important to know:

- the rainfall characteristics of the area for which the pavement is being constructed;
- the permeable pavement area and the non-permeable contributing area supplying runoff to the pavement area;

- the materials being used and their properties, such as the void ratio of the base course materials and the infiltration rate of the paver materials; and
- For a reuse system, the water demand is needed in order to estimate the required volume of the base course. However, in this case, the design thickness obtained by the mechanistic design approach was considered. The amount of stormwater that can be harvested with that mechanistic design thickness was assessed in the hydraulic design.

The design thickness (D) of the pavement base course depends on the volume of water harvested from the storage (SH), the pavement area (A), and the void ratio (VR) of the base course.

$$D = \frac{SH}{A \times VR} \qquad (11.5)$$

The pavement area and void ratio of the material depend on the design requirements. The volume of water harvested from the storage depends on the runoff volume and storage ratio of the pavement system. Each city has a hydrological effectiveness curve, which determines the efficiency of unit runoff ($L/s/m^2$) into the pavement in relation to the storage as a percentage of mean annual runoff volume (% MARV, which is denoted as the storage ratio). The hydrological effectiveness curve can assess a storage system's discharge (infiltration) performance, which considers EIA (equivalent impervious catchment area), historical rainfall data, storage, outflow, bypass, and overflow (DPLG, 2010). An example of a hydrological effectiveness curve for the city of Adelaide, Australia, is shown in Figure 11.3.

$$SH = \%MARV \times AAR \qquad (11.6)$$

where
% MARV is dependent on the efficiency of the system and AAR = average annual runoff, which is distinct for each area and dependent on rainfall data (AvgR).

$$AAR = AvgR \times CA \qquad (11.7)$$

where CA is the total catchment area, which is the sum of the permeable paving area and any non-permeable contributing catchment area.

All these factors were considered in the pavement hydraulic design software package, *DesignPave v2.0*, which was developed by the authors and is being implemented in the Australian pavement industry. In this research, the software was used to calculate the % of stormwater harvested from the

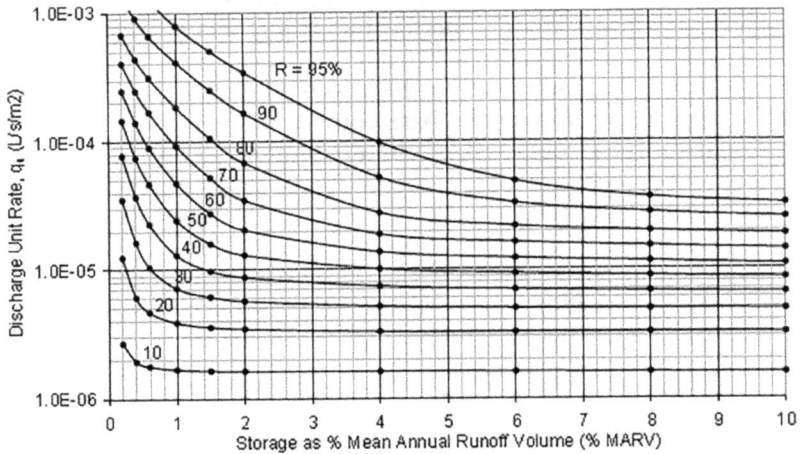

Figure 11.3 Hydrological effectiveness curve for Adelaide for a rainfall intensity ranging from 500 to 600 mm per annum (DPLG, 2010).

runoff (which determines the amount of water harvested in the pavement) for the design thickness that was considered.

The amount of water harvested (HW) depends on the hydrological effectiveness (E) and the average annual runoff volume (AAR).

$$HW = E \times AAR \tag{11.8}$$

If the design thickness is larger and the runoff volume is high, the pavement can collect and store more stormwater. But for a low runoff value zone, a thicker pavement would not be useful, and thus, the hydraulic design would suggest a suitable design thickness for the pavement considering its use and location.

11.2.3 Estimating carbon footprint of a pavement

To analyse the carbon footprint of the pavement materials used, the life cycle of the raw aggregates was assessed (Figure 11.4). The carbon footprint of the paver materials was also assessed by reviewing the available literature. Figure 11.4 shows how estimated carbon footprint values vary across the literature depending on the assumptions made under various conditions. In this chapter, the average of the values shown in Figure 11.4 was considered for the carbon footprint assessment. Rahman et al. (2020a) found that use of recycled aggregates in a pavement instead of natural aggregates, while maintaining the design safety, could reduce the energy consumption and

	Energy (MJ/t)	CO_2 (kg/m^2)*
[A] Rahman et al. 2020	48	11
[B] Hammond and Jones 2008	100	13
Average considered	**74**	**12**

	CO_2 (kg/m^2) (80 mm pavers)
[B] Hammond and Jones 2008	26.42
[A, C] Rahman et al. 2020, Rahman et al. 2020	35.97
[D] Sing et al. 2020	35.31
[E] Beecham 2020	41.8
Average considered	**35**

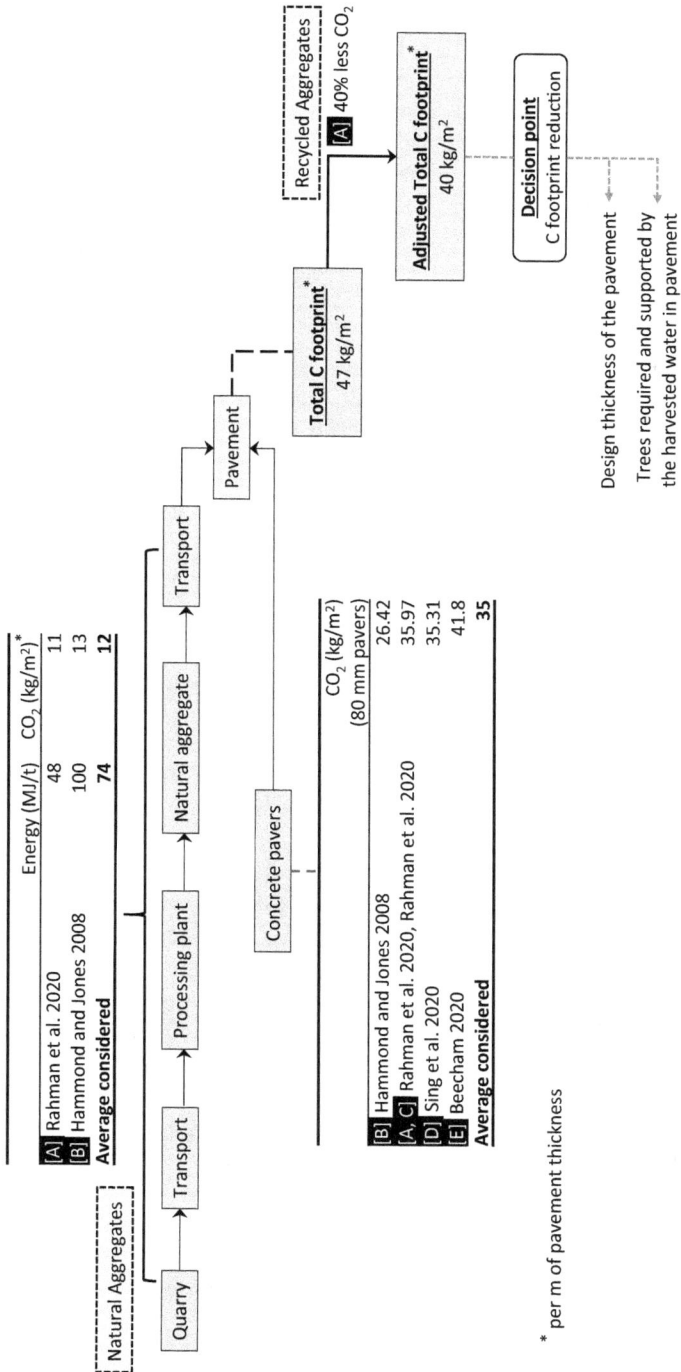

* per m of pavement thickness

Natural Aggregates

Quarry → Transport → Processing plant → Transport → Natural aggregate → Transport → Pavement

Concrete pavers

Recycled Aggregates

Pavement

Total C footprint*
47 kg/m^2

[A] 40% less CO_2

Adjusted Total C footprint*
40 kg/m^2

Decision point
C footprint reduction

Design thickness of the pavement

Trees required and supported by the harvested water in pavement

Figure 11.4 Life cycle carbon footprint of pavement materials and a schematic diagram of an approach to reduce this footprint through the use of recycled aggregates and tree plantings. (a) Rahman et al., 2020a. (b) Hammond and Jones, 2008b. (c) Rahman et al., 2020b. (d) Singh et al., 2020. (e) Beecham, 2020.

CO_2 production by 40%. That reduction potential was also considered in the integrated pavement system here to reduce its carbon footprint.

In this chapter, a 4,000 m² rectangular car park (200 m×20 m) was considered as a case study for the carbon footprint and sequestration assessment.

A permeable pavement system was considered with trees planted along the sides. The permeable pavement was designed to provide harvested stormwater for passive irrigation of the adjacent trees. The key features of the permeable pavement system are consistent with the guidelines of the Concrete Masonry Association of Australia (CMAA, 2010) and are as follows:

- 80-mm think pavers with openings along narrow joints and with a design surface infiltration rate of 0.00009 m/s.
- Uniform granular base course material with a 40% void ratio and a base course permeability of 0.5 m/s.
- Layer system: paver, bedding sand, base course, subgrade (Figure 11.2a), with no subgrade infiltration.
- No impermeable contributing catchment upstream of the permeable pavement carpark.

The trees were designed to be planted 5 m apart, to allow for the maximum tree foliage by their mature stage. A total of 88 trees are assumed to be accommodated in two rows and two columns.

The following assessment was made for three different Australian rainfall conditions using climate data from the cities of Adelaide, Brisbane, and Sydney:

- The permeable pavement design thickness required for each city to harvest the estimated water demand for the trees.
- The water demand of the trees allowing for season and various stages of growth from juvenile to fully mature. In a South Australia study, Teskey and Sheriff (1996) estimated a typical water demand for the selected trees of approximately 2.88 L/d/m² tree, which is close to the value recommended in a Western Australian study by GWA (2019) of 3 L/d/m² tree. It should be noted that the study areas from which these two values derive have similar climates. Figure 11.5 shows the variation of irrigation water demand for the trees according to season and tree size.

  As the permeable pavement in this case is required to provide irrigation water for the trees, the average water supply required in excess of the trees' natural ability to uptake water from soil was considered here. The pavement design thickness was aimed to support the water demand requirements of mature stage trees (16 m² on average), which results in a value of approximately 1541 kL/year for 88 adult trees.

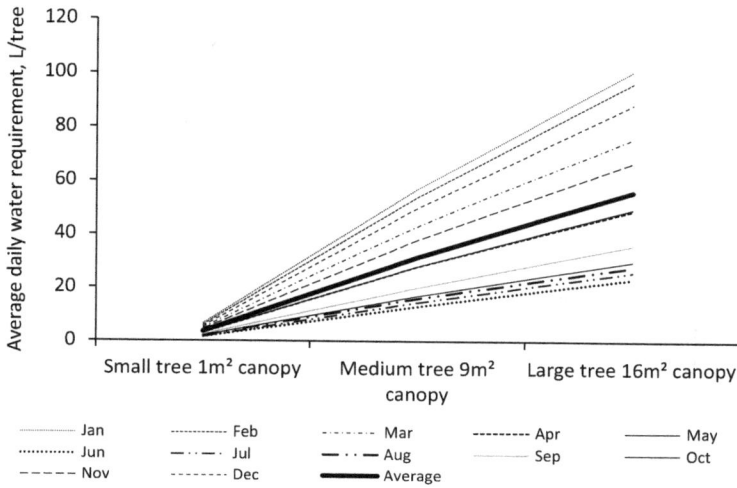

Figure 11.5 Variation of irrigation water demand with tree size and Australian season of the year. (Adapted from GWA, 2019.)

- The time required for the trees to sequestrate the carbon footprint of the pavement was also assessed. The carbon sequestration rate of the plants depends on the mass of the tree (Cook & Knapton, 2009; Tak & Kakde, 2020), and the rate of CO_2 removed is estimated as per the ratio of molecular weights for CO_2 to carbon, i.e. 44/12 (Tak & Kakde, 2020).

$$\text{Carbon storage} = \text{Total Biomass} \times 50\% \qquad (11.9)$$

$$CO_2 \text{ removed} = \text{Carbon storage} \times {44}/{12} \qquad (11.10)$$

Equation (11.9) is derived from Lieth and Whittaker (1975) and Razakamanarivo et al. (2011). The biomass variation with tree age was considered for the calculation of the sequestration rate for two common Australian native tree species, namely eucalyptus and acacia. Zhang et al. (2012) estimated the non-linear variation of biomass with the growing age for these two native tree species over a typical period of 20 years, as shown in Figure 11.6.

Typically, a tree requires 20 years to reach its fully mature stage after which time the biomass is assumed to be constant. Also, it was assumed that the trees reach a medium stage (with a canopy size of $9\,m^2$) at around 10 years (Beecham et al., 2010). Depending on the size of the tree and the rate of sequestration per m^2 of tree, the estimated

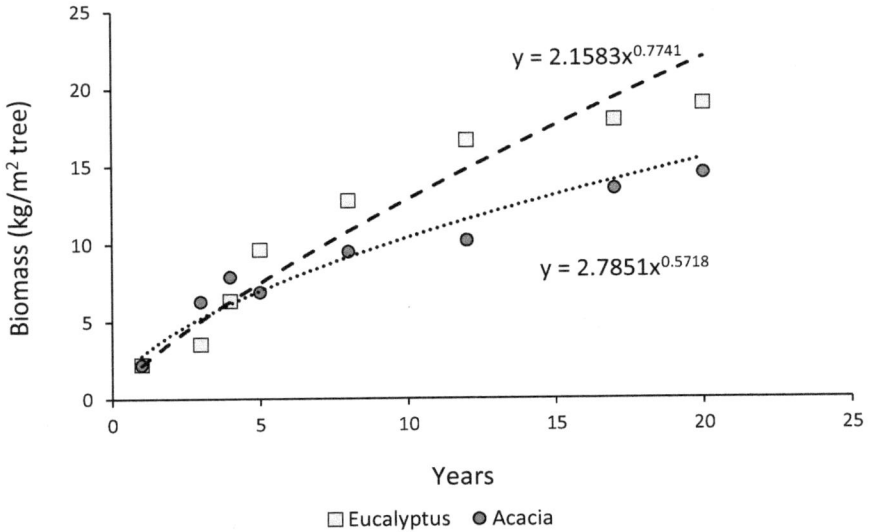

Figure 11.6 Biomass (kg/m^2) in two Australian native tree species in relation to their age. (Adapted from Zhang et al. 2012.)

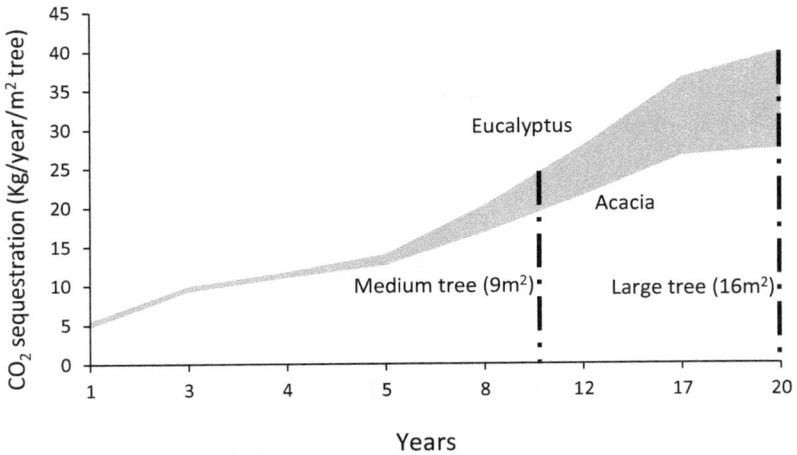

Figure 11.7 Estimated CO_2 sequestration rate (kg/year/m^2 of tree canopy) by two native Australian tree species over 20 years (rate is constant thereafter).

CO_2 sequestration for the two selected native Australian tree species over a 20 year is shown in Figure 11.7. While the total carbon footprint for the pavement was considered for its design life of 30 years, the CO_2 sequestration rate from 20 to 30 years of time was also considered as constant.

11.3 RESULTS: NEUTRALIZING THE CARBON FOOTPRINT

The design thickness of the impermeable CBP increases with the wheel load (equivalent standard axle (ESA) for the car park) on the pavement. For a wheel load as small as 50,000 ESA and for a subgrade CBR of 4, the mechanistic design thickness is estimated as 190 mm (Figure 11.8). The thickness was calculated using the recently developed Australian standard design software (*DesignPave v2.0*). The base course thickness increases to 250 mm for a 300,000 ESA. The design thickness range selected here is assessed for the carbon footprint and for estimating the potential for stormwater harvesting if the impermeable pavements are replaced by permeable pavements. A 250-mm pavement thickness is considered as a reasonable maximum thickness considering the potential wheel load exposure of the car park. The aim of the analysis is then to see which design thickness serves the purpose of neutralizing the carbon footprint from the pavement.

Since the carpark is to contain permeable pavers rather than an impermeable pavement, a hydraulic design was also undertaken and this shows that the range of base course thicknesses can be as low as 100 mm for stormwater harvesting, depending on the demand and runoff volume in that area. As mentioned earlier, for this hypothetical case, the design was repeated for three different cities in Australia, having different rainfall and hydrological effectiveness zones. Therefore, the potential stormwater harvesting rate with pavement thicknesses ranging from 100 to 250 mm for the cities of Adelaide, Brisbane, and Sydney was estimated (Figure 11.9) using the hydraulic design method available in the *DesignPave v2.0* software

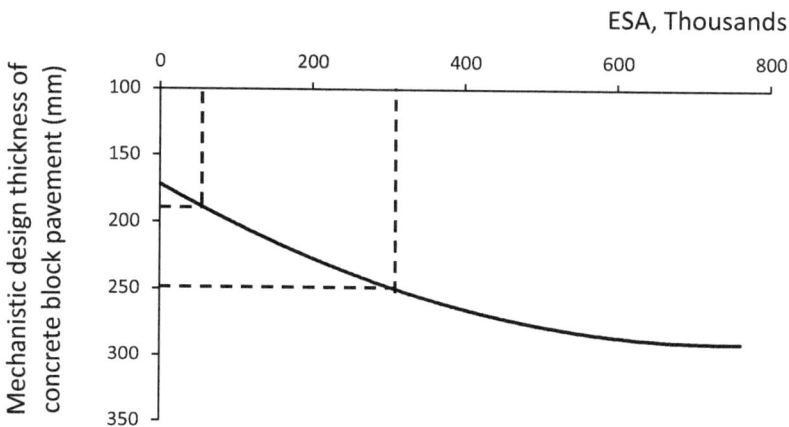

Figure 11.8 Mechanistic design thickness (mm) for concrete block pavement, indicating the thickness range for ESA values ranging from 50,000 to 3,00,000 for car parks with a subgrade with CBR = 4.

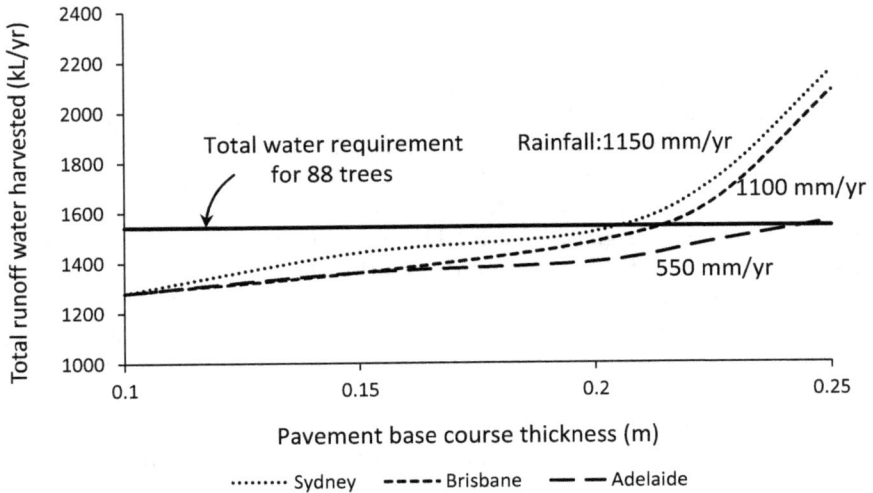

Figure 11.9 Stormwater harvesting potential (kL/year) for various design thicknesses compared to the total water requirement for 88 trees.

package. Because of the different hydrological effectivenesses and the runoff volumes for each city, the harvesting rate varied along with the design thickness. The average annual rainfall for Sydney and Brisbane were similar, and thus, the harvesting rate followed a similar pattern but the hydrological effectivenesses were different. For Adelaide, the harvesting rate was relatively lower, because of the lower annual rainfalls, even for higher water demands or pavement volumes.

As mentioned in the methods section, the total irrigation water requirement for 88 trees in their mature stage would be 1541 kL/year. From Figure 11.9, it can be assessed that a pavement thickness between 200 and 250 mm would be required to supply the irrigation water requirements for the trees planted to neutralize the carbon footprint. The decision point of the pavement thickness is therefore dependent on the amount of stormwater to be harvested or required. With the change in the tree numbers, the water requirement would change, and thus, the design thickness requirement would also change.

From Figure 11.4, the pavement carbon footprint without using recycled aggregates in the basecourse is approximately 47 kg/m^3 (for 1 m pavement thickness), with a reduced footprint of 40 kg/m^3 when using recycled aggregates. For design thicknesses of 100–275 mm, the total carbon footprint therefore ranged from 143 to 147 tonnes, respectively (Figure 11.10).

Considering the CO_2 sequestration rate specified in Figure 11.7, the estimated total CO_2 sequestration by 88 planted trees over the period of 50 years is shown in Figure 11.10. The size variation of trees with age was

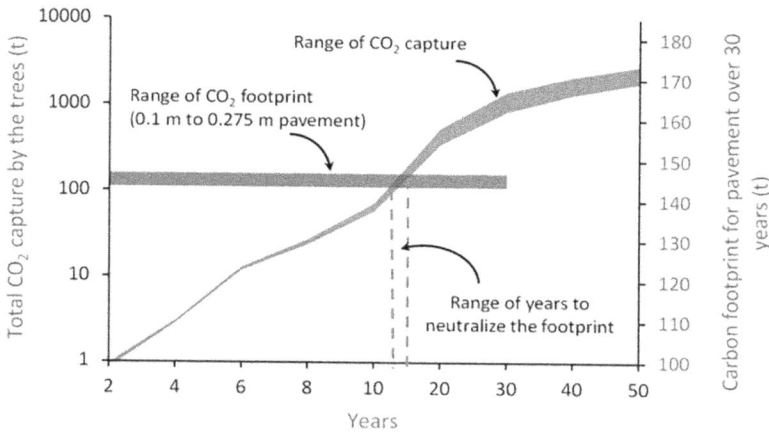

Figure 11.10 Estimated range of CO_2 sequestration for two Australian native tree species (88 trees) compared to the CO_2 footprint for a range of pavement design thicknessess, indicating the number of years required for carbon neutralization.

considered while calculating the total sequestration rate, adopting the relationship provided in Figure 11.6. If a constant footprint is considered for the pavement over its design life (20–30 years), it would require 12–15 years for 88 trees to neutralize the total carbon footprint of the 4,000 m² car park pavement (Figure 11.10).

This analysis demonstrates that pavement sustainability may be achieved by neutralizing the carbon footprint generated by the materials used in the pavement and that the integrated pavement system (recycled aggregate + permeable pavement + planted trees) is expected to reach a net-zero carbon state within its design life. If the benefits of recycling stormwater were also considered in the estimation, the net-zero state could be achieved even earlier.

11.4 CONCLUSIONS

This study has assessed the potential to reach to a stage where the carbon footprint generated by a carpark pavement could potentially be neutralized by adopting a carbon-neutral construction strategy. The total carbon footprint (in the form of CO_2) was calculated for a traditional carpark pavement system, and this was compared with the use of alternative recycled construction materials. Furthermore, an integrated pavement system was considered that included trees planted around the pavement for additional carbon sequestration. In addition, the irrigation water requirement for the trees was designed to be supplied by the stormwater harvested in the permeable pavement.

This analysis provides a robust framework for the sustainability assessment of carpark constructions but further research is required, particularly to account for the growth of different tree species planted in other climatic environments as well as in different soil conditions.

REFERENCES

Abey, S. T., & Kolathayar, S. (2020). Embodied energy and carbon emissions of pavements: a review. In *Advances in energy and built environment* (pp. 167–173). Springer Singapore. https://doi.org/10.1007/978-981-13-7557-6_14

Austroads. (2017). *Guide to Pavement Technology Part 2: Pavement Structural Design*. Austroads.

Azam, A. M., Cameron, D. A., & Rahman, M. M. (2015). Permanent strain of unsaturated unbound granular materials from construction and demolition waste. *Journal of Materials in Civil Engineering, 27*(3). https://doi.org/10.1061/(ASCE)MT.1943-5533.0001052

Beecham, S. (2020). Using green infrastructure to create carbon neutral cities: an accounting methodology. *Chemical Engineering Transactions, 78*, 469–474. https://doi.org/10.3303/CET2078079

Beecham, S., Lucke, T., & Myers, B. (2010). Designing porous and permeable pavements for stormwater harvesting and reuse. *First European IAHR Congress*, International Association for Hydro-Environment Engineering and Research, Edinburgh, UK.

Boussinesq, J. V. (1885). *Application des potentiels à l'étude de l'équilibre et du mouvement des solides élastiques*. Gauthier-Villars, Paris, France.

Cai, X., Wu, K., Huang, W., Yu, J., & Yu, H. (2020). Application of recycled concrete aggregates and crushed bricks on permeable concrete road base. *Road Materials and Pavement Design*, 1–16. https://doi.org/10.1080/14680629.2020.1742193

CCAA. (2008). *Use of Recycled Aggregates in Construction*. https://www.ccaa.com.au/imis_prod/documents/Library%20Documents/CCAA%20Reports/RecycledAggregates.pdf

CMAA (2010). *Permeable Interlocking Concrete Pavements Design and Construction Guide*. Concrete Masonry Association of Australia (CMAA), Sydney, Australia.

Cook, I., & Knapton, J. (2009). Assessment of embodied carbon in conventional and permeable pavements surfaced with pavers. *9th International Conference on Concrete Block Paving*, Buenos Aires, Argentina. https://www.cmaa.com.au/Technical/Manuals/DownloadManual/51?ManualName=assessment-of-embodied-carbon-in-conventional-and-permeable-pavements-surfaced-with-pavers.pdf

DPLG (2010). *Water Sensitive Urban Design Technical Manual for the Greater Adelaide Region*. https://www.sa.gov.au/__data/assets/pdf_file/0011/11540/WSUD_chapter_6.pdf

Edwards, J. M., & Valkering, C. P. (1974). Structural design of asphalt pavements for road vehicles - the influence of high temperatures. *Highways and Road Construction, 42*(1770), 4–9.

GWA (2019). *Citrus Irrigation Recommendations in Western Australia.* Department of Primary Industries and Regional Development, Government of Western Australia (GWA). Retrieved 20 Aug 2021 from https://www. agric.wa.gov.au/water-management/citrus-irrigation-recommendations-western-australia?page=0%2C2

Hammond, G. P., & Jones, C. I. (2008a). Embodied energy and carbon in construction materials. *Proceedings of the Institution of Civil Engineers - Energy, 161*(2), 87–98. https://doi.org/10.1680/ener.2008.161.2.87

Hammond, G. P., & Jones, C. I. (2008b). *Inventory of Carbon and Energy (ice)-Version 1.6a.* Department of Mechanical Engineering, University of Bath

Houben, L. J. M., Molenaar, A. A. A., Fuchs, G. H. A. M., & Moll, H. O. (1984). Analysis and design of concrete block pavements. *2nd International Conference on Concrete Block Paving.*

Kang, D.-H., Gupta, S. C., Ranaivoson, A. Z., Siekmeier, J., & Roberson, R. (2011). Recycled materials as substitutes for virgin aggregates in road construction: I. Hydraulic and mechanical characteristics. *Soil Science Society of America Journal, 75*(4), 1265–1275. https://doi.org/10.2136/sssaj2010.0295

Kumar, K., Kozak, J., Hundal, L., Cox, A., Zhang, H., & Granato, T. (2016). In-situ infiltration performance of different permeable pavements in a employee used parking lot - a four-year study. *Journal of Environmental Management, 167,* 8–14. https://doi.org/10.1016/j.jenvman.2015.11.019

Kuruppu, U., Rahman, A., & Rahman, M. A. (2019). Permeable pavement as a stormwater best management practice: a review and discussion. *Environmental Earth Sciences, 78*(10), 1–20. https://doi.org/10.1007/s12665-019-8312-2

Lei, B., Li, W., Luo, Z., Tam, V. W. Y., Dong, W., & Wang, K. (2020). Performance enhancement of permeable asphalt mixtures with recycled aggregate for concrete pavement application. *Frontiers in Materials, 7*(253). https://doi.org/10.3389/fmats.2020.00253

Lieth, H., & Whittaker, R. H. (1975). *Primary Productivity of the Biosphere.* Springer Verlag, New York.

Liu, Y., Li, T., & Yu, L. (2020). Urban heat island mitigation and hydrology performance of innovative permeable pavement: a pilot-scale study. *Journal of Cleaner Production, 244*(8), Article 118938. https://doi.org/10.1016/j.jclepro.2019.118938

Mahapatra, G., & Kalita, K. (2018). Effects of interlocking and supporting conditions on concrete block pavements. *Journal of the Institution of Engineers (India): Series A, 99*(1), 29–36. https://doi.org/10.1007/s40030-018-0267-x

Mohammadinia, A., Disfani, M. M., Narsilio, G. A., & Aye, L. (2018, 2018/04/20/). Mechanical behaviour and load bearing mechanism of high porosity permeable pavements utilizing recycled tire aggregates. *Construction and Building Materials, 168,* 794–804. https://doi.org/10.1016/j.conbuildmat.2018.02.179

Monrose, J., Tota-Maharaj, K., & Mwasha, A. (2021). Assessment of the physical characteristics and stormwater effluent quality of permeable pavement systems containing recycled materials. *Road Materials and Pavement Design, 22*(4), 779–811. https://doi.org/10.1080/14680629.2019.1643397

Mullaney, J., Lucke, T., Trueman, S. J., & Bai, S. H. (2016). The growth and health of street trees planted in permeable pavements. *Acta Horticulturae, 1108,* 77–82. https://doi.org/10.17660/ActaHortic.2016.1108.10

Odemark, N. (1949). *Investigations as to the Elastic Properties of Soils and Design of Pavements According to the Theory of Elasticity.* Statens Vaginstitut, Meddelande, Stockholm, Sweden.

Rahman, M. A., Imteaz, M. A., Arulrajah, A., Piratheepan, J., & Disfani, M. M. (2015). Recycled construction and demolition materials in permeable pavement systems: geotechnical and hydraulic characteristics. *Journal of Cleaner Production, 90*, 183–194. https://doi.org/10.1016/j.jclepro.2014.11.042

Rahman, M. M., Beecham, S., Iqbal, A., Karim, M. R., & Rabbi, A. T. Z. (2020a). Sustainability assessment of using recycled aggregates in concrete block pavements. *12*(10), 4313. https://www.mdpi.com/2071-1050/12/10/4313

Rahman, M. M., Beecham, S., McIntyre, E., & Iqbal, A. (2018). Mechanistic design of concrete block pavements. *Geotechnics and Transport Infrastructure*, Melbourne, Australia.

Rahman, M. M., Hora, R. N., Ahenkorah, I., Beecham, S., Karim, M. R., & Iqbal, A. (2020b). State-of-the-art review of microbial-induced calcite precipitation and its sustainability in engineering applications. *Sustainability, 12*(15), 1–43. https://doi.org/10.3390/su12156281

Razakamanarivo, R. H., Grinand, C., Razafindrakoto, M. A., Bernoux, M., & Albrecht, A. (2011). Mapping organic carbon stocks in eucalyptus plantations of the central highlands of madagascar: a multiple regression approach. *Geoderma, 162*, 335–345.

Sanicola, O., Lucke, T., & Devine, J. (2018). Using permeable pavements to reduce the environmental impacts of urbanisation. *International Journal of Geomate, 14*(41), 159–166. https://doi.org/10.21660/2018.41.Key3

Shackel, B. (1980). Progress in the evaluation and design of interlocking concrete block pavements, *Biennial Conference of the Concrete Masonry Association of Australia*, Brisbane, Australia.

Shackel, B. (1986). Computer-aided design and analysis of concrete segmental pavements, *Workshop on Interlocking Concrete Pavements*, Melbourne, Australia.

Shackel, B. (1990). *Design and Construction of Interlocking Concrete Pavements.* Elsevier Applied Science.

Shackel, B. (1992). Computer methods for segmental concrete pavement design, *Concrete Segmental Paving Workshop, Proceedings of Australian Road Research Board Conference*, Perth, Australia.

Shackel, B. (2000). The development and application of mechanistic design procedures for concrete block paving, *Japan Interlocking Block Pavement Engineering Association (JIPEA) World Congress*, Tokyo, Japan.

Shackel, B., & Arora, M. G. (1978). The evaluation of interlocking block pavements - an interim report. *Proceedings of Conference of Concrete Masonry Association of Australia*, Sydney, Australia.

Shackel, B., & Lim, D. O. O. (2003). Mechanism of paver interlock. *7th International Conference on Concrete Block Paving (PAVE AFRICA 2003)*, Sun City, South Africa.

Singh, A., Vaddy, P., & Biligiri, K. P. (2020). Quantification of embodied energy and carbon footprint of pervious concrete pavements through a methodical life-cycle assessment framework. *Resources, Conservation and Recycling, 161*, 104953. https://doi.org/10.1016/j.resconrec.2020.104953

Tak, A. A., & Kakde, U. B. (2020). Analysis of carbon sequestration by dominant trees in urban areas of thane city. *International Journal of Global Warming*, *20*(1), 1–11.

Teskey, R. O., & Sheriff, D. W. (1996). Water use by pinus radiata trees in a plantation. *Tree Physiology*, *16*(1–2), 273–279. https://doi.org/10.1093/treephys/16.1-2.273

UNFCCC (2021). The paris agreement. *United Nations Framework Convention on Climate Change*. Retrieved 25 Aug 2021 from https://unfccc.int/process-and-meetings/the-paris-agreement/the-paris-agreement

Zhang, H., Guan, D., & Song, M. (2012). Biomass and carbon storage of eucalyptus and acacia plantations in the pearl river delta, south china. *Forest Ecology and Management*, *277*, 90–97. https://doi.org/10.1016/j.foreco.2012.04.016

Chapter 12

Communal rainwater systems

An alternative to individual household systems

Thulo Ram Gurung
KBR

Ashok K. Sharma
Victoria University

Cara D. Beal and Rodney A. Stewart
Griffith University

CONTENTS

DOI: 10.1201/9781003368335-12

12.1 INTRODUCTION

12.1.1 Individual rainwater tanks

Individual household rainwater tanks are already an established decentralised feature in many parts of the world. Various countries have implemented mandatory regulation requiring the installation of rainwater tanks in new buildings; for example, new buildings in Catalonia, Spain, with certain garden sizes and new dwellings in Belgium with roof area greater than 100 m² are mandated to install rainwater tanks (Domènech and Saurí, 2011). In Queensland, Australia, before the Queensland Development Code (QDC) Mandatory Part (MP) 4.2 was made part mandatory, the installation of rainwater tanks was a popular approach to achieve potable water savings target outlined in the code (Beal et al., 2012). The mandatory regulation resulted in the increase of rainwater tanks from 24% to 36% (ABS, 2013) in the South East Queensland region between 2007 and 2010.

The increase in individual rainwater tanks (IRWT) in Australia has resulted in a plethora of studies conducted, including their mains water savings potential (e.g. Beal et al., 2012), cost-effectiveness (e.g. Hall, 2013), energy efficiency of pumps (e.g. Talebpour et al., 2014), tank optimisation and reliability (e.g. Imteaz et al., 2011, Sharma et al., 2015), and water quality issues (e.g. Ahmed et al., 2014). Although IRWT are a popular approach to achieving potable water savings, the lack of householders' skills and knowledge in maintaining these tanks may result in their inefficiency, which may potentially contribute to health risks. Furthermore, with population growth in cities around the world resulting in urban densification through the subdivision of large land plots to smaller residential areas, space constraints may limit the storage capability for an individual allotment.

12.1.2 Communal rainwater systems

An alternative configuration to the traditional IRWT is the implementation of communal rainwater systems (CRWS), which have been reported to reduce demand for mains drinking water by around 60% (Coombes et al., 2000). As depicted in Figure 12.1, such systems collect, store, and treat rainwater across multiple households within a residential development in a single storage and treatment facility. The treated water is then redistributed to the household for either potable or non-potable purposes. The option of providing rainwater as a potable source is particularly suited for an off-grid community, which has difficulty in accessing the mains water supply. CRWS are intended to be plumbed for internal household use, and a continuous supply is required to avoid disruption to the intended use. As such systems are climate dependent, they will be unable to fully achieve 100% reliability. Hence, a supplementary source is required, such as on-site bore water and, if accessible, mains water top-up.

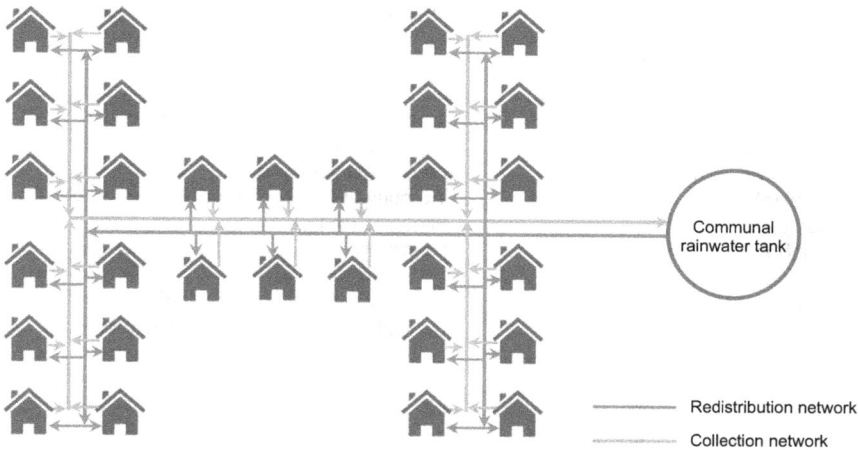

Figure 12.1 Schematic of a typical communal rainwater system.

While there is an extensive literature source on IRWT, there is a lack of knowledge on the design and life cycle costing of CRWS as it is a relatively new and emerging configuration. Table 12.1 summarises some examples of CRWS in Australia.

Limited studies of rainwater use in multiple and single residential dwellings have reported the former to be more economical than the latter. For instance, IRWT had a payback period of 21 years, against 5 years for a multi-dwelling rainwater tank in Brazil (Ghisi and Ferreira, 2007; Ghisi and Olvieira, 2007); 45 years against 31 years in Spain (Domenech and Sauri, 2011); and in Australia, 23 years (Willis et al., 2013) against between 9 and 22 years, depending on the city's location (Zhang et al., 2009).

The cost-effectiveness of decentralised water systems utilised in multi-dwelling units, as compared to their use in single dwellings, has also been reported in similar studies (e.g. Friedler and Hadari, 2006; Mourad et al., 2012). Similarly, an investigation into the cost-effectiveness of rainwater systems of different spatial configurations in Spain showed that a neighbourhood-scale rainwater system for new developments was the most cost-effective (Farreny et al., 2011). Furthermore, Roebuck et al. (2011) reported that individual household rainwater tanks are not financially viable when all costs are included, and recommended further work on the cost benefits of installing rainwater tank systems on a larger scale.

The outcomes of these previous studies have highlighted the cost-effectiveness of using rainwater tanks in multi-dwelling developments, whereby the costs of installing, operating, and maintaining the system are shared by a number of families (e.g. economies of scale). Traditional centralised water or sewage network system utilises such economy of scale to keep the costs

Table 12.1 Examples of developments adopting communal rainwater systems in Australia

Scheme	Building type	Summary of scheme
Fitzgibbon Chase Potable Roof water (PotaRoo) Scheme, Fitzgibbon, QLD (DILGP, 2012)	1,230 residential homes	Rainwater collected from a total roof catchment of 110,000 m² in satellite communal scale plants. The rainwater is then pumped into an 800 kilolitres (kL) centralised storage and treatment facility for redistribution to consumers. Treated water will initially be distributed through the non-potable reticulation line (supplied by stormwater). Over a period of monitoring, the water will be supplemented into the mains water grid.
Capo di Monte, Mount Tambourine, QLD (Cook et al., 2013)	46 residential homes (off-grid)	Rainwater collected from a total roof catchment of 10,700 m². Rainwater is stored in two 200-kL tanks; 100 kL assigned for residential potable use and the remaining 100 kL for firefighting purposes. Collected rainwater is treated to potable standards in on-site water treatment facility for use in kitchens, bathrooms and laundry. Wastewater is also harvested and treated for non-potable purposes. Bore water is used as a supplementary top-up for both alternative sources.
Green Square North Towers, Brisbane, QLD (Cook et al., 2014)	12-storey commercial building	Rainwater collected from a roof catchment area of 1,600 m² and gravity fed to a 110 m³ basement tank. Water is then pumped to two smaller tanks of 21 and 28 kL each on the roof, from where it is used for toilet flushing and irrigation. Mains water is provided as a backup supply.
Kogarah Town Square, Kogarah, NSW (Sydney Water, 2007)	193 apartments, commercial buildings	Rainwater is collected from the roof and upper level terraces in an underground tank. The collected water is screen-filtered and used for toilet flushing, car washing, and water features in the town square. Stormwater is also collected and used for irrigation purposes.

of providing the services low. However, the diseconomies of scale prevalent with pipe collection systems may potentially increase the costs of such systems after an optimal household scale has been reached (Booker, 1999; Clark, 1997; Fane et al., 2002). CRWS, which also utilise pipe systems to collect and distribute the rainwater, would also be likely to experience such diseconomies of scale in the collection network.

This chapter explores the design, financial, and management aspects of CRWS through:

- Guidance on the design concept of a CRWS from various literature sources and an example site, as no known specific guidelines on its design exists.
- Financial assessment of a CRWS based on the principles of economy of scale through the sharing of the capital and ongoing expenses amongst a number of households.
- Management of the CRWS through formal arrangements in place for its maintenance and operation.

12.2 STRUCTURE AND DESIGN OF COMMUNAL RAINWATER SYSTEMS

12.2.1 Design summary of communal rainwater systems

The primary function of the CRWS is to collect, transport, treat, and recirculate the treated rainwater back to the residential households for either potable or non-potable use. The design of the CRWS is provided in detail in Gurung and Sharma (2014) and also serves as a method in determining the most economical household scale for the subsequent section. The main steps in the design are as follows:

1. Develop a representative housing layout using local information for the development, such as size of lot, roof area, road width, and housing density.
2. Design the various scales of housing layouts (e.g. 4, 8, 16, etc., lots) based on the collated information.
3. Select the location of the communal tank considering the topography of the development (see Figure 12.2).
   a. For a flat topography, the tank should be situated in the centre to minimise the costs of laying the collection and distribution pipes.
   b. In a sloped topography, the tank should be placed on the lower side of the gradient to utilise the slope of the land for rainwater collection.
4. Plan the layout of the collection and distribution systems for all households.
5. Determine the application type of the rainwater system, that is, either for potable or for non-potable purposes, and collect water demand information for the associated end-uses.

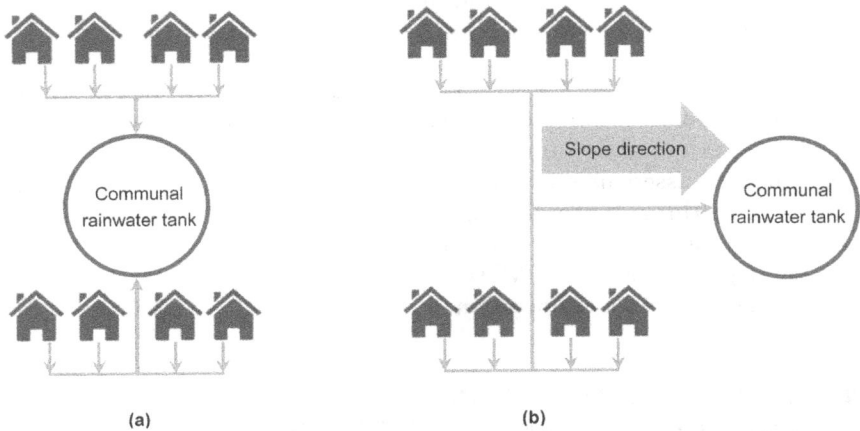

Figure 12.2 Examples of communal rainwater tank location in (a) a flat topography and (b) a sloping topography.

6. Estimate the peak flow rates for both the rainwater collection and distribution systems using local guidelines.
7. For each layout, estimate the sizes of the storage tanks, collection and recirculation pipes, and the pumping capacity. Interim storage tanks should be considered if the main storage tanks are determined to be placed deep underground as it can be uneconomical to construct and operate.

12.2.2 Design example of a CRWS

In a CRWS, rainwater collected from a community is normally transferred via gravity through a pipe network to a storage tank, where it is then further treated and stored before being redistributed through a distribution network to the connected household (see Figure 12.3). Thus, there are a number of physical features to consider in the design of a CRWS, namely:

- Rainwater collection pipes – size of gravity collection pipe system.
- Storage tanks – size of storage tanks and, if required, interim tanks.
- Distribution system network – sizes of distribution pipes and recirculation pumps.
- Water treatment facility – the type of on-site treatment required for the rainwater.

As CRWS are relatively new in Australian cities, there is a lack of how such systems are likely to perform. Hence, the design of the CRWS in this section

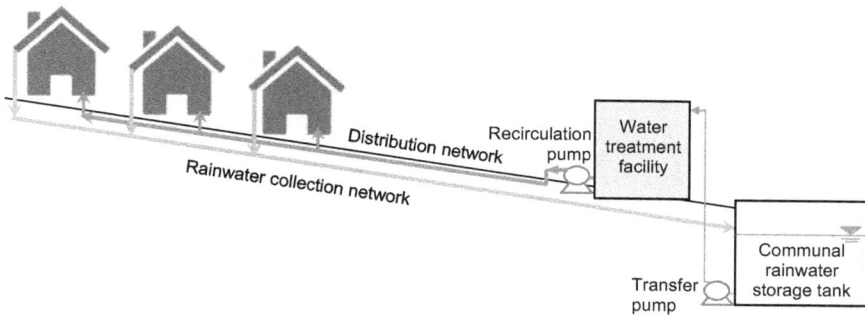

Figure 12.3 Features of a typical communal rainwater system.

makes references to available literature sources and a well-known communal system in Capo di Monte (CDM) located at Mt Tamborine, Queensland.

12.2.2.1 Rainwater collection pipes

Rainwater collection pipes convey rainwater from roofs, through downpipes and in-ground connecting pipes, to the storage tanks. Peak flow rates from each roof are required to adequately size the collection pipes and can be estimated using the rational method (DERM, 2007), if the roof area and rainfall intensity are known. A rearranged Manning's equation can then be used to size the gravity collection pipes, depending on the chosen pipe material. The rearranged Manning's equation is as shown in Equation (1).

$$ d = \left(\frac{4^{5/3}.Q.n}{\pi.S^{1/2}} \right)^{3/8} \tag{1} $$

where d is the diameter of the pipe (m), Q is the peak flow rate (m³/s), n is Manning's roughness coefficient, and S is the slope of the pipe.

In CDM, an in-ground connecting pipe connects the downpipes of each property to the main collection pipe. The size of the in-ground pipe limited the rainfall intensity entering the collection pipes to approximately 2 mm/5 min. Using the rational method, this intensity produced a peak flow rate of 1.28 L/s per dwelling, for a roof size of 220 m² with 100% connectivity and a loss coefficient of 0.875.

Local guidance recommends that a minimum pipe velocity of 0.6 m/s be maintained to ensure that pipes are self-cleansing (DERM, 2007). Hence, in the CDM area, for a slope of 0.5%, a 100-mm-diameter polyvinyl chloride (PVC) pipe, with Manning's roughness of 0.01 (Rossman, 2010), is required to produce a minimum pipe full velocity of 0.6 m/s (DERM, 2007).

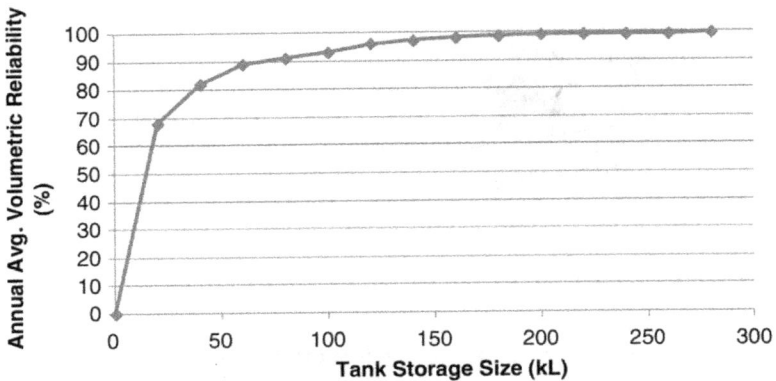

Figure 12.4 Volumetric reliability of various tank sizes for the CRWS at Capo di Monte (Cook et al., 2012).

12.2.2.2 Storage tanks

12.2.2.2.1 Main storage tank

The volume of the required storage tanks can be determined using a water balance tool, such as UVQ (Mitchell and Diaper, 2006). The sizing of the tank is dependent on a number of factors, namely historical rainfall data, size of the roofs, the intended water usage, and reliability of the supply. Since CRWS is a climate-dependent water supply source, a tank reliability of 100% is not achievable, and a supplementary source (e.g. on-site bore water) may be required.

In the CDM site, rainwater is used mainly for indoor potable purposes, and hence, has little seasonal variation, unlike water used for outdoor purposes. A reliability analysis of the CRWS at this site was conducted using a 20-year period rainfall data obtained from the Bureau of Meteorology, and with the maximum rainfall intensity restricted to 2 mm/5 min due to the flow limitation of the connectors to the main collection system. Figure 12.4 shows the results of a reliability analysis of various tank storage sizes at CDM. A tank with a storage volume of 160 kL would provide a reliability of 98% over the 20-year period, while half of this tank volume (80 kL) would provide a reliability of 90%.

In CDM, rainfall did not have to meet all the water demand as groundwater was used as a backup supply in times of low rainfall. The presence of a backup supply is especially important, as it reduces the need for large storage tanks, while still providing a reliable supply of water to the end users. For instance, a storage volume of 240 kL produced a reliability of 99.5% in CDM. This is a large increase in storage volume for only a marginal gain in reliability, which will affect the construction costs of the CRWS. Furthermore, space constraints may limit the use of such large tanks.

12.2.2.2.2 Interim tank and pump

In the event that collection pipes run deep underground (e.g. due to the flat topography of the development site), a smaller interim tank may be required for the temporary storage of rainwater before it is pumped to the main storage tank. This is to avoid costly deep excavations for the larger storage tanks. Sizing of the interim tanks can again be done using a similar water balance tool as used for the main storage tank (such as UVQ) with the required size of the pump based on Bernoulli's equation (Swamee and Sharma, 2008).

12.2.2.3 Distribution system network

Local planning guidelines and hydraulic methods for pipe sizing can be used to determine water distribution pipe sizes and the required pumping power. From local guidelines, the minimum pipe diameter for potable mains is 100 mm (GCCC, 2008). Pipe velocities are not to exceed 2.5 m/s to prevent pipe failures, and pressures at residential properties are to be maintained between 22 and 80 m (SEQ Code, 2013).

In a simple and small CRWS (e.g. where there are few households and ground-level variation is minimal), checks for mains water velocity can be done by dividing the peak supply flow by the area of the pipe. The peak supply flow for the study area can be estimated using peak day and peak hour factors provided by the local guidelines (e.g. SEQ Code, 2013). The minimum pressure for potable water at the individual dwellings and the head required for the distribution pumps can be calculated using Bernoulli's equation (Swamee and Sharma, 2008), with the Darcy-Weisbach formula (Swamee and Sharma, 2008) used to calculate the head losses in the pipe network. The power required for the distribution pumps can be calculated by using the pump equation (Swamee and Sharma, 2008) and assuming suitable values of pump and motor efficiencies. The required size of the pumps can then be determined by using the pump curve of the selected pump manufacturer.

However, a more complex and interconnected CRWS supply network would require the use of a water distribution network software, such as EPANET (Rossman, 2000), to check for failure criteria and pumping requirements. Such distribution software requires the user to be familiar with the hydraulics of the modelled CRWS water supply network so that pipes and pumps are sized accordingly.

12.2.2.4 Water treatment

The application of a CRWS provides the opportunity to use the collected rainwater for either non-potable or potable purposes. If the CRWS is used as a non-potable source, the downpipes should have screening and first

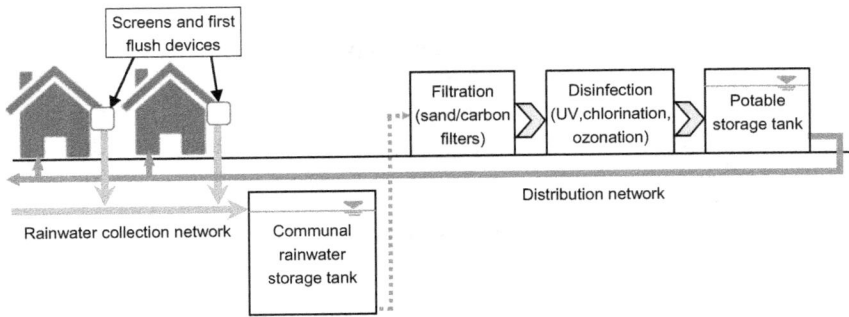

Figure 12.5 Typical rainwater treatment configuration.

flush systems installed at individual dwellings before the rainwater enters the collection network. Such systems would be sufficient for improving the rainwater quality for non-potable applications. However, if the rainwater water is to be used for potable purposes due to the potential presence of faecal indicators in rainwater tanks, the rainwater needs to be treated before being supplied to the households (Ahmed et al., 2014). Rainwater will require physical, chemical, and disinfection processes to achieve Australian Drinking Water Guidelines (NHMRC and NRMCC, 2011) or any other local drinking water guidelines objectives for safe use of the rainwater for potable purposes.

The water treatment unit at CDM consists of a transfer pump, which pumps the rainwater from the storage tank through a sand and carbon filter, ultraviolet (UV) sterilisation system, and chlorination processes, to a holding tank, where it is then distributed to the households. The treatment processes at CDM are more than adequate to treat rainwater to potable standards, as the filtration and UV disinfection ensure the sufficient destruction of protozoa, while the addition of chlorine prevents bacterial regrowth in the distribution system (Cook et al., 2015). However, care has to be taken during the chlorination process as the presence of a high organic content in the rainwater may result in the formation of disinfection by-products (DBPs), which can cause health issues to the consumer (Knight et al., 2010). UV sterilisation is generally sufficient for rainwater to be treated to potable standards; as it is easy to use, has a short retention time, does not require the addition of chemicals, and has a low chance of forming DBPs. However, unlike chlorination, it provides no residual disinfection in the distribution systems (Cook et al., 2015). Additionally, upstream filtration is necessary to ensure the effectiveness of the UV disinfection system, as the presence of large particles in the rainwater may make it difficult to disinfect. Figure 12.5 shows the basic components of a rainwater treatment facility.

Additionally, other options of disinfection are available, including ozonation, and adding hydrogen peroxide. Further treatment beyond the disinfection process to reduce trace contaminants includes membrane filtration methods, such as microfiltration, nanofiltration, and reverse osmosis. However, such complex water treatment processes are only to be included in the treatment process if the presence of highly contaminated water source, such as stormwater, is part of the supply system.

12.2.3 Economies of scale and financial assessment of communal rainwater systems

Economies of scale is a well-established feature of the traditional centralised water or sewage pipe systems, whereby the sharing of the costs of large infrastructure systems produces cost advantages to both the customers and the service provider. Similarly, in a CRWS, an economy of scale would be present due to the multiple connected households. However, with an ever-increasing number of households connected to the CRWS, the lengths of the collection pipes would eventually result in a diseconomy of scale. This section investigates the economies of scale of CRWS to determine the optimal number of households where minimum costs occurs and compares against an IRWT system.

12.2.3.1 Economies of scale of communal rainwater systems – example application

The conceptual design of the CRWS is based on the method described in the previous section and in further detail in Gurung and Sharma (2014). Housing layouts ranging from 4 to 576 households that are connected to a communal rainwater tank were developed for an area, assumed to have a flat topography, in South East Queensland. Local information was sourced and used to characterise the CRWS to estimate the costs of the individual components in the system. Lot dimensions of 16 m by 25 m were adopted for the study to obtain the approximate required lengths of the collection and distribution pipes. The collection pipes were assumed to have a slope of 0.5% to produce the minimum pipe full velocity. The average roof area for a single dwelling was estimated to be 220 m² and was assumed to be fully connected to the CRWS. From an end-use study of the local area (Beal et al., 2011), the potable water demand was estimated as 82.3 litres per person per day (L/p/d), while a household occupancy rate of 2.6 (OESR, 2012) was used. A 94% supply reliability was used to size the rainwater tank, with the remaining demand provided by bore-water top-up. Unit rates for the different components within the system (e.g. pumps, pipes, tanks) were sourced from local distributors.

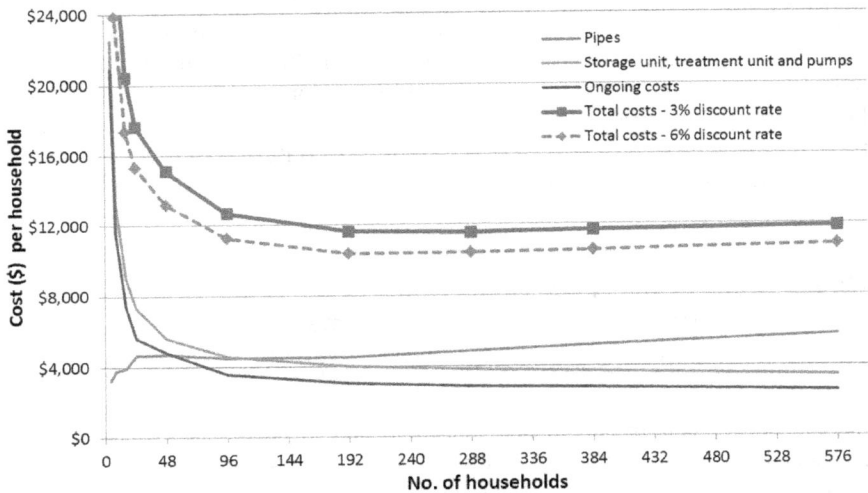

Figure 12.6 Cost per household for the various components in a CRWS (3% discount rate) and the total costs per household for both the 3% and 6% discount rates.

The net present value (NPV) method was used to conduct the economic life cycle cost (LCC) estimation of the various CRWS configurations. The NPV is expressed in Equation (2).

$$\text{NPV} = \frac{F}{(1+i)^n} \qquad (12.2)$$

where F is the future cost, i is the discount rate, and n is the analysis period.

An analysis period of 50 years and a discount rate of 3% were used to conduct the LCC estimation. A 50-year period was chosen as anything beyond this period will make the present value of the future investment insignificant. A sensitivity analysis was also conducted using a 6% discount rate due to the uncertainty of the rates used. The units used for the economical assessment, henceforth, are all in Australian dollars ($).

The results of the NPV are plotted in Figure 12.6 and show that the most economic dwelling scale was in the range of 192 to 288 households, with the lowest NPV of $11,543 occurring for development of 288 households. Beyond this number of dwellings, the cost of the communal system on a per household basis increases due to the diseconomy of scale from increasing pipe costs, which are not sufficiently counterbalanced by the economies of scale of other components in the system. For lower number of households, (<100) treatment and storage units are the dominant capital cost components, while the diseconomy of pipes affects the capital costs components of the higher-scale households, in agreement with observations by both Booker (1999) and Clark (1997).

The sensitivity analysis conducted for a discount rate of 6% showed little difference in the overall costs of the CRWS when compared to the baseline 3%: $10,390 against $11,543 (see Figure 12.6). This was due to the construction costs of the CRWS incurring a large capital investment, as it was assumed that the CRWS will be built within the first year. The discount rates only had an influence on the ongoing costs, which were small in comparison with the capital costs. For instance, at the optimal household layout, the costs of the main capital units (i.e. pipes, treatment units, and storage units) made up 75% of overall costs, with the rest made up of ongoing costs. Incidentally, the sensitivity analysis had a minimal effect on the optimal housing scale, which was similar to the results of the financial analysis conducted on a communal sewer model (Clark, 1997).

12.2.3.2 Comparison against individual rainwater tanks

A comparison of CRWS against IRWT was done to understand the cost-effectiveness of both systems. IRWT in Queensland, Australia, are normally only connected to non-potable appliances under the now part-mandated QDC MP 4.2. As the communal systems in this example supply to potable appliances, it is unrealistic to compare the CRWS and IRWT configurations on a per dwelling basis due to the different water demands for the two systems. Instead, the cost-effectiveness for both systems was expressed and compared as a levelised cost, which assumes that the unit cost of water as well as the discount rate is constant over the period of analysis (Hall, 2013). The levelised cost equation is expressed using Equation (3).

$$\text{Levelised cost } (\$/\text{kL}) = \frac{C + \sum_{t=1}^{n} A_t/(1+i)^t}{\sum_{t=1}^{n} Y_t/(1+i)^t} \tag{12.3}$$

where C is the capital cost, A is the annual costs, i is the discount rate, Y is the annual yield of water (kL), t is the year, and n is the period of analysis.

The levelised costs for the CRWS for all household layouts were calculated for both discount rates over the 50-year analysis period. The results shows that the optimal housing layout for the levelised cost was unchanged when compared against the NPV, with the optimal levelised costs still occurring at the 192 to 288 household range. The levelised costs for the 3% and 6% discount rates are $6.11/kL and $8.97/kL, respectively.

The levelised cost method was used to compare the cost-effectiveness of the IRWT against the CRWS. The NPV of the IRWT for non-potable use, with a demand of 54 L/p/d, was obtained using similar parameters to the communal system (i.e. roof area, household occupancy). A 5 kL rainwater

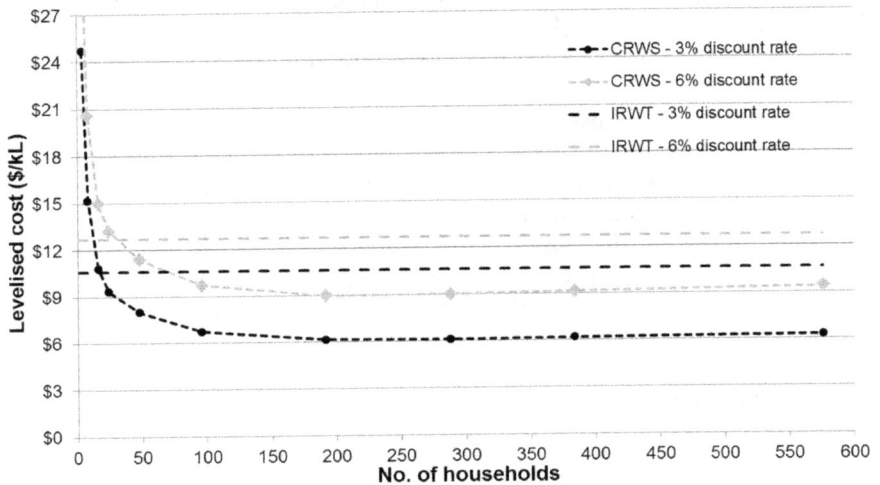

Figure 12.7 Levelised costs ($/kL) of a CRWS and an IRWT for 3% and 6% discount rates on a per household basis.

tank, as recommended in the QDC MP4.2, was used and produced a volumetric reliability of 90%. A NPV of $12,576 was obtained for a 3% discount rate, and $9,232 for a 6% discount rate, over the 50-year analysis period.

While the NPV for an IRWT is higher than the NPV for the optimal communal household level for both discount rates, converse results are obtained for their levelised costs. A levelised cost of $10.59/kL and $12.69/kL was obtained for the IRWT for the 3% and 6% discount rates, respectively. The levelised costs for the IRWT were higher than the levelised costs of the CRWS, and highlight the cost-effectiveness of the latter configuration over the former. In fact, it is more economical to implement CRWS than IRWT for a development with more than 24 dwellings. Figure 12.7 presents the levelised costs for both the CRWS and IRWT.

12.2.3.3 Other influencing factors

The method presented in this section provides a means to determine the cost-effectiveness of CRWS. However, variable factors, such as land topography, climatic conditions, rainwater end-uses, connected roof area, household occupancy, level of treatment required, local water guidelines, will vary the final costs for different locations. To understand the influence of local factors on costs, an analysis was done to determine the effects of a sloped topography on the optimal household configuration. A 0.5% slope, similar to the gradient of the pipes, was assumed. The unit costs of the

Table 12.2 Cost differences between a flat and sloped topography for the optimal household layout

Components	Flat topography	Sloped topography	Percent difference
Capital costs			
Pipes	$4,856	$3,819	21%
Storage unit, treatment unit, and pumps	$3,826	$3,531	8%
Ongoing costs			
Maintenance	$2,078	$1,990	4%
Operation	$553	$485	12%
Replacement	$230	$211	8%
Total NPV	$11,543	$10,036	13%

pipes were reduced by 16% as a result of the parallel configuration of the collection and recirculation pipes (Booker, 1999). Table 12.2 shows the comparison of the results of the flat and sloped topography for the optimal household scale on an individual household basis.

The sloped topography resulted in an overall reduction in all costs components. Shallower pipe depths with a parallel configuration reduced the capital costs of the pipes by 21%. The removal of the interim storage tank and its pump reduced the costs of the storage unit, treatment unit, and pumps by 8%, and the ongoing costs by between 4% and 12%. The overall NPV reduced from $11,543 to $10,036, a reduction of 13%. The levelised cost for the sloped topography configuration was $5.31/kL, which was approximately half of the IRWT configuration ($10.59/kL).

This simple analysis demonstrates how local factors affect the overall costs and the costs of the various components of a CRWS. Hence, a single financial assessment will not provide conclusive evidence on the final costs of a CRWS. Instead, individual financial assessments are required for CRWS at different locations by taking into consideration the local varying factors, for example, climatic conditions/rainfall pattern and rainwater end-uses, which will affect the size and, hence, costs of the storage tanks.

12.3 MANAGEMENT OF COMMUNAL RAINWATER SYSTEMS

As discussed in the previous sections, CRWS provides an advantage over IRWT by potentially providing lower levelised costs of using the rainwater through an economy of scale, and with the added benefit of using the rainwater for either potable or non-potable purposes. In the case of IRWT, the lack of time, motivation, and skills of householders to maintain the tanks may result in issues with ongoing maintenance that may lead to increased failure rates of systems components and poor water quality (Mankad et al., 2012). Such failure and maintenance issues would reduce the positive impacts of IRWT on mains water savings.

On the other hand, CRWS provide the opportunity of sub-contracting the operation and maintenance to industry specialists who have the skills and knowledge to operate the system efficiently, allowing the householders to adopt a more passive role in mains water savings (Cook et al., 2015). Regular monitoring of the CRWS as part of the management system would enable the early identification and fixing of any faults, enabling the system to run efficiently and maximise its mains water savings potential. On their part, householders should cooperate with their neighbours to utilise the rainwater from the CRWS in a responsible and formal manner to limit the cases of conflicts, for instance, complaints against individuals using more water than allocated or necessary.

The management arrangement for a CRWS requires the roles and responsibilities of the management entity to be defined in the early stages. This includes the responsibilities of outsourcing the operation and management of the CRWS to service providers and a payment model that ensures financial sustainability of the CRWS. Other decisions to be made by the management entity include setting up a risk management plan to identify and mitigate potential risks, such as water quality and health risks (Cook et al., 2015). Hence, the management entity must serve as a professional body that coordinates the needs and requirements of the various stakeholders of the CRWS, including the householders and regulatory officials. This type of agreement could be formalised under a Body Corporate arrangement.

12.4 CONCLUSIONS

A CRWS can provide a decentralised alternative to the popular traditional individual household rainwater tanks, offering the option to supply to either potable or non-potable appliances. While information on CRWS is limited, there have been a number of CRWS case studies that have been implemented successfully. Such case studies provide a good point of information and reference to serve as a guideline in the design of a CRWS, which subsequently requires a good knowledge of local water guidelines and hydraulic methods. CRWS may offer cost savings over the traditional individual rainwater systems, due to the involved economies of scale, as multiple households share the costs of the systems. The final costs may vary for different locations due to a number of factors, such as topography, climatic conditions, and rainwater end-uses. In addition, CRWS provides an opportunity to put in place formal agreements for a responsible entity to address the maintenance challenges usually experienced for IRWT.

REFERENCES

ABS (2013). *4602.0.55.003- Environmental Issues: Water use and Conservation*, March 2013. Australian Bureau of Statistics.

Ahmed, W., Brandes, H., Gyawali, P., Sidhu, J.P.S. and Toze, S. (2014). Opportunistic pathogens in roof-captured rainwater samples, determined using quantitative PCR. *Water Research*, 53, 361–369.

Beal, C., Stewart, R.A., Huang, T.T. and Rey, E. (2011). South East Queensland residential end use study. *Journal of the Australian Water Association*, 38 (1), 80–84.

Beal, C.D., Sharma, A., Gardner, T. and Chong, M. (2012). A desktop analysis of potable water savings from internally plumbed rainwater tanks in South-East Queensland, Australia. *Water Resources Management*, 26 (6), 1577–1590.

Booker, N. (1999). Estimating the economic scale of greywater reuse systems. *Program Report FE-88*. CSIRO Molecular Science.

Clark, R. (1997). *Water Sustainability in Urban Areas: An Adelaide and Regions Case Study, Report Five: Optimum Scale for Urban Water Systems*, Department of Environment and Natural Resources, Adelaide, South Australia.

Cook, S., Sharma, A., Gurung, T. R., Chong, M. C., Umapathi, S., Gardner, T., Carlin, G. and Palmer A. (2012). Performance of cluster scale rainwater harvesting systems: analysis of residential and commercial development case studies. *Urban Water Security Research Alliance Technical Report No. 68*.

Cook, S., Sharma, A. and Gurung, T. R. (2014). Evaluation of alternative water sources for commercial buildings: a case study in Brisbane, Australia. *Resources, Conservation and Recycling*, 89, 86–93.

Cook, S., Sharma, A., Gurung, T.R., Neumann, L.E., Moglia, M. and Chacko, P. (2015). Cluster scale rainwater harvesting. In: A. Sharma, D. Begbie and T. Gardner (Eds.) *Rainwater Tank Systems for Urban Water Supply: Design, Yield Health Risk, Economics and Social Perceptions*. IWA Publishing, London, UK

Coombes, P.J., Argue, J.R. and G. Kuczera (2000). Figtree place: a case study in water sensitive urban development (WSUD). *Urban Water*, 1 (4), 335–343.

DERM (2007). *Queensland Urban Drainage Manual (QUDM)*, Volume 1, Second Edition, Department of Environmental and Resource Management, Queensland Government.

DILGP (2012). *Fitzgibbon Chase Guide Book*. Department of Infrastructure, Local Government and Planning, Queensland Government, Australia, June 2012.

Domenech, L. and Sauri, D. (2011). A comparative appraisal of the use of rainwater harvesting in single and multifamily buildings of the Metropolitan Area of Barcelona (Spain): social experience, drinking water savings and economic cost. *Journal of Cleaner Production*, 19 (6–7), 598–608.

Fane, S.A., Ashbolt, N.J. and White, S.B. (2002), Decentralised urban water reuse: the implications of system scale for cost and pathogen risk. *Water Science Technology*, 46 (6), 281–288.

Farreny, R., Gabarrell, X. and Rieradevall, J. (2011). Cost-efficiency of rainwater harvesting strategies in dense Mediterranean neighbourhoods. *Resources, Conservation and Recycling*, 55 (7): 686–694.

Friedler, E. and Hadari, M. (2006). Economic feasibility of on-site greywater reuse in multi-storey buildings. *Desalination*, 190(1–3), 221–234.

GCCC (2008). *Gold Coast Planning Scheme Policies, Policy 11: Land Development Guidelines, Section 4: Water Reticulation e Design Requirements*. Gold Coast, Queensland, Gold Coast City Council.

Ghisi, E. and Ferreira, D.F. (2007). Potential for potable water savings by using rainwater and greywater in a multi-storey residential building in southern Brazil. *Building and Environment*, 42 (7), 2512–2522.

Ghisi, E. and de Oliveira, S.M. (2007). Potential for potable water savings by combining the use of rainwater and greywater in houses in southern Brazil. *Building and Environment*, 42 (4), 1731–1742.

Gurung, T.R. and Sharma, A. (2014). Communal rainwater tank systems design and economies of scale. *Journal of Cleaner Production*, 67, 26–36.

Hall, M. (2013). Review of rainwater tank cost effectiveness in South East Queensland. *Urban Water Security Research Alliance Technical Report No. 105*, Queensland Government, Australia.

Imteaz, M.A., Ahsan, A., Naser, J. and Rahman, A. (2011). Reliability analysis of rainwater tanks in Melbourne using daily water balance model. *Resources Conservation and Recycling*, 56(1), 80–86.

Knight, N., Shaw, G., Sadler, R. and Wickramasinghe, W. (2010) Disinfection By-product (DBP) Formation and Minimisation in Southeast Queensland Drinking Water. *Urban Water Security Research Alliance, Technical Report*.

Mankad, A., Tucker, D. and Greenhill, M.P. (2012). Mandated versus retrofitted tank owners: psychological factors predicting maintenance and management. *Urban Water Security Research Alliance Technical Report No. 51*.

Mitchell, G. and Diaper, C. (2010). UVQ User Manual. *CMIT Report No. 2005-282*, CSIRO.

Mourad, K. A., Berndtsson, J. C. and Berndtsson, R. (2011). Potential fresh water saving using greywater in toilet flushing in Syria. *Journal of Environmental Management*, 92 (10), 2447–2453.

NHMRC and NRMMC (2011). *Australian Drinking Water Guidelines Paper 6 National Water Quality Management Strategy*. National Health and Medical Research Council, National Resource Management Ministerial Council, Commonwealth of Australia, Canberra.

OESR (2012). *Population and Dwelling Profile; South East Queensland, Office of Economic and Statistical Research*, Queensland Government, April 2012.

Roebuck, R.M., Oltean-Dumbrava, C. and Tait, S. (2011). Whole life cost performance of domestic rainwater harvesting systems in the United Kingdom. *Water and Environment Journal*, 25 (3), 355–365. doi: 10.1111/j.1747–6593.2010.00230.x.

Rossman, L.A. (2000). *EPANET2, Users Manual*. US Environmental Protection Agency. Cincinnati, OH.

SEQ Code (2013). *South East Queensland Water Supply and Sewerage Design and Construction Code*, South East Queensland Code, Queensland.

Sharma, A.K., Begbie, D. and Gardner, T. (2015). *Rainwater Tank Systems for Urban Water Supply – Design, Yield, Energy, Health risks, Economics and Community perceptions*, IWA Publishing.

Swamee, P.K. and Sharma, A.K. (2008). *Design of Water Supply Pipe Networks*. John Wiley and Sons, New Jersey.

Sydney Water (2007). Best practice guidelines for water conservation in commercial office buildings and shopping centres, Part 3: Alternative water sources.

Talebpour, M.R., Sahin, O., Siems, R. and Stewart, R.A. (2014). Water and energy nexus of residential rainwater tanks at an end use level: case of Australia. *Energy and Buildings*, 80, 195–207.

Willis, R.M., Stewart, R.A., Giurco, D.P., Talebpour, M.R. and Mousavinejad, A. (2013). End use water consumption in households: impact of socio-demographic factors and efficient devices. *Journal of Cleaner Production*, 60, 107–115.

Zhang, Y., Grant, A., Sharma, A., Chen, D. and Chen, L. (2009). Assessment of rainwater use and greywater reuse in high-rise buildings in a brownfield site. *Water Science & Technology*, 60(3), 575–581.

Chapter 13

Climate change and geo-infrastructures

Assessment and challenges

Susanga Costa
Deakin University

Dilan Robert
RMIT University

Jayantha Kodikara
Monash University

CONTENTS

13.1 INTRODUCTION

Climate change is no longer a topic limited to debate within scientific community, but a phenomenon experienced by ordinary people in day-to-day life. While the drivers of global warming are still being debated by some in the political arena, the evidence of extreme weather events, rising sea level, ocean acidification, and sea ice loss cannot be ignored. From bush fires in Australia, and cyclones in the USA to floods in South Asia and droughts in Africa, the world has seen the aggressive face of climate change across the globe. In the four decades from 1975 to 2014, frequency of natural disasters recorded around the world has tripled (Thomas and Lopez, 2015), revealing that our infrastructure needs more holistic and resilient approach against climate change-driven scenarios.

DOI: 10.1201/9781003368335-13

Extreme weather events induced by climate change in particular and environmental factors in general have a significant effect on the performance of shallow ground behaviour and geo-infrastructures placed in this zone. The impact of climate change on infrastructure is multi-dimensional. In regions where an increased precipitation is expected, more infrastructure damage and soil erosion are likely. Increased ambient and ground temperatures will accelerate evaporation and lead to desiccation crack formation. Decrease in precipitation will affect natural water bodies and cause water pollution. Rise in sea level, as we have seen in the recent past, could cause coastal erosion, wetland flooding, saltwater intrusion into aquifers and agricultural lands, and loss of habitat for fish and wildlife.

The effect of climate change on built environment as well as natural resources has stimulated the worldwide scientific community to investigate new adaptation methods to deal with current and expected future climatic changes. Recent research contains investigations into the impact on landslides, mining excavations, buildings, dams, road pavements, and slopes (Alvioli et al. 2018; Bagui et al. 2022; Choudhury et al. 2022). Questions that need to be answered include whether the existing infrastructure will be able to survive the impact and whether our design practices for new infrastructure are sufficiently robust (Vardon 2015). Majority of these studies, however, focus on developing mitigating measures for extreme climate events. While addressing the influence of extreme events is of clear importance, changes in average climatic conditions and their consequences cannot be undermined despite the lesser attention paid to date.

Impact of climate change on infrastructure can be determined either by vulnerability or by the magnitude of the change. Geo-infrastructure becomes more vulnerable to deterioration in areas where ground consists of predominantly expansive soils. Moisture profiles in the ground are important particularly for those structures placed at shallow ground. Extreme shrink-swell soil movements and seasonal cycles of shrinkage and swelling caused by change in soil moisture can damage buried service pipes, building foundations, road pavements, and embankments (Rajeev et al. 2012). For example, in Victoria, Australia, many houses were damaged due to uneven settlement during a major drought that spanned from 1996 to 2009 (Karunarathne et al. 2016). This is because the strength of geomaterials, including natural soils, rocks, and mixtures of artificially manufactured crushed rocks, depends, to varying degrees, on the moisture content and is sensitive to environmental factors. Based on published research, Table 13.1 summarizes many ways how different aspects of climate change affect infrastructure systems.

This chapter aims to encapsulate the current state of knowledge in soil-atmospheric interaction in relation to climate change and geo-infrastructures. Firstly, the predictions of expected global climate changes are summarized followed by a discussion on theoretical knowledge required

Table 13.1 Climate change impact on different geo-infrastructure types

Climate change event	Impact on geo-infrastructure	References
Extreme precipitation	• Swelling of soils leading to increased pipe loading • Water accumulation on the road • Increased landslide • Increased embankment failure	Meyer et al. (2010); Hunt and Watkiss (2011); Camp et al. (2013); Lambert et al. (2013); McNutt (2013); Bollinger et al. (2014); Miara et al. (2017); Kundzewicz et al. (2018);
Drought or decrease in precipitation	• Instability of road substructure that will lead to closed or blocked roads and higher maintenance cost • Consolidation, oxidation of soil, differential settlement • Instability of buried service pipes	
Sea-level rise (SLR)	• Increase in sea level caused inundation of road infrastructure and consequences associated with flood	
Extreme events including flooding, storms, hurricanes	• Damages to pavement structure • Increased landslides • Soil erosion and sedimentation • Movement of trees and roots due to extreme wind leading to increased pipe loading • Damages to infrastructures due to debris	
Temperature rise	• Formation of desiccation cracks • Thermal expansion of pipes • Asphalt melting and rutting • Drying-induced excessive ground settlement	

Source: Modified after Rathnayaka et al. (2021).

to assess and predict climate loading on geo-infrastructure. Main focus of the work is to identify the problems created by the changing climate for key geo-infrastructure and to explore the mitigation and adaptation measures.

13.2 SUMMARY OF PREDICTED CLIMATE CHANGE AND LAND USE

There is a strong connection between climate change and land use. More than 20% of total global anthropogenic greenhouse gas emissions are from different forms of land use (e.g. agriculture, deforestation, soil erosion, and salinization). The special report on climate change and land published by the Inter-governmental Panel for Climate Change (IPCC, 2019) recognizes that climatic changes over land areas are more aggravated than global mean (i.e. average for both land and sea). Compared to the pre-industrial era (1850–1900), averaged land surface air temperature for the period 1999–2018 has increased more than the global mean surface temperature. Between these two periods, mean land surface air temperature rose by 1.53°C compared to a 0.87°C rise in global mean air temperature. Moreover, surface air temperature will continue to rise over the 21st century under all assessed emission scenarios with an alarming 2°C increase that is predicted between 2081 and 2100 under the worst emission scenario (IPCC 2014). Consequently, there will be more hot extreme days and fewer cold extreme days in the future.

Other significant impact from climate change is in the variation of precipitation patterns across the globe. High-latitude regions (e.g. North America, Northern parts of Asia, Antarctica, etc.) will experience an increase in precipitation. Mid-latitude dry regions (e.g. Europe, some parts of South America, Australia, etc.) will receive less precipitation, while mid-latitude wet regions (e.g. the Pacific) receive more precipitation. This means prolonged droughts for some regions and more frequent floods for others. Global ocean warming and sea-level rise is another key climatic factor that impacts coastal infrastructure. According to IPCC (2014), global mean sea level will continue to rise over the next 80 years in the range of 0.45–0.82 m under the least conservative emission scenario.

A major concern related to the impact of climate change on geo-infrastructure is the increased intensity and frequency of extreme weather and climate events. Experience in the last 50 years indicates that natural hazards such as droughts, heat waves, heavy downpours, hurricanes, tornadoes, wildfires, and floods will be more frequent in future. Extreme events can cause catastrophic damages to critical infrastructure incurring mammoth human and financial costs. During the 20-year period from 2000 to 2019, 7,348 natural disasters were reported around the world. It cost 1.23 million lives and US$2.97 trillion in economic losses (CRED 2020). This is a dramatic increase in comparison with the preceding 20-year period (1980–1999), during which only 4,212 natural disasters were reported. Human cost was 1.23 million lives, and economic losses were estimated at US$1.63 trillion. CRED (2020) attributes this sharp increase in natural disasters to climate change. Failure of geo-infrastructure under natural hazards exacerbates the damage and interrupts community life. On the other hand, by adapting new design guidelines to build climate-resilient geo-infrastructure, they can be used to mitigate the impact of extreme events.

13.3 THEORETICAL BACKGROUND OF SOIL–ATMOSPHERE INTERACTION

Our structures are built on the ground, and hence, understanding the interaction between soil and atmosphere is paramount to evaluate the climate change impact on geo-infrastructure as well as to develop mitigation

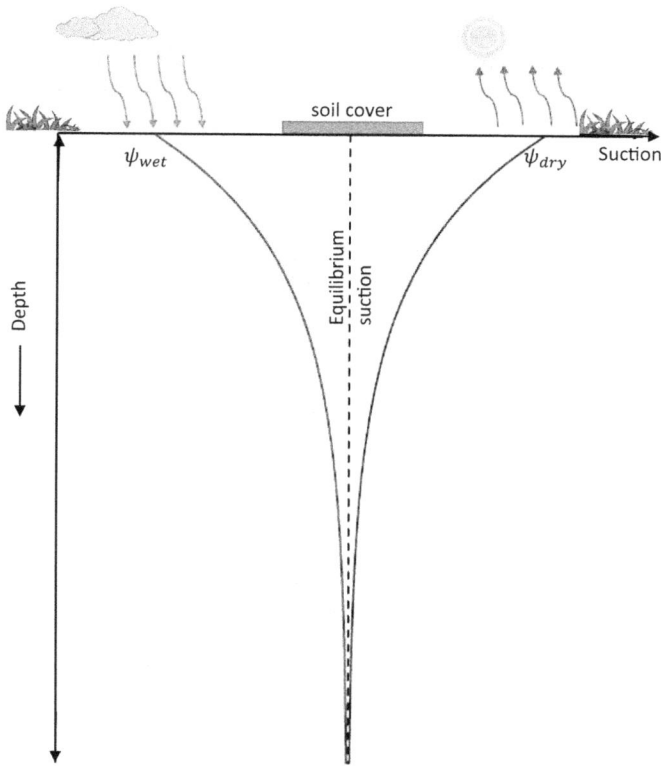

Figure 13.1 Seasonal suction variation within the reactive zone.

measurements. Ground responds to seasonal climatic changes through variations in soil suction and moisture content. Figure 13.1 shows a simplified suction profile near ground surface. In most areas, soil is unsaturated close to the ground surface. Suction is maximum at the ground surface at any given time and gradually decreases with the depth and reaches zero at the water table. Suction is related to moisture content; thus, it is influenced by rainfall and amount of evaporation. During dry season, when evaporation is greater than the rainfall, suction at the ground surface reaches its highest (ψ_{dry}) value. During wet season, when there is more rainfall, suction reduces and reaches a minimum (ψ_{wet}). Under covered areas, however, suction profile across the depth remains reasonably constant. The depth over which the change in suction takes place is known as "reactive zone" (or "vadose zone"). Demarcating this depth, however, is not straightforward. Suction variation below this depth is usually considered negligible for geotechnical engineering purposes, and the value of suction at this depth is taken as the equilibrium suction (Vann and Houston 2021). Change in suction becomes crucial if the soil in the reactive zone is expansive. Soil scientists and researchers in agronomy, hydrology, and climatology have widely studied soil–atmosphere interaction (Cui et al. 2013). The main focus of these studies, however, was on soil water evaporation and transpiration rates. In order to analyse the impact of climate change on geo-infrastructure from geotechnical engineering perspective, coupling of thermal, hydrological, and mechanical processes is necessary.

In geotechnical engineering, several models have been developed incorporating coupled mass and heat flow. Wilson et al. (1994) presented the following equations for the one-dimensional mass and heat transfer at soil–atmosphere boundary.

Mass (water and water vapour) transfer:

$$\frac{\partial h}{\partial t} = C_w \frac{\partial}{\partial y}\left(k_w \frac{\partial h_w}{\partial y}\right) + C_v \frac{\partial}{\partial y}\left(D_v \frac{\partial P_v}{\partial y}\right) \tag{13.1}$$

Heat transfer:

$$C_h \frac{\partial T}{\partial t} = \frac{\partial}{\partial y}\left(\lambda \frac{\partial T}{\partial t}\right) - L_v \left(\frac{P + P_v}{P}\right) \frac{\partial}{\partial y}\left(D_v \frac{\partial P_v}{\partial y}\right) \tag{13.2}$$

where C_w is the modulus of volume change with respect to liquid phase, C_v is the modulus of volume change with respect to vapour phase, D_v is the diffusion coefficient of the water vapour through soil (kg.m/(kN.s)), P_v is the partial pressure in the soil due to water vapour (kPa), k_w is the coefficient of hydraulic conductivity (m/s), h_w is the hydraulic head, C_h is the volumetric specific heat (J/(m^3.°C)), L_v is the latent heat of vaporization for water (J/kg), T is the soil temperature, and λ is the thermal conductivity (W/(m.°C)).

More recently, An et al. (2017) conducted a two-dimensional numerical analysis to determine volumetric water content and temperature in soil inside an embankment. They presented the following coupled equations for soil–atmosphere interaction (An et al. 2017).

Mass transfer:

$$C_\varphi \frac{\partial \varphi}{\partial t} + C_{\varphi T} \frac{\partial T}{\partial t} = \nabla.\left[K_\varphi \nabla \varphi\right] + \nabla.\left[K_{\varphi T} \nabla T\right] + \rho_l \nabla K_w \tag{13.3}$$

Heat transfer:

$$C_T \frac{\partial T}{\partial t} + C_{T\varphi} \frac{\partial \varphi}{\partial t} = \nabla.\left[K_T \nabla T\right] + \nabla.\left[K_{T\varphi} \nabla \varphi\right] \tag{13.4}$$

where C_φ is the volumetric isothermal capacity of moisture (kg/m^3.K), $C_{\varphi T}$ is the volumetric thermal capacity of moisture (kg/m^3.K), C_T is the volumetric thermal capacity of the structure (J/(m³.K)), $C_{T\varphi}$ is the volumetric thermal capacity of the structure (J/(m³.K)), φ is the matric suction head (m), K_{φ} is the isothermal moisture diffusivity (m²/s), $K_{\varphi T}$ is the thermal moisture diffusivity (m²/s.K), K_T is the thermal soil structure diffusivity (W/m.K), $K_{T\varphi}$ is the thermal soil structure diffusivity (W/m), T is the absolute temperature (K), ρ_l is the liquid density (kg/m³), and K_w is the saturated hydraulic conductivity (m/s).

In expansive soils, substantial swelling and shrinkage can be seen in response to moisture change. To assess the settlement and potential damages to structures caused by this swell-shrink movement, it is necessary to couple mechanical response of soil along with mass and heat flow. Hemmati et al. (2012) showed how thermo-hydro-mechanical modelling can be undertaken for soil–atmosphere interaction on vegetated land.

The aforementioned equations and associated models have produced promising results. Numerical predictions of volumetric water content and soil temperature using Equations (13.3) and (13.4) were in close agreement with field measurements (An et al. 2017). Costa et al. (2023) presented data on numerically determined soil suction, volumetric water content, and soil temperature under both recorded and predicted climate conditions. They used the software VADOSE/w, which is built on Equations (13.1) and (13.2). In another study by Soltani et al. (2019), soil suction was predicted using the software SVFlux, which is also governed by Equations (13.1) and (13.2).

Despite these remarkable advancements in understanding of soil–atmosphere interaction, more robust constitutive models are also necessary to explain the dynamic soil properties inside the unsaturated zone near ground surface. Generally, design standards for geo-infrastructure assume linear elasticity and limit equilibrium for the fully plastic condition. In real life, however, soil does not behave neither in perfectly elastic nor in perfectly plastic manner (Potts 2002). Hence, it is important to use a suitable constitutive model that is capable of capturing actual unsaturated soil behaviour in the field.

Jayasundara et al. (2019) laid out four criteria to select an appropriate constitutive model to describe unsaturated soil behaviour under climate loading. These criteria are briefly noted below:

i. *The unsaturated constitutive model should comply with the parameters that can be readily acquired with proximal soil*: Existing constitutive models for unsaturated soils use state variables such as suction, specific volume, and degree of saturation. However, accurate determination of these parameters in the field is challenging, thereby placing a limit to using the model. Alternatively, a framework such as MPK (Kodikara 2012) can be utilized. MPK framework uses specific volume, mean net stress, and specific water volume as state variables, which are relatively easier to be determined in the field.

ii. *The unsaturated constitutive model should be able to smoothly transit and explain saturated soil behaviour:* Although mostly unsaturated, near surface soil can be occasionally saturated, for example, in the aftermath of a heavy rainfall. Thus, the selected constitutive model should be able to smoothly describe saturated soil behaviour as well.

iii. *All coefficients in the unsaturated constitutive model should be easily obtainable through less sophisticated and less time-consuming experiments:* The available constitutive models occupy a large number of material parameters that demand time-consuming tests for determination.

iv. *The selected model should capture significant phenomenological observations of unsaturated soils:* For example, the environmentally stabilized state of soil (Kodikara et al. 2020) achieved after several wet-dry cycles. The selected model should be able to describe such observations.

Soil response to the future climate loading only be estimated using reliable numerical or empirical models. Hence, it is essential to develop rigorous finite element codes using appropriate constitutive models and boundary conditions to simulate the thermal-hydro-mechanical response of unsaturated zone.

13.4 IMPACT OF CLIMATE CHANGE ON GEO-INFRASTRUCTURE

The impact of climate change on geo-infrastructure can be discussed in two fronts: change in mean climatic factors and change in extreme events. Assessment of climate resilience of existing geo-infrastructure and planning for new geo-infrastructure should consider both these aspects. The following sections discuss the main concerns for geo-infrastructure arising from future climate predictions and suggest some adaptive measures.

13.4.1 Temperature rise and desiccation cracks

The vast volume of literature published on desiccation cracking phenomenon in clay is evidence to how problematic the cracks can be for geotechnical applications. Numerous studies have been conducted to understand the mechanisms and to simulate crack patterns (Tang et al. 2021; Cuadrado et al. 2022). Expected climate changes in most parts of the world are set to intensify this problem with rise in temperature and decline in precipitation. Landfill clay covers, unsealed road pavements, earth embankments, and foundations of lightweight structures will be among the worst affected.

Prolonged periods of hot extreme days, as expected in many regions, will wither soil into excessive shrinkage forming larger and deeper cracks than what has been observed during normal dry periods in the past. While any form of shrinkage cracks is unwelcomed, minute cracks are more likely to self-heal during wetting cycles. With larger and deeper cracks on the other hand, sidewalls will collapse during wetting, further widening the cracks.

Desiccation cracking is an unavoidable phenomenon in clay soils when exposed to climate. One option to control desiccation cracking in geo-infrastructure is by preventing the moisture flow (evaporation). This can be achieved by covering the soil surface with geosynthetics. Another option is to modify the soil used in construction to mitigate desiccation cracking potential. This demands the development of new composite materials, innovative designs to minimize exposure, and novel construction methods to enhance structural integrity. Fibre reinforcement and stabilization using novel additives (e.g. reclaimed waste materials, enzymes, geopolymers) have provided hope for mitigating desiccation cracks (Chaduvula et al. 2017; Xie et al. 2020; Tang et al. 2021). These findings should be investigated further and bring them to practice.

13.4.2 High evaporation and suction increase

Higher surface air temperature driven by climate change can lead to higher subsurface temperature and evaporation. Changes in soil moisture and suction will have an impact on geotechnical properties of soil posing new challenges to geotechnical designs. Cherukuvada (2008) predicted soil suction for six locations in Australia under expected climate change between 2070 and 2080 (see also Costa et al. 2023). Table 13.2 shows the approximate maximum suction predicted for each location. According to the values shown in Table 13.2, suctions near ground surface can reach as high as 12,000 kPa. Very high suctions are not usually found in the field by geotechnical engineers. In most situations, suctions in unsaturated soils dealt in the field are below 500 kPa (Wilson et al. 1997). Thus, the climate change induced suction rise in future will create problems in design of shallow geo-infrastructures.

Table 13.2 Predicted soil suction for six locations in Australia 2080

Location	Predicted suction (kPa)
Mildura	11,500
Mt Isa	6,000
Horsham	4,000
Mt Gambier	3,250
Weipa	600
Latrobe Valley	1,400

Source: Cherukuvada (2008).

The impact of increased suction will be of twofold on geo-infrastructures. An enhancement in shear strength can be expected with higher suctions, which will be a positive (perhaps temporarily) consequence. This is particularly important to stability of slopes and embankments where common cause of failure is due to porewater pressure rise. Modelling of mudslides in Italy as conducted by Dehn et al. (2000) reported a significant reduction in slope displacement due to the lowered water table. A slight improvement in resilient modulus can also be expected with increased suction (Edris 1976). However, negative consequences in terms of large cyclic pore pressure fluctuations, shrink-swell, and desiccation cracking will dominate the overall impact on geo-infrastructures due to increase in suction.

Many countries still rely on earth structures that were constructed prior to the era of modern soil mechanics. For example, design standards for shallow foundations (AS2870) do not account for a detailed characterization of soil moisture/suction variation during service life of the structure. The change in material properties and the progressive deformation caused by the strain softening have a serious effect in determining the lifespan of these structures. Results from Cherukuvada (2008) and Costa et al. (2023) also suggest a substantial increase in amplitude of pore pressure fluctuations due to climate change in the coming decades. Recent finite element modelling indicates that the annual total magnitude of pore pressure fluctuations is significant in controlling the rate of progressive failure (Kovacevic et al. 2001; Rouainia et al. 2020). It was also found in these analyses that an increase in amplitude of porewater pressure fluctuations reduces the number of cycles that the material will survive before collapse. Hence, the pore pressure fluctuations will have a critical impact on the long-term serviceability of geo-infrastructures in future.

13.4.3 Soil moisture depletion and temperature

Soil moisture content is an important factor that influences the climate system. It directly affects evaporation from non-vegetated as well as vegetated surfaces and controls the precipitation storage. Moisture depletion will affect geo-infrastructures as well as the ecosystem. Costa et al. (2023) reported a significant decline in volumetric water content in soil by 2070–2080 with severe fluctuations throughout the year. Such a decrease in water content could dry the soil substantially, causing excessive desiccation cracking problems in plastic soils with vegetation growth becoming more vulnerable.

Drier and warmer surface soil layers can give rise to flash floods and more frequent bushfires in arid regions. Hotter climates will generate higher evapotranspiration leading to unstable atmospheric conditions, which will trigger intense, shorter-duration rainfall (Kurylyk 2019). This will cause aggravated flooding issues deteriorating road pavements and embankments. On the other hand, because of the correlation between lightning and intense, shorter-duration rainfall (convective precipitation events), more lightning strikes can be expected causing bushfires in regions with higher combustibility (Kurylyk 2019).

While climate changes cause soil moisture to deplete, dryness in soil responds back to climate by changing the atmospheric temperature (Seneviratne et al. 2013; Zhou et al. 2021). Soil moisture controls the evapotranspiration and surface energy fluxes, eventually influencing the atmospheric temperature. Therefore, the soil moisture decrease induced by atmospheric temperature rise will further increase the temperature in return (Vogel et al. 2018). This is known as soil moisture feedback in climate science. This phenomenon coupled with high-intensity rainfall events will aggravate shrinkage and swelling in geo-infrastructures. The adverse effects of shrinkage cracks and uneven settlement/heave are well known to geotechnical engineering community (Costa et al. 2018). Hence, it is important to alleviate the soil moisture loss from surface using barriers or other means.

13.5 RECENT RESEARCH ON THE EFFECT OF CLIMATE CHANGE ON GEO-INFRASTRUCTURE

Despite the significance the issue of climate change has gained in the recent past, only a handful of research literature has been published on the effect of climate change on geo-infrastructure. However, there is a considerable number of technical reports assessing the impact of climate change on infrastructure systems in general. These are mainly produced by governments and other relevant agencies to help them with future planning and budget allocations. Assessing the risk, development of adaptation measures, maintenance cost prediction, and life cycle analysis are the main foci of published literature. Some of those recent notable research literature related to geo-infrastructure are listed in Table 13.3.

Every country and economy rely heavily on road and rail system. This is reflected by the fact that bulk of the studies related to climate change effects on geo-infrastructure focus on transport infrastructure. Commonly noted geo-infrastructure elements in road systems are embankments, both natural and anthropogenic slopes, stabilized or non-stabilized subgrades, flexible pavements, unsealed pavements, retaining walls, and foundations. Slope failure, landslip, and debris flow have been identified as the major concerns related to both road and rail transportation systems. For flexibles pavements, rutting, bleeding, excessive deformation, depressions, and ground settlement have been identified as the common problems related to climate change effects (Maadani et al. 2021).

Table 13.3 Recent research on impact of climate change on geo-infrastructure

Reference	Scope and comments
Pavements	
Maadani et al. (2021)	A state-of-the-art review on the effect of extreme weather events on flexible pavements in North America.
Piryonesi and El-Diraby (2021)	Pavement condition prediction under climate change using machine learning tools.
Qiao et al. (2019)	Life cycle cost analysis for a climate-resilient asphalt road design mix.
Mulholland and Feyen (2021)	Estimate of increased maintenance cost due to adverse climatic changes on railway and road infrastructure.
Kwiatkowski et al. (2020)	Relationship between atmospheric freezing-thawing temperature and porous asphalt pavement temperature.
Wang et al. (2019)	A framework for long-term adaptation planning for UK road systems under climate change.
Gudipudi et al. (2017)	Pavement performance assessment under predicted climate conditions.
Ortega et al. (2020)	Assessing the criticality of transport network in Spain under climate change projections.
Zapata et al. (2007)	Development of a mechanistic pavement design guide to incorporate climate change.
Slopes and landslides	
Occhiena and Pirulli (2012)	Investigating the connection between an increase in rockfalls and climate change.
Crozier (2010)	Modelling landslide response to predicted climate.
Dehn et al. (2000)	Assessment of a landslide under climate change projections.
Robinson et al. (2017)	Impact of extreme rainfall events on landslides under projected climate.
Shallow foundations/reactive soils	
Sun et al. (2017)	Calculating new Thornthwaite Moisture Index (TMI) for the changed climate and estimating the associated ground movement.
Gerard et al. (2016)	Relationship between climate conditions and evaporation in soil.
Embankment	
An et al. (2018)	Hydro-thermal modelling of an embankment coupled with soil–atmosphere interaction.

There is a void in research literature regarding studies on the impact of climate change on other types of geo-infrastructure such as pile foundations, clay covers and barriers, natural slopes, and rehabilitated mine slopes. Understanding the impact on these infrastructure elements and developing adaptation measures must be undertaken before it becomes too late and too costly.

With respect to engineering solutions to alleviate the effects of climate change, it is necessary to develop sustainable additives and construction techniques to make geomaterials more climate resilient. There have been advancements in developing composite materials by mixing problematic soils with novel, sustainable additives. Promising results have been obtained with geopolymers, fibres, fly ash, bio-enzymes, and reclaimed and recycled waste materials (Cardoso et al. 2016; Xie et al. 2020; Karami et al. 2021). Most of these additives bring extra benefits to the composite in mechanical properties such as dry density, unconfined compressive strength, CBR, tensile strength, and shear strength. However, thorough investigations are necessary on hydraulic properties, durability, and workability of the modified materials.

Another requirement to build climate-resilient geo-infrastructure is more accurate, real-time, low-cost condition monitoring. It is important to continuously monitor the ground movement in critical infrastructure (e.g. dams, slopes) under extreme weather events. While there have been recent developments in inventing new instruments and monitoring techniques, there is an urgent need to produce more accurate instruments and automated monitoring techniques.

13.6 CONCLUSIONS

There is a clear consensus that climate change impacts geo-infrastructures in one way or another. The effects are mostly unfavourable and demand engineered solutions. The existing knowledge on soil–atmosphere interaction is capable of assessing soil response to climate change to an extent. However, the selection of an appropriate constitutive model and the determination of associated state variables remains a challenge.

Key problems on geo-infrastructure are related to temperature rise and desiccation cracks, high evaporation and suction increase, soil moisture depletion, and temperature. These problems are multi-dimensional and require thermo-hydro-mechanical coupling to produce theoretical analysis. In practice, however, mitigating the sensitivity of soil and rock to climate conditions is the key to make geo-infrastructure more resilient and high performing. Future research must focus on novel additives that can increase tensile strength, reduce porosity, and mitigate cracking potential and erodibility. Attention must also be given to develop robust instruments for condition monitoring and field data collection.

REFERENCES

Alvioli, M., Melillo, M. Guzzetti, F., Rossi, M., Palazzi, E., Hardenberg, J., Brunetti, M.T., Peruccacci, S., 2018. Implications of climate change on landslide hazard in Central Italy. *Science of the Total Environment* 630, 1528–1543.

An, N., Hemmati, S., Cui, Y., 2017. Numerical analysis of a soil volumetric water content and temperature variations in an embankment due to soil-atmosphere interaction. *Computes and Gotechnics* 83, 40–51.

An, N., Hemmati, S., Cui, Y.J., Maisonnave, C., Charles, I., Tang, C.-S. 2018. Numerical analysis of hydro-thermal behaviour of Rouen embankment under climate effect. *Computers and Geotechnics* 99, 137–148.

Bagui, S.K., Das, A., Sharma, R., Pandey, Y., 2022. Pavement design considering changing climate temperature. In: Ghosh, C., Kolathayar, S. (eds) *A System Engineering Approach to Disaster Resilience. Lecture Notes in Civil Engineering*, vol 205. Springer, Singapore. https://doi-org.ezproxy-b.deakin.edu.au/10.1007/978-981-16-7397-9_13

Bollinger, L.A., Bogmans, C.W.J., Chappin, E.J.L., Dijkema, G.P.J., Huibregtse, J.N., Maas, N., Schenk, T., Snelder, M., van Thienen, P., de Wit, S., Wols, B., Tavasszy, L.A., 2014. Climate adaptation of interconnected infrastructures: a framework for supporting governance. *Regional Environmental Change* 14, 919–931. https://doi.org/10.1007/s10113-013-0428-4

Camp, J., Abkowitz, M., Hornberger, G., Benneyworth, L., Banks, J.C., 2013. Climate change and freight-transportation infrastructure: current challenges for adaptation. *Journal of Infrastructure Systems* 19, 363–370. https://doi.org/10.1061/(ASCE)IS.1943-555X.0000151

Cardoso, R., Silva, R.V., Brito, J., Dhir, R., 2016. Use of recycled aggregates from construction and demolition waste in geotechnical applications: a literature review. *Waste Management* 49, 131–145.

Choudhury, B.U., Nengzouzam, G., Islam, A. (2022) Runoff and soil erosion in the integrated farming systems based on micro-watersheds under projected climate change scenarios and adaptation strategies in the eastern Himalayan mountain ecosystem (India). *Journal of Environmental Management* 309, 114667.

Costa, S., Cherukuvada, M., Islam, T., Kodikara, J., 2023. Impact of climate change on shallow ground hydro-thermal properties. *Bulletin of Engineering Geology and the Environment* 82, 16. https://doi.org/10.1007/s10064-022-03046-7.

Costa, S., Kodikara, J., Barbour, S.L., Fredlund, D.G., 2018. Theoretical analysis of desiccation crack spacing of a thin, long soil layer. *Acta Geotechnica* 13(1), 39–49.

CRED   2020. Human cost of the disasters: an overview of the last 20 years. *Published by the Centre for Research on the Epidemiology of Disasters and United Nations Office for Disaster Risk Reduction.*

Crozier, M.J. 2010. Deciphering the effect of climate change on landslide activity: a review. *Geomorphology* 124, 260–167.

Cuadrado, A.C., Najdi, A., Ledesma, A., Olivella, S., Prat, P.C., 2022. THM analysis of a soil drying test in an environmental chamber: The role of boundary conditions. *Computers and Geotechnics* 141. https://doi.org/10.1016/j.compgeo.2021.104495.

Cui YJ, Ta AN, Hemmati S, Tang AM, Gatmiri B, 2013. Experimental and numerical investigation of soil-atmosphere interaction. *Engineering Geology* 165, 20–28.

Dehn, M., Burger, G., Buma, J., Gasparetto, P., 2000. Impact of climate change on slope stability using expanded downscaling. *Engineering Geology* 55, 193–204.

Edris, E.V., Lytton, R.L., 1976. Dynamic properties of subgrade soils including environmental effects. *Research Report No. TTI-2-18-74-164-3.* Texas Transportation Institute, College Station, Texas.

Gerard, P., Douzane, M., Mpawenayo, R., Debaste, F., 2016. Influence of climatic conditions on evaporation in soil samples. *Environmental Geotechnics*. doi. org/10.1680/jenge.15.00069.

Gudipudi, P.P., Underwood, B.S., Zalghout, A., 2017. Impact of climate change on pavement structural performance in the United States. *Transportation Research Part D* 57, 172–184.

Hemmati, S., Gatmiri, B., Cui, Y.J., Vincent, M., 2012. Thermo-hydro-mechanical modelling of soil settlements induced by soil-vegetation-atmosphere interactions. *Engineering Geology* 139–140, 1–16.

Hunt, A., Watkiss, P., 2011. Climate change impacts and adaptation in cities: a review of the literature. *Climatic Change* 104, 13–49. https://doi.org/10.1007/s10584-010-9975-6

IPCC 2014. *Climate Change 2014: Synthesis Report. Contribution of Working Groups I, II and III to the Fifth Assessment Report of the Intergovernmental Panel on Climate Change* [Core Writing Team, R.K. Pachauri and L.A. Meyer (eds.)]. IPCC, Geneva, Switzerland, 151 pp.

IPCC 2019. Summary for Policymakers. In: *Climate Change and Land: an IPCC Special Report on Climate Change, Desertification, Land Degradation, Sustainable Land Management, Food Security, and Greenhouse Gas Fluxes in Terrestrial Ecosystems* [P.R. Shukla, J. Skea, E. Calvo Buendia, V. Masson-elmotte, H.- O. Pörtner, D. C. Roberts, P. Zhai, R. Slade, S. Connors, R. van Diemen, M. Ferrat, E. aughey, S. Luz, S. Neogi, M. Pathak, J. Petzold, J. Portugal Pereira, P. Vyas, E. Huntley, K. Kissick, M. elkacemi, J. Malley, (eds.)]. In press.

Jayasundara, C., Deo, R.N., Kodikara, J. 2019. An integrated conceptual approach for the monitoring and modelling of geo-structures subjected to climate loading. *Physics and Chemistry of the Earth* 114, 102798. https://doi.org/10.1016/j.pce.2019.08.006

Karami, H., Pooni, J., Robert, D., Costa, S., Li, J., Setunge, S., 2021. Use of secondary additives in fly ash based soil stabilization for soft subgrades. *Transportation Geotechnics* 29. https://doi.org/10.1016/j.trgeo.2021.100585

Karunarathne, A.M.A.N, Gad, E.F., Disfani, M.M., Sivanerupan, S., Wilson, J.L., 2016. Review of calculation procedures of Thornthwaite Moisture Index and its impact on footing design. *Australian Geomechanics* 51(1), 85–95.

Kodikara, J.K., 2012. New framework for volumetric constitutive behaviour of compacted unsaturated soils. *Canadian Geotechnical Journal* 49, 1227–1243.

Kodikara, J., Jayasundara, C., Zhou, A.N., 2020. A generalised constitutive model for unsaturated compacted soils considering wetting/drying cycles and environmentally-stabilised line. *Computers and Geotechnics* https://doi.org/10.1016/j.compgeo..2019.103332.

Kovacevic, K., Potts, D.M., Vaughan, P.R., 2001. Progressive failure in clay embankments due to seasonal climate changes. *Proceedings of the 15th International Conference Soil Mechanics Geotechnical Engineering*, CRC Press, vol. 3, 2117–2130.

Kundzewicz, Z.W., Krysanova, V., Benestad, R.E., Hov, Ø., Piniewski, M., Otto, I.M., 2018. Uncertainty in climate change impacts on water resources. *Environmental Science & Policy* 79, 1–8. https://doi.org/10.1016/j.envsci.2017.10.008

Kurylyk, B.L., 2019. Engineering challenges of warming. *Nature Climate Change* 9, 807–808.

Lambert, J.H., Wu, Y.-J., You, H., Clarens, A., Smith, B., 2013. Climate change influence on priority setting for transportation infrastructure assets. *Journal of Infrastructure Systems* 19, 36–46. https://doi.org/10.1061/(ASCE)IS.1943-555X.0000094

Maadani, O., Shafiee, M., Egorov, I., 2021. Climate change challenges for flexible pavement in Canada: an overview. *Journal of Cold Climate Engineering* DOI: 10.1061/(ASCE)CR.1943-5495.0000262.

McNutt, M., 2013. Climate change impacts. *Science* 341, 435–435. https://doi.org/10.1126/science.1243256

Meyer, M.D., Amekudzi, A., O'Har, J.P., 2010. Transportation asset management systems and climate change: adaptive systems management approach. *Transportation Research Record* 2160, 12–20. https://doi.org/10.3141/2160-02

Miara, A., Macknick, J.E., Vörösmarty, C.J., Tidwell, V.C., Newmark, R., Fekete, B., 2017. Climate and water resource change impacts and adaptation potential for US power supply. *Nature Climate Change* 7, 793–798. https://doi.org/10.1038/nclimate3417

Mulholland, E., Feyen, L., 2021. Increased risk of extreme heat to European roads and railways with global warming. *Climate Risk Management* 34, 100365. https://doi.org/10.1016/j.crm.2021.100365

Occhiena, C., Pirulli, M., 2012. Analysis of climatic influences on slope microseismic activity and rockfalls: case study of the Matterhorn Peak (Northwestern Alps). *Journal of Geotechnical and Geoenvironmental Engineering.* https://doi.org/10.1061/(ASCE)GT.1943-5606.0000662.

Ortega, E., Martin, B., Aparicio, A., 2020. Identification of critical section of the Spanish transport system due to climate scenarios. *Journal of Transport Geography* 84, 102691. https://doi.org/10.1016/j.jtrangeo.2020.102691

Qiao, Y., Santos, J., Stoner, A.M.K., Flinstch, G., 2019. Climate change impact on asphalt pavement construction and maintenance. *Journal of Industrial Ecology.* https://doi.org/10.1111/jiec.12936.

Rajeev, P., Chan, D., Kodikara, J., 2012. Ground-atmosphere interaction modelling for long-term prediction of soil moisture and temperature. *Canadian Geotechnical Journal* 49, 1059–1073.

Rathnayaka, B., Siriwardana, C., Amaratunga, D., Haigh, R., Robert, D., 2021. Climate change impacts on built environment: a systematic review. *Proceedings of International Conference on Structural Engineering and Construction Management*, Kandy, 17–19 December.

Robinson, J.D., Vahedifard, F., AghaKouchak, A., 2017. Rainfall-triggered slope instabilities under a changing climate: comparative study using historical and projected precipitation extremes. *Canadian Geotechnical Journal* 54, 117–127.

Rouainia, M., Helm, P., Davies, O., Glendinning, S., 2020. Deterioration of an infrastructure cutting subjected to climate change. *Acta Geotechnica* 15, 2997–3016.

Scaringi, G., Loche, M., 2022. A thermal-hydro-mechanical approach to soil slope stability under climate change. *Geomorphology* 401, 108108. https://doi.org/10.1016/j.geomorph.2022.108108

Seneviratne, S.I., Wilhelm, M., Stanelle, T., Hurk, B.V.D., Hagemann, S., et al., 2013. Impact of soil moisture-climate feedbacks on CMIP5 projections: first results from the GLACE-CMIP5 experiment. *Geophysical Research Letter* 40, 5212–5217.

Soltani, H., Muraleetharan, K. K., Bulut, R., Zaman, M., 2019. Prediction of soil suction using measured climatic data. *Environmental Geotechnics*. https://doi.org/10.1680/jenge.15.00064.

Sun, X., Li, J., Zhou, A., 2017. Assessment of the impact of climage change on expansive soil movements and site classification. *Australian Geomechanics* 52(3), 39–50.

Tang, C.-S., Zhu, C., Cheng, Q., Zeng, H., Xu, J.-J., Tian, B.-G., Shi, B., 2021. Desiccation cracking of soils: A review of investigation approaches underlying mechanisms and influencing factors. *Earth-Science Reviews* 216. https://doi.org/10.1016/j.earscirev.2021.103586.

Thomas, V., Lopez, R., 2015. Global increase in climate-related disasters. *ADB Economics Working Paper Series, Asian Development Bank No.* 466.

Vann, J.D., Houston, S.L., 2021. Field soil suction profiles for expansive soils. *Journal of Geotechnical and Geoenvironmental Engineering* 147(9). https://doi.org/10.1061/(ASCE)GT.1943-5606.0002570

Vardon, P.J., 2015. Climatic influence on geotechnical infrastructure: a review. *Environmental Geotechnics EG* 3, 166–174.

Vogel, M.M., Zscheischler, J., Seneviratne, S.I., 2018. Varying soil moisture–atmosphere feedbacks explain divergent temperature extremes and precipitation projections in central Europe. *Earth System Dynamics* 9, 1107–1125, https://doi.org/10.5194/esd-9-1107-2018, 2018.

Wang, T., Qu, Z., Yang, Z., Nichol, T., Dimitriu, D., Clarke, G., Bowden, D., 2019. How can the UK road system be adapted to the impacts posed by climate change? By creating a climate adaptation framework. *Transportation Research Part D* 77: 403–424.

Wilson, G.W., Fredlund, D.G., Barbour, S.L., 1994. Coupled soil-atmosphere modelling for soil evaporation. *Canadian Geotechnical Journal* 31, 151–161.

Wilson, G.W., Fredlund, D.G., Barbour, S.L., 1997. The effect of soil suction on evaporative fluxes from soil surfaces. *Canadian Geotechnical Journal* 34, 145–155.

Xie, Y., Costa, S., Zhou, L., Kandra, H., 2020. Mitigation of desiccation cracks in clay using fibre and enzyme. *Bulletin of Engineering Geology and the Environment* 79, 4429–4440 https://doi.org/10.1007/s10064-020-01836-5.

Zapata, C.E., Andrei, D., Witczak, M.W., Houston, W.N., 2007. Incorporation of environmental effects in pavement design. *Road Materials and Pavement Design* 8(4), 667–693, DOI: 10.1080/14680629.2007.9690094.

Zhou, S., Williams, A.P., Lintner, B.R., Berg, A.M., Zhang, Y. et al., 2021. Soil moisture-atmosphere feedbacks mitigate declining water availability in drylands. *Nature Climate Change* 11, 38–44.

Chapter 14

Risk-based asset management decision-making in life management of concrete water reservoirs

H.S. Richards, G. Chattopadhyay, and Harpreet Singh Kandra
Federation University

N. Islam
Swan Hill Rural City Council

CONTENTS

14.1 INTRODUCTION

The water sector in Australia and many countries around the world has been facing significant challenge with asset management and cost-effective maintenance for reliable supply of ever-increasing demand. Concrete water storage reservoirs are cost-effective solutions for any water supply network for meeting the demand during peak period and balancing demand and supply. However, there are challenges due to degradation in concrete structure, over time, that could lead to water quality problems, if not detected well in advance. In addition, lack of redundancy in the system could increase the risk of not meeting supply commitments when necessary inspections, maintenance, and major repairs and renewals are undertaken. Therefore, risk-based decision-making in management of concrete water reservoirs is being embraced by major networks around Australia.

The ISO and Australian and New Zealand standards have defined risk as "effect of uncertainty on objectives [1, Section 14.3.1]." In particular,

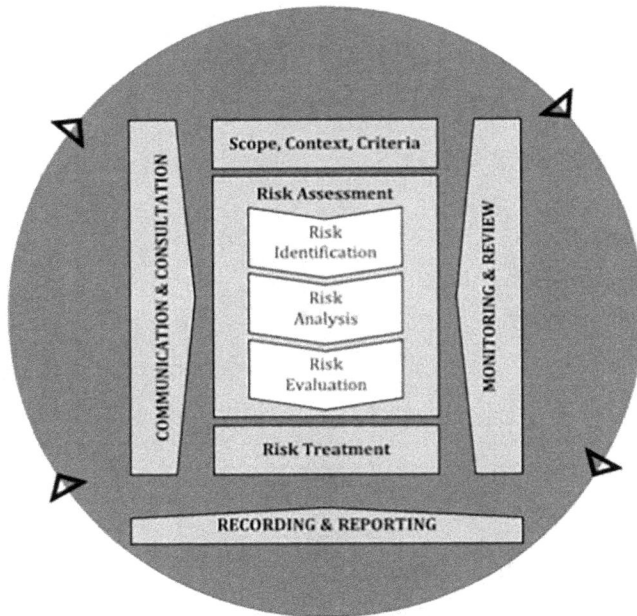

Figure 14.1 Risk management process. (Adapted from [1, Fig. 4 of clause 6.1].)

AS/ISO 31000:2018 mentions risk as the effect of uncertainty on objectives (ISO31000, 2014). Risk is usually expressed in terms of risk sources, potential events, their consequences, and their likelihood.

The risk assessment and management processes are iterative processes and broadly include establishing the context of risk assessment; risk analysis and evaluation; and preparation of risk management plan. The assessment of risk is done as a combination of the likelihood of its occurrence and consequences. Risk assessment is followed by formulation of risk management strategies aimed to reduce either the likelihood of risks or the impact of its occurrence. AS/ISO 31000:2018 (2014) states that the risk management process involves the systematic application of policies, procedures, and practices to the activities of communicating and consulting, establishing the context and assessing, treating, monitoring, reviewing, recording, and reporting risk, as shown in Figure 14.1.

Risk assessment literature in the water sector ranges widely from ecological risk assessment (e.g. Chen 1993) to health risk assessment (e.g. Yong 1997; Oesterholt 2007); assessing impacts of contamination from faecal matter (e.g. Lieverloo 2006) and different chemicals (e.g. Pinar 2006; Sun 2006; Evens, 2007); comparative risk assessments between different options (e.g. Regan 1997); risks from the use of reclaimed water (e.g. Zhao 2006); assessments to develop water quality criteria (Morgan 2000); and assessing safety of water infrastructure with focus on tampering and terrorism (Lambert 1997; Ezell, 2000). Risk assessments have also been undertaken in relation to the operation of facilities to manage process risk in field facilities and plants (e.g. Schlechter 1996; Pal 2003); fire and safety risks (Slye Jr. 2001); and for disaster management (e.g. as in pipeline industry). Operational risk management work is also being carried in the business sector to manage network risks (e.g. Soh 1995; Monton 1997) and financial risks (Kalapodas 2006; Renzo 2007).

In this study, an analysis of asset data critical to the potable water supply has been undertaken. Inspection and maintenance history of a balance and service concrete reservoir in regional Australia constructed 40 years ago has been used for analysis. Risk profile has been developed through the application of techniques using available internal data and industry-wide information.

This chapter has developed a risk-based decision-making for renewal of existing reservoir and feasibility analysis for construction of a secondary reservoir. Analysis has been carried out for continuing to operate current reservoir, as-is, without mitigation of risks, which was then compared with a scenario wherein a secondary reservoir is constructed. This analysis also investigated necessary improvements in dealing with data quality and further enhancement of inspection and maintenance strategies for reducing risks and costs and further enhancing performance of the network.

14.2 METHODS

14.2.1 Case study details 'concrete reservoir'

Concrete reservoir constructed 40 years ago has been selected as a case study. The region is a key centre for industries in the regional Australia, with a population of approximately 66,000 people and a vibrant community of mineral processing, energy generation, and transport facilities.

14.2.2 Issues with concrete reservoirs

Concrete water reservoirs, such as the subject reservoir, generally have reinforced cast in-situ concrete walls and floor with precast concrete columns supported by a metal-sheeted roof. Challenges associated with concrete water reservoirs are:

- floor joint defects, leaks,
- degradation of concrete structure and
- movements because of natural activities and disasters.

Figure 14.2 Leak from reservoir drainage pipework.

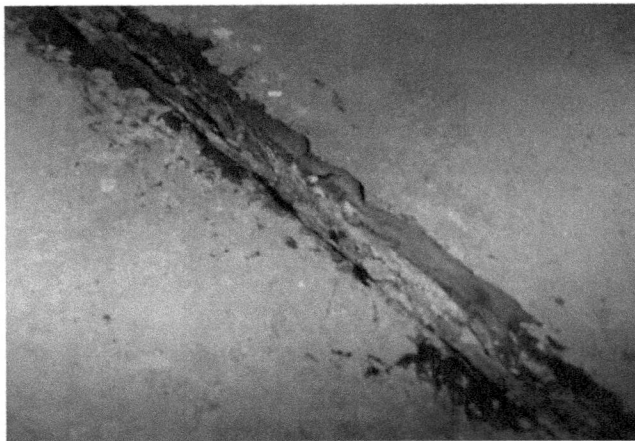

Figure 14.3 Floor joint defects.

Leaks are a common occurrence in these concrete reservoirs, and possible reasons may be due to wall softening and degradation, which may also result from the aggressive nature of the water and its quality, if water is not meeting requirements of the Australian Drinking Water Guidelines and Water Quality Management Plan. Identifying the source of the leak is generally a challenge with concrete reservoirs and needs to be determined before any remedial works can be undertaken (Figures 14.2 and 14.3).

Level of sediment accumulated over years also poses challenge in accurate detections. Horizontal cracking, leakages, and deflection might occur, if wall's vertical bending moments are above design limits.

Visual inspection, diagnostic testing, and design evaluations are used for the structural integrity (Whittaker et al. 2020). Services of professional divers are utilised to perform visual inspection of the interior defects, including floor joints lifting out of the floor rebates. Diagnostic testing is used for finding depth of carbonation for any corrosion and sulphate-affected deterioration. It can also include assessing chloride content, sulphate content, and depth of carbonation, cover meter, and taped cover at breakout, electrode potentials, and resistivity. At times, excavation may be needed to determine the reasons for leak(s).

14.2.3 Analysis

This research is based on study and analysis of asset data, conditions, and performance data critical to the potable water supply for the case study reservoir. Inspection and maintenance history of a balance and service concrete reservoir constructed 40 years ago has been used for analysis. Demand and changing risk profile analysis over a 20-year planning horizon was conducted to better understand the risk to the business and customers.

Given that the life of concrete reservoirs is generally long, data on failure and periods of non-availability is traditionally not collected consistently over entire life of the asset and available scantly. The major reasons for the reservoir being out of service might be leakage and lining failure or degradation of internal wall. The time needed to get these fixed as downtime is analysed for calculating the availability of this vital asset.

Mathematically, the availability of an item is the proportion of time for which it can perform to its specified requirements, i.e., has not failed or is unavailable. Operational availability in classical term is defined accordingly and calculated using Equations (1) and (1a), as below:

$$A_O = \frac{\text{MTBMA}}{\text{MTBMA} + \text{MDT}} \tag{14.1}$$

$$\text{MTBMA} = \frac{1}{\lambda + f_{\text{PM}}} \tag{14.1a}$$

where

MTBMA is the mean time between maintenance activities (expressed in Equation (14.1a))
MDT is the mean down time
λ is the failure rate assuming all failures are repaired, and
f_{PM} is the frequency of preventative maintenance

Table 14.1 Condition ratings and definitions

Rating	Definition
0	Not rated
1	Excellent
2	Good
3	Fair
4	Poor
5	Very poor

Statistics on concrete reservoir down times for wall and/or floor remedial works has been collected. This did not consider any reservoir works where the reservoir could be kept offline for additional time. Repairs conducted with the reservoir online were also excluded. As the concrete reservoirs are differently sized, a prorated down time can take the geometry into consideration.

In this case study, flow out of the reservoir through the common inlet was unknown. Data was available only for inflow and the reservoir volume. Using the principles of mass balance, the outlet (kL) (a proxy measure for demand) was determined using Equation (2).

$$\text{Outlet } (kL) = \left(\text{Inlet } (kL)_{\text{End Hour}} - \text{Inlet } (kL)_{\text{Start Hour}}\right) +$$
$$\left(9,100 \ (kL)\right)^*$$
$$\left[\left(\frac{\text{Reservoir Level Percent}_{\text{Start Hour}} - \text{Reservoir Level Percent}_{\text{End Hour}}}{100}\right)\right] \tag{14.2}$$

Multicriteria condition rating has been developed considering condition of coatings, structural roof access and security, joints and seals, and pipework for overall condition of concrete water reservoirs. These ratings, as outlined in Table 14.1, have been developed by the management of these capital-intensive assets.

14.2.4 Quantification of risks

Risk has been quantified and expressed in financial terms ($) as the product of the total consequences ($) and the likelihood (probability). For the quantitative version of risk matrix, likelihood is staged from rare to almost certain. The associated probabilities for these adopted in the quantitative risk matrix are provided in Table 14.2.

Table 14.2 Risk matrix qualitative likelihoods and equivalent probabilities

Qualitative likelihood	Associated probability
Almost certain	≤ 1
Likely	≤ 0.1
Possible	≤ 0.01
Unlikely	≤ 0.001
Rare	≤ 0.0001

Table 14.3 Reservoir downtimes

	Reservoir downtime	Adjusted downtime
Minimum (days)	7	4
Maximum (days)	30	31
Median (days)	9	8
Average (days)	14	11
Standard deviation (days)	9	10
Standard error	3	4

Table 14.4 Functional failures

Description of functional failure adopted for risk analysis	Primary consequence types
Any leak led to scouring down hill	Environmental and reputation
Partial collapse of wall/roof resulting in struggle to maintain safe drinking water	Water quality and reputation
Multiple leaks	Safety and reputation
Structural engineers state that reservoir cannot be operated following on from current external inspection regime and internal ROV/diver footage	Business disruption and reputation

Consequences from the risk matrix are quantified in financial terms ($) for each consequence category: Safety, Water Quality, Reputation, Environment, Business Disruption, Legal and Financial, and then aligned with the financial consequences for each consequence severity (Negligible, Low, Medium, High, Severe) with a sensitivity between each band. For example, the low consequences range from $0 to $50,000 as per the financial consequence. The user can choose to take a value in-between $0 and $50,000 based on perceived (or known) sensitivity.

An estimation has been made for the reservoir downtimes, as listed in Table 14.3. Definitions for functional failure which would cause the reservoir to become non-operational as provided in Table 14.4 for maintenance and inspection regimes for "Do Nothing Scenario".

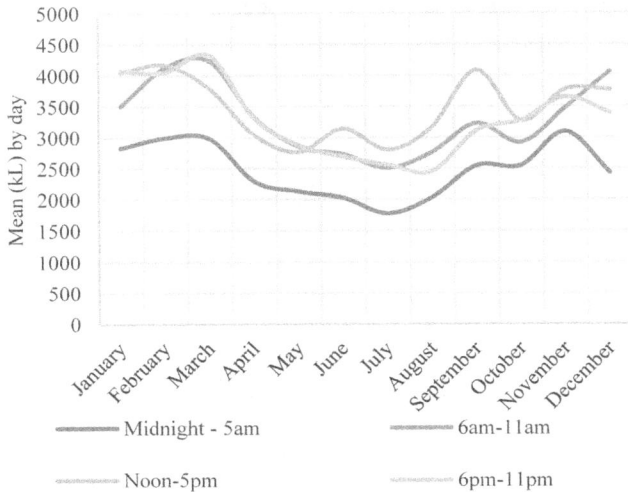

Figure 14.4 Mean demand (kL) by month.

14.3 RESULTS AND DISCUSSIONS

Analysis of demand suggests that the mean daily demand exceeds the capacity of the reservoir in both a season-based and month-based analysis. The data, as shown in Figure 14.4, suggest that mean daily demand as expected is high in spring and summer (10,277 kL in winter to 1,47,44 kL in summer). In particular, the months of January and February record the maximum demand (14,454 and 15,337 kL, respectively). The mean daily demand is low in autumn and winter (11,485 and 10,277 kL, respectively), lowest in particular in the months of July and August (9,641 and 10,422 kL, respectively).

The standard deviation for the daily demand by seasons varies from 1,301 to 1,687 kL (14%–19% of reservoir capacity, respectively). The standard deviation for the daily demand by month varies from 581 to 1,829 kL (6%–20% of reservoir capacity, respectively, as in Figure 14.5). These may be due to the quality of inlet data and step changes of 300 kL from raw data. The peak day demand and consequently reserve storage have been calculated using Equations (3) and (4).

$$\text{Peak Day Demand} = 12,577\,\text{kL} \times 2 = 25,154\,\text{kL} \tag{14.3}$$

$$\text{Reserve Storage} = \frac{25,154\,\text{kL}}{3} = 8.3\,\text{ML} \tag{14.4}$$

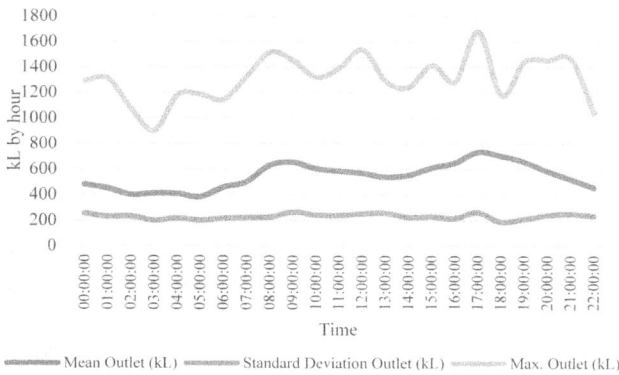

Figure 14.5 Hourly mean demand (kL).

Table 14.5 Likelihood of failure

Time frame	Qualitative	Estimated expected frequency
0–5 years	Unlikely	1 in 500
5–10 years	Less possible	1 in 100
10–15 years	Possible	1 in 50
15–20 years	Likely	1 in 10

Table 14.6 Profile of risk cost over 20 years in 5-year periods for not mitigating

End year	Likelihood	Annualised for period	Risk cost per event/year ($AUD)	Risk cost for period ($AUD)	Cumulative risk cost ($AUD)
5	0.002	0.0004	36,400	1,76,665	176,665
10	0.01	0.002	1,82,000	8,40,453	1,017,118
15	0.02	0.004	3,64,000	1,599,324	2,616,442
20	0.1	0.02	1,820,000	7,608,511	10,224,953

The likelihood of functional failure was assessed for each 5-year period and presented in Table 14.5.

Following the demand analysis, consequences of operations were assessed. The greatest consequence for water network in Australia was assessed to be reputation (adverse media reference), business disruption (inability to supply) followed by water quality and environmental and financial losses. A discount rate of 3% is used, and risk cost of no mitigations planned is provided in Table 14.6. Using the semi-quantitative risk matrix, the consequences were estimated to be $91 million AUD per event.

Table 14.7 Nominal cost secondary reservoir

Size	Construction concrete	Epoxy-lined tank
5 ML	$2.2M	$670K
10 ML	$3.3M	$870K

Alternatively, costs of constructing a secondary reservoir have also been estimated as shown in Table 14.7 for comparison with the base case with not mitigation.

Based on the estimated construction values, capital expenditure for construction of a secondary reservoir prior to 15 years is a better option compared with the "Do Nothing" scenario. Other alternatives to meet the demand in addition to construction of secondary reservoir, e.g. pump system, were also considered. High-level reasoning for discarding this option is due to the fact that costs of pump operation and maintenance over the life cycle of the reservoir are higher in comparison with the life of alternative reservoir.

14.4 CONCLUSIONS

The demand analysis in this case study showed that the reservoir is under-sized and does not meet the national code (WSAA Water Supply Code of Australia 3.1) or the requirements of the Strategic Water Plan. The risk-based asset management proposed construction of a secondary reservoir.

There has been a leak at the reservoir for over two years. It is recommended that leaks need to be quantified and monitored, through the installation of a flow meter or simply through monitoring the rate of flow on a weekly basis. This includes recording data on date, time, reservoir level, leak flow rate, and weather conditions. If the leakage worsens, this may influence decisions on accelerated capital investment. In addition, findings proposed the reservoir to be cleaned regularly using preventive maintenance plan. The development of an inspection specification or guideline, similar to that used by Water Corporation, undertaken by qualified personnel/agency, is recommended to ensure consistent inspection regime, data quality, and document management.

Risk-based decision-making in asset management of concrete reservoirs is recommended, including data quality issues to be resolved, and further analysis be undertaken to determine the risk of the high lift pump system. Reliability of the high lift pumps can adversely impact on likelihood of functional failures and associated costs.

ACKNOWLEDGEMENT

The authors would like to thank Federation University Australia and Water networks for the kind support and valuable inputs in this research.

REFERENCES

Abunada, M., N. Trifunovic, M. Kennedy and M. Babel, "Optimization and reliability assessment of water distribution networks incorporating demand balancing tanks," *Procedia Engineering*, vol. 70, pp. 4–13, 2014.

Aqualift, "Storage Tank Cleaning Intervals 101," Aqualift, Sydney, 2019.

Benitez, P, F. Rodrigues, S. Talukdar, S. Gavilan, V. Humberto and E. Spacone, "Analysis of correlation between real degradation data and a carbonation model for concrete structures," *Cement and Concrete Composites*, vol. 95, pp. 247–259, 2019.

Breysse. D., "Nondestructive evaluation of concrete strength: An historical review and a new perspective by combining NDT methods," *Construction and Building Materials*, vol. 33, pp. 139–163, 2012.

Cox, L. A., "What's wrong with risk matrices?" *Risk Analysis*, vol. 28, no. 2, pp. 497–512, 2008.

Dujim, N. J., "Recommendations on the use and design of risk matrices," *Safety Science*, vol. 76, pp. 21–31, 2015.

Gulati, R., *Maintenance and Reliability Best Practices*, South Norwalk: Industrial Press, 2013.

International Organization for Standardization, ISO31000: Risk Management Guidelines, Geneva, 2018.International Organization for Standardization, ISO 55000: Asset Management-Overview, principles and terminology. Geneva, 2014.

ISO. ISO 55000 Asset Management-Overview, principles and terminology. Geneva, 2014.

Modarres, M., M. P. Kaminskiy and V. Krivtsov, *Reliability Engineering and Risk Analysis A Practical Guide*, Boca Raton, FL: CRC Press, 2017.

O'Connor, P. and A. Kleyner, *Practical Reliability Engineering*, Chichester: John Wiley & Sons, Ltd., 2011.

Rehman. S. K. U., Z. Ibrahim, S. A. Memon and M. Jameel, "Nondestructive test methods for concrete bridges: a review," *Construction and Building Materials*, vol. 107, pp. 58–86, 2016.

Smith, D., *Reliability, Maintainability and Risk*, Oxford: Butterworth-Heinemann, 2011.

Water Corporation, "Inspection Guidelines for the Condition Assessment of Concrete Structures," Water Corporation, Perth, 2018.

Water Services Association of Australia, "Water Supply Code of Australia Version 3.1," Water Services Association of Australia, Melbourne, 2011.

Whittaker, G., H. Karampour and J. Wagner, "Structural condition assessment and renewals planning of drinking water concrete reservoirs," *Concrete in Australia*, vol. 46, no 4, pp. 51–58, 2020.

Chapter 15

Review of clogging processes in stormwater treatment filters

Harpreet Singh Kandra
Federation University Australia

David McCarthy
Monash University

Ana Deletic
Queensland University of Technology

CONTENTS

15.1 INTRODUCTION

Urbanization, including clearing of natural vegetation, compaction of soils, introduction of impervious surfaces, and changes in drainage pathways, causes significant changes to the hydrology of urban systems, as shown in Figure 15.1. These changes result in an increase of stormwater runoff

DOI: 10.1201/9781003368335-15

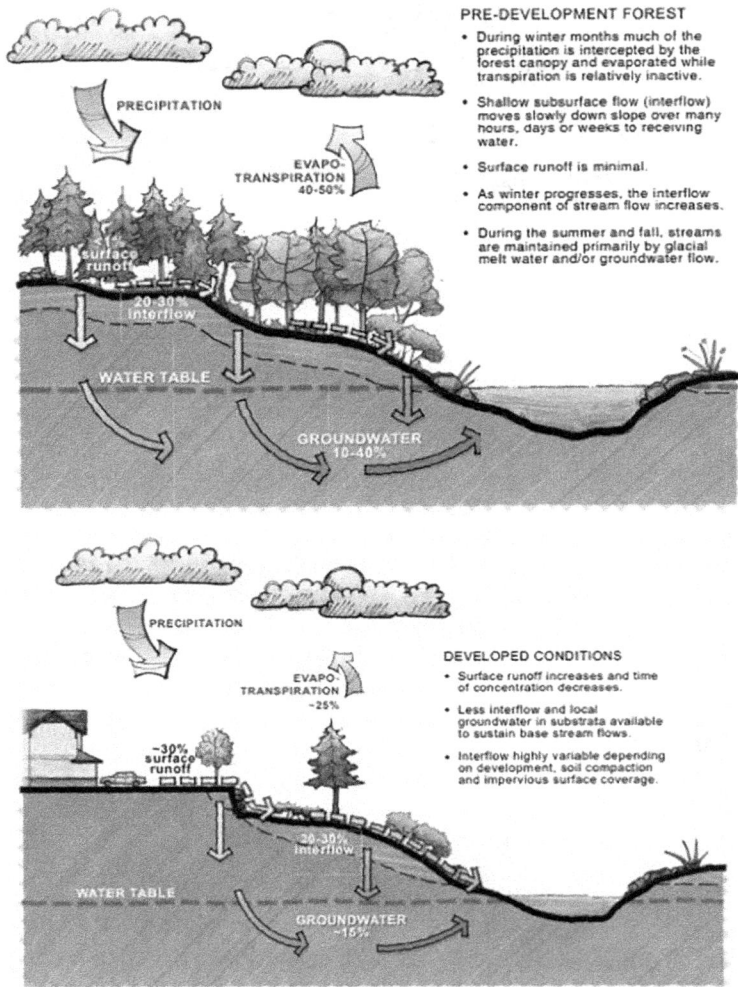

PRE-DEVELOPMENT FOREST

• During winter months much of the precipitation is intercepted by the forest canopy and evaporated while transpiration is relatively inactive.

• Shallow subsurface flow (interflow) moves slowly down slope over many hours, days or weeks to receiving water.

• Surface runoff is minimal.

• As winter progresses, the interflow component of stream flow increases.

• During the summer and fall, streams are maintained primarily by glacial melt water and/or groundwater flow.

DEVELOPED CONDITIONS

• Surface runoff increases and time of concentration decreases.

• Less interflow and local groundwater in substrata available to sustain base stream flows.

• Interflow highly variable depending on development, soil compaction and impervious surface coverage.

Figure 15.1 Comparison of pre-urban and an urban catchment.

Source: Lancaster et al. (2005).

volumes, flow frequency, and peak flood discharges (Walsh, 2000; Roesner et al., 2001; Hatt et al., 2004; Line and White, 2007). Increased peak flows have been the core focus of urban drainage design, to protect urban dwellers against flooding by providing hydraulically effective transport of surface runoff from urban areas into local receiving waters (Butler and Davies, 2004; Rauch et al., 2005). This traditional approach to managing

Table 15.1 Water quality characteristics of different water sources as reported in several literatures

Quality parameter	Stormwater	Potable supply	Untreated grey water	Untreated wastewater	Treated wastewater
BOD5 (mg/L)	3–73		90–290	100–500	8–80
Suspended solids (mg/L)	13–1622		45–330	100–500	11–250
Total dissolved solids (mg/L)	44–208	500	284–1700	250–850	520–4940
Cadmium (µg/L)	0.2–46	2	<10		0–2
Copper (mg/L)	0.005-0.56	1	0.018-0.39	0.001-0.2	0.001-0.12
Iron (mg/L)	2.4–7.3	0.3	0.094–4.37	0.3	0.03–1.6
Lead (mg/L)	0.007–2.04	0.01	<0.05–0.15	0.05	0–0.03
Manganese (mg/L)	0.04–0.11	0.1	0.014-0.075	0.0003	0.02–0.08
Sodium (mg/L)	12–116	180	29–230	70–300	41–1540
Zinc (mg/L)	0.026–2.4	3	<0.01–0.44	0.055	0.0–0.26
Total phosphorus (mg/L)	0.049–2.14		0.6–27.3	4–30	
Total nitrogen (mg/L)	0.50–12.6		2.1–31.5	20–85	6.1–44.2
Nitrate (mg/L)	0.1–6.2	50	<0.1	5–30	0.1–19.5

Source: Adapted from Mitchell et al. (2002).

stormwater peak flows eventually changes patterns and volume of infiltration, evapotranspiration, and surface and subsurface flows, causing further changes to flow regimes, thereby degrading our precious ecosystems (Burton and Pitt 2002; Walsh et al., 2005).

Urbanization also influences the water quality of our receiving water bodies. A survey conducted by the United States EPA found that urban stormwater to be the most important source of pollution in the coastal waters (Burton and Pitt, 2002). Commonwealth of Australia (2002) also recognizes stormwater as a major source of pollution in streams and coastal waters of Australian cities. The resulting degradation affects the aquatic life and its diversity because of the changes in photosynthesis processes, oxygen depletion (by decomposition of waste), and toxicity changes. It also affects usability of water as bacteria and viruses are introduced in stormwater and visual amenity of waterways is also affected. Besides, these pollutants in stormwater also affect groundwater resources (Burton and Pitt, 2002).

Approaches to treat and manage volume of stormwater flow and its composition target to remove, reduce, retard, and/or prevent urban stormwater runoff quantity and pollutants from reaching the receiving waters (Strecker et al., 2001). The key pollutants in urban stormwater include sediments and suspended solids, nutrients, metals, oils and surfactants, organic matter, and microorganisms. A comparison of characteristics of these contaminants in different waters is given in Table 15.1.

Treatment of stormwater often leads to operational issues, of which clogging is of significant concern. This chapter reviews current knowledge on clogging processes that occur in filter media used for urban stormwater treatment. The first section provides background information on urban stormwater flows and pollutants followed by a discussion on urban stormwater management approaches. The subsequent sections discuss the types of clogging processes and factors influencing clogging in filter media.

Several large compilations of worldwide data sets of urban water quality suggest that characteristics of urban stormwater are site-specific (Duncan, 1999; Smullen et al., 1999; Fuchs et al., 2004). The composition of stormwater is influenced by many factors, such as the extent of urbanization, type of collection surface (e.g., roads, roofs, car parks), atmospheric conditions, population density, waste disposal and sanitation practices, soil type, climatic conditions (i.e., dry and wet weather flows), and the presence of construction activities (Duncan, 1999; Wong, 2000; Marsalek and Chocat, 2002; Goonetillekea et al., 2005). The variations in characteristics of stormwater in space and time have also been documented by different studies (Smullen et al., 1999; Francey et al., 2010).

Several studies suggest that total suspended solids (TSS) in water often carry the majority of other stormwater pollutants in an attached form. For instance, in an analysis of highway stormwater runoff characteristics undertaken by Han et al. (2006) for over 3 years, it was found that suspended solids were associated with most particulate-bound metals such as chromium, copper, nickel, lead, and zinc. Other studies (Liebens, 2002; Pitt et al., 2004; Stead-Dexter and Ward, 2004; Gnecco et al., 2005; and Weiss et al., 2006) have also shown that these metals are largely associated with sediments in stormwater runoff. Taylor et al. (2005) found that the proportion of nitrogen in particulate form was higher during storm events. Similarly, in study of seven urban catchments in South Eastern Australia, Francey et al. (2010) found that the presence of TSS and total phosphorous has a strong correlation.

Therefore, given that pollutants are associated with sediment, it is recommended that the removal of suspended sediment from urban stormwater flows could assist in improving downstream water quality. This is particularly the case for nutrients and heavy metals, which can cause eutrophication and toxicity to fish, respectively (EPA, 1986; Burton and Pitt, 2002).

15.2 URBAN STORMWATER TREATMENT AND MANAGEMENT

Strategies to treat and manage can be both structural measures – a physical device – and non-structural measures (Department of Water and Swan River Trust, 2007). To prevent or minimize pollutants from entering stormwater runoff and/or reduce the volume of stormwater requiring management, non-structural controls such as institutional and pollution prevention practices are often designed. Such non-structural measures do not involve any fixed, permanent facilities, and they usually employ change management measures through government regulation (e.g., planning and environmental laws), persuasion, and/or economic instruments.

These measures, based on integrated approaches, are termed different in various parts of the world, such as sustainable urban drainage systems (SUDS), best (or better) management practices (BMPs), low-impact design (LID), and water-sensitive urban design (WSUD) (Mikkelsen et al., 1996; Wong, 2006; Fletcher et al., 2007). The aim of these strategies is to control stormwater flows and pollutants by removing, reducing, retarding, or preventing urban stormwater runoff quantity and pollutants from reaching receiving waters (Strecker et al., 2001). Additionally, there is a focus on the integration of urban water cycle management and urban planning and design, based on sustainability principles such as water conservation, waste management, and environmental protection (Lloyd et al., 1998).

Access to non-potable water supplies through the use of treated stormwater could potentially reduce pressures on the existing potable urban water supply systems and also assist in better water security in a world affected by climate change. Several authors have investigated the use of treated urban stormwater for a range of non-potable urban water uses and have found that it is of better quality as compared with untreated sewage or industrial wastewater discharge (Mitchell et al., 2002; Brown and Farrelly, 2007). Furthermore, stormwater treatment and harvesting systems can also assist with flood protection and restore flow regimes and water quality in water bodies (Fletcher et al., 2007).

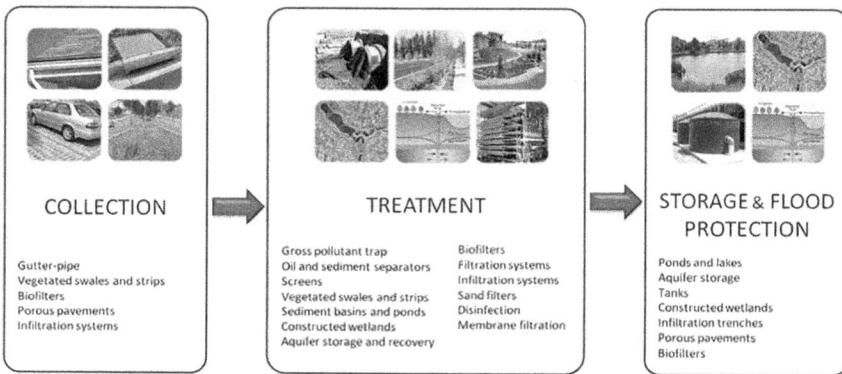

COLLECTION	TREATMENT		STORAGE & FLOOD PROTECTION
Gutter-pipe	Gross pollutant trap	Biofilters	Ponds and lakes
Vegetated swales and strips	Oil and sediment separators	Filtration systems	Aquifer storage
Biofilters	Screens	Infiltration systems	Tanks
Porous pavements	Vegetated swales and strips	Sand filters	Constructed wetlands
Infiltration systems	Sediment basins and ponds	Disinfection	Infiltration trenches
	Constructed wetlands	Membrane filtration	Porous pavements
	Aquifer storage and recovery		Biofilters

Figure 15.2 Technologies used for stormwater management.

Source: CWSC (2010).

15.2.1 Overview of structural systems

Structural stormwater treatment technologies, known as water-sensitive urban design (WSUD) systems, employed for urban stormwater runoff management could be categorized in several ways, with their main distinction being that some are waterbody based (e.g., wetlands or ponds), whereas others are filter media based (e.g., biofilters or infiltration systems). A brief overview of commonly used stormwater treatment technologies is provided below (CWSC, 2010).

15.3 NON-FILTRATION-BASED SYSTEMS

15.3.1 Gross pollutant traps, oil separators, and sediment separators/traps

These systems have high hydraulic loading rates of 50–100 m/h as the aim is to drain the stormwater as quickly as possible (Allison et al., 1998). However, there is hardly any removal of key pollutants such as solids, nitrogen, phosphorous, and heavy metals possible (CWSC, 2010).

- *Gross pollutant traps* are generally provided for limited treatment before stormwater is discharged into the receiving waters. They range from simple screens to structures that use various combinations of screens, stalling flow, settlement, floatation, and flow separation. These systems are not effective in the removal of fine or dissolved pollutants (Hatt et al. 2004a).
- *Oil separators* are used to remove hydrocarbons that are derived in urban stormwater from motor vehicles and roads. They are usually installed at the sources of spills (e.g., petrol stations and airports), before the oil emulsifies in large volumes of urban runoff (Wong, 2006b). The main purpose is to remove free-floating oil to minimize the effects on receiving waters and surrounding environment. In certain cases, they may also be provided to reduce load on downstream treatment systems such as biofilters or wetlands.
- *Sediment separators* prevent discharge of coarse sediment to downstream treatment measures. They range from simple earthen or concrete basin designs to complex structures using vortices and secondary flows. Sediment traps are seldom effective in the removal of fine or dissolved pollutants (Hatt et al. 2004a).

Wetlands are used in stormwater management either as standalone facilities or in combination with other WSUD systems, such as stormwater detention ponds. These systems have very low hydraulic loading rates of 0.003–0.03 m/h (Wadzuk et al. 2010; Yi et al. 2010). These are relatively shallow vegetated waterbody treatment systems that are designed to detain water for treatment on a periodic or permanent basis. They are designed to trap sediment, nutrients, bacteria, and toxins, and promote oxygen recovery (Ellis, 1993). Water treatment is performed as a combination of sedimentation, filtration, and biological nutrient uptake.

Wetlands, whether natural or constructed, additionally offer landscape amenity, recreational opportunities, habitat provision, and flood retention (Mitchell et al., 2007). The role of vegetation in wetland performance is critical, and they can be quite effective in the removal of fine sediment and dissolved pollutants (Hatt et al. 2004a). Several studies report variable performance of these systems dependent on construction features and climatic conditions. These systems have long life but also need a lot of space and maintenance (CWSC, 2011).

Stormwater management ponds are designed to intercept the runoff before it reaches a stream and detain the water for a longer period with a hydraulic loading rate is 0.02–0.1 m/h. This allows more time for the sediment and attached nutrients to settle out. These are artificial open water bodies, usually deeper than 1.5 m, that treat stormwater mainly by sedimentation and detention, with some nutrient uptake from emergent vegetation along the margin (Wong, 2006). System type and design may vary depending on the nature/size of pollutants treated by these systems (sedimentation basins and ponds) and other features offered such as urban lakes. The treatment train provided by lakes includes sedimentation, biological uptake, and exposure to UV disinfection. Lake's design and maintenance is important to avoid the risk of algal blooms, and pre-treatment is a requirement to avoid this risk (Wong, 2006b).

Vegetated swales are open, grassed surfaces that both collect and treat stormwater by filtration before discharging it to drainage system or receiving water. These technologies aim to reduce runoff velocity and retain coarse sediments (Hatt et al. 2004a). Depending on the end-use objectives (such as providing water for reuse, ground water recharge, or discharge to receiving waters), they may either promote or discourage infiltration. They can also be adopted as a pre-treatment measure preceding other WSUD technologies. These systems have low hydraulic loading rates of 0.02–0.06 m/h, which implies large areas are required for treatment.

It can therefore be concluded from that the non-filtration-based systems have very low hydraulic loading rates, thereby reducing their capacity of flow attenuation. These systems are not very effective in the removal of pollutants such as nutrients, heavy metals, and microorganisms. Further, a lot of variation has been observed in their pollutant removal performances.

15.4 FILTRATION/INFILTRATION-BASED SYSTEMS

Biofiltration systems are also known as bio-retention systems, biofilters (especially for large-scale applications), rain gardens (at the household scale), or tree pits biofilters (at street scale). Biofilters are vegetated buffers on top of a filtration medium (e.g., sandy loam, sand, and/or gravel) and facilitate flow attenuation, sediment, and pollutant removal (CWSC, 2011).

Stormwater runoff from close catchments flows across the vegetation before percolating through the soil media (FAWB, 2022). By flowing through the vegetation and the filter media, stormwater is subjected to several physical, chemical, and biological processes. As stormwater enters the dense vegetation in a biofilter, its velocity diminishes, enhancing sedimentation of particulates. Physical filtration continues as stormwater percolates down through the filter media, which also enhances binding of dissolved contaminants. Additionally, the vegetation and especially the soil microbial community take nutrients (nitrogen) and, to a lesser extent, other contaminants (heavy metals), ensuring both their growth and survival (CWSC, 2011). Depending on the design of the system, effluents reaching the bottom of the filter media might either infiltrate the underlying soil or be collected in a drainage pipe for conveyance to the receiving waters.

Biofiltration systems have higher hydraulic loading rates of 0.1–0.2 m/h but there is considerable variability in the field performance of these systems, which is likely due to different design characteristics. Pre-treatment may be required to reduce clogging risks of these systems as operational problems such as overflowing could result because of clogging (i.e., reduced hydraulic performance) (Hatt et al., 2012). However, the presence of plants is one of the main reasons why clogging is maintained in biofilters.

Infiltration systems are similar to filtration systems but allow water to percolate in the underlying soil. They are usually located near the point of discharge. These systems have been found as advantageous towards achieving better receiving water quality as they maintain the natural local hydrology and water table levels and reduce pollution discharges to receiving waters (Duchene et al., 1992).

Depending on the nature and location of storage, these may be differentiated as infiltration trenches, infiltration basins, and leaky well systems. These systems have low hydraulic loading rates of 0.01–0.08 m/h and solid removal rate of 36%–50%. However, nitrogen removal rates are low, and leaching of NOx has been observed (Landphair et al., 2000; Birch et al., 2005).

Porous pavements promote infiltration at high flow-through rates. Depending on their construction, all these pavements could be grouped into monolithic structures that include porous concrete and porous pavement (asphalt) and modular structures that include porous pavers or modular lattice structures with a gap in between each paver (Yong, 2008).

The treatment processes that occur in these systems include sedimentation, straining, adsorption, and biological degradation, etc. Their infiltration rates can range between 50 mm/h and up to 40 m/h depending on the pavement design (Bean et al., 2007). There is a huge variation in the performances amongst different designs of porous pavements but consistent for a given design.

Non-vegetated filters enable stormwater treatment by letting stormwater runoff flow through a porous non-vegetated medium (CWSC, 2011). These may be further categorized depending on the type of filter media used such as sand, coarse gravel, engineered media, and their combinations. These systems are effective in the removal of sediment and adsorbed pollutants (Hatt et al. 2004a). Like the biofiltration systems, pollutant removal in these systems results from sedimentation and filtration. These filters may require more maintenance as compared with biofiltration systems, to ensure that the top layer remains porous and does not clog with accumulated sediments. This is because of the much higher infiltration rate of these systems and non-availability of vegetation. Maintenance involves the removal of the top layer of filter media, where contaminants such as oils and sediments are retained (Hatt et al. 2004a).

These systems can be engineered, for example, by using soils containing naturally occurring and/or bio-engineered microorganisms that degrade toxic pollutants and organic materials that remove nutrients as stormwater infiltrates through the soil (Hatt et al. 2004a).

While sand-based systems have low hydraulic loading rates of 0.03–0.13 m/h and lower nutrient removal rates, engineered media-based systems can be designed to have high loading rates with better treatment performance, as listed in Table 2.2 (Schang et al., 2010; Poelsma et al., 2010). Additionally, the space requirements for the high-flow-rate systems are considerably less. For instance, Schang et al. (2010) state that the enviss™ systems can be sized seven times smaller than biofilters for a given impervious catchment area. On the contrary, systems such as biofilters need to wait for plant establishment periods, require water to maintain plant health (which can be an issue during extensive dry periods), and have very low infiltration rates because of which they are not suited for confined urban environments. High-flow-rate systems that could fit into highly urbanized areas could be beneficial for both discharge applications and possibly reuse scenarios, especially if their treatment performance could exceed current technologies (Schang et al., 2010; Bratieres et al., 2012).

However, high infiltration performance and non-availability of any vegetation make these systems more prone to clogging. The treatment performance of such systems may be lower as compared to filters with lower hydraulic performances because of shorter residence time within the filter bed. Therefore, better understanding of clogging and treatment processes would help design better systems that could maintain high flows while treating stormwater contaminants effectively.

In summary, compared to the other stormwater treatment systems, filtration-based systems have been found to be more acceptable as they have been found capable to effectively treat both particulate and dissolved pollutants, including both microorganisms and chemicals (Fletcher et al., 2004). Additionally, these filtration-based systems can also be engineered for site-specific requirements, depending on the objectives of treatment, water quality standards for end-use, space availability, and site-specific pollutant characteristics.

For these reasons, filtration-type stormwater treatment technologies are gaining traction around the globe: Denmark (Warnaars et al., 1999), France (Barraud et al., 1999; Raimbault et al., 1999; Le Coustumer and Barraud 2007), Germany (Pagotto et al., 1999; Zimmer et al., 1999; Dierkes et al., 2002a), Japan (Fujita, 1994; Akagawa et al. 1997; Nozi et al., 1999), Malaysia (Nawang 1999), Singapore (Tan et al., 2003), Sweden (Holmstrand, 1984), Switzerland (Mikkelsen et al., 1997), the UK (Kelso et al., 2000), the USA (Lindsey et al., 1992; Pitt et al., 1999), and Australia (Argue, 1999; Coombes 2001; Lloyd, 2004).

15.5 OPERATIONAL ISSUES WITH FILTRATION-BASED SYSTEMS

Literature acknowledges that filtration-based stormwater filters have some operational challenges. The acceptance of filtration-based systems to manage urban stormwater is impeded due to a range of concerns, including maintenance and longevity issues (Ellis and Marsalek, 1996). Numerous studies have highlighted that clogging of these systems is a key operational issue affecting their longevity and performance.

Field studies to evaluate the performance of infiltration-based systems, such as infiltration basins and trenches, have been undertaken by Veenhuis et al. (1988), Schueler et al. (1992), Lindsey et al. (1992), Warnaars et al. (1999), and Bardin et al. (2001). These studies observed a decline in hydraulic performance and/or failure of these systems over time. For instance, a field survey of several infiltration systems, conducted by Lindsey et al. (1992), showed that only 38% of infiltration basins were functioning as designed after 4 years of operation, with 31% considered to be clogged. Similarly, field-based investigations of permeable pavements show a reduction in permeability by a factor of 10–100 due to clogging (Pratt, 1995; Illgen et al., 2007; Yong et al., 2008). Clogging has been also observed in laboratory-based studies for stormwater treatment using soil-based filter media (Hatt et al. 2008) and gravel stormwater filters (Siriwardene et al., 2007).

For instance, Figure 15.3 (left) shows the picture of ponded biofilter at Banyan Reserve stormwater treatment system (Melbourne) after a storm

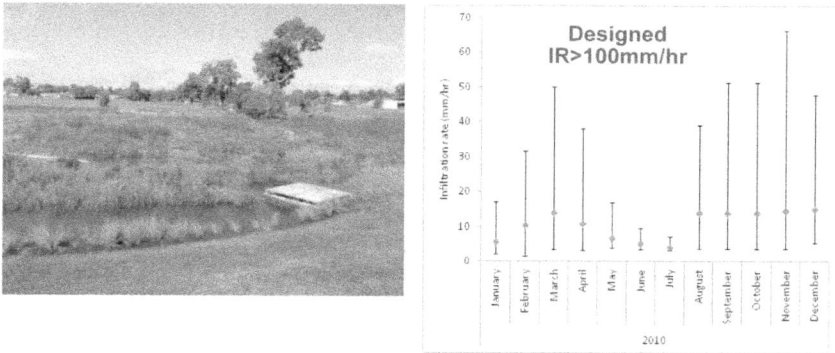

Figure 15.3 Biofilter after event in April 2010 (left) and hydraulic performance results compared with designed IR of 100 mm/h (right).

event (Hatt et al., 2012). Figure 15.3 (right) presents the results of hydraulic performance, over a period of 12 months, as observed for this biofilter, as compared with the design infiltration rate (IR). By contrast, the designed hydraulic performance was more than 100 mm/h, and the observed hydraulic performance of the system dropped significantly to around 20 mm/h.

15.6 CLOGGING OF FILTRATION/INFILTRATION-BASED WSUD SYSTEMS

Clogging of infiltration systems has been defined as the process of decrease in permeability of a filtration system and occurs due to the accumulation of materials associated with treatment/sediment removal processes (Bouwer, 2002; Perez-Paricio, 2001). This trapped material could accumulate in the surface and/or subsurface of the filter, thereby preventing infiltration (Siriwardene et al., 2007; Hatt et al., 2008). Therefore, either the infiltration rate (also called as hydraulic performance or hydraulic conductivity or permeability) of the system diminishes or the piezometric head increases, depending on the boundary conditions.

The phenomenon of clogging has also been observed and studied in other filtration systems and for other influents, such as fine sand clogging by septic tank effluent (Spychala and Blazejewski, 2003), sewage flow in vertical flow constructed wetlands (Winter and Goetz, 2003), and municipal wastewater filtration using different types of geotextiles (Yaman et al., 2006).

The development of a clogging layer leads to problems such as increased overflows and long periods of ponding, which eventually result in a range of concerns: from public safety to aesthetics, health, hygiene, and reduction in operational efficiency (Le Coustumer et al., 2007; Knowles et al., 2011).

Advanced clogging may eventually necessitate remediation of the clogged media, thus limiting the asset lifetime of the system; or the system may just become a wetland.

In order to take clogging into account in the design of a treatment system, therefore, conservative values of hydraulic performance are generally used, and a safety factor is generally applied to the measured hydraulic conductivity. For instance, New Jersey Department of Environmental Protection, 2004, suggests a safety factor of 2. However, such safety factors are chosen arbitrarily since no study has been undertaken to quantify the actual evolution of the hydraulic conductivity over time.

Clogging has been observed in potable water treatment and supply and has been effectively managed in these systems by backwash/back-flushing processes. Backwashing operations involve reversing the flow and increasing the velocity at which water passes back through the filter. This, in effect, blasts the clogged particles off the filter. However, de-centralized stormwater treatment and harvesting systems have limited infrastructure and resources and normally operate without backwashing.

Based on review of literature on clogging processes, it has been found that most of the field studies have simply reported the existence and importance of clogging. However, these studies have failed to go beyond quantifying the extent or rate of clogging, and none have been able to explain the nature of the clogging process. This is because clogging was generally not the desired outcome or focus of the research conducted. In some studies, where clogging was observed, insufficient detail was recorded to evaluate the conditions leading to clogging for the applied dosing regime. A greater understanding of the conditions that cause clogging may therefore lead to improvements in operation and design for optimal treatment efficiency and improved system reliability.

15.7 CLOGGING PROCESSES

Clogging, as discussed above, occurs because of the accumulation of materials removed during the treatment/sediment removal processes. In general, clogging occurs because of three main processes: chemical, biological, and physical (CITE). Each clogging process is discussed below.

15.7.1 Chemical clogging processes

Chemical clogging processes are caused by the precipitation of calcium carbonate, gypsum, phosphate, and other chemicals within the filtration media. It is during the filtration of influent that chemical reactions may occur within the filter bed and can eventually lead to continuous deposition of matter on the surfaces of the pores (Rice, 1974; Salli et.al, 1993;

Rinck-Pfeiffer, 2000; Larroque and Franceschi, 2011). For instance, many natural and altered hydro-geologic systems are characterized by chemical disequilibrium, which is caused by the changes in temperature, pressure, and oxidation/reduction potential. The resulting chemical reactions then result in either precipitation of a mineral phase or transformation of one mineral phase into another (Baveye et al., 1998).

Clogging is known to depend on chemical factors that control the colloidal stability of the particles (Mays and Hunt, 2007). The factors affecting chemical clogging include the characteristics of influent (such as pH, its ionic strength, fraction of organic compounds, mineralogical composition) and operational conditions (such as temperature and pressure changes that assist precipitation/dissolution). For instance, Schubert (2002) states that high loads of biodegradable substances in the river water can lead to chemical clogging beneath the infiltration areas due to strong changes in redox potential and pH values, which may cause precipitation of substances (e.g., $FeCO_3$) in the pores of the soil.

For instance, Rinck-Pfeiffer et al. (2000), while investigating bore clogging at a South Australian recycled water aquifer storage and recovery site, found that chemical processes had a role to play. But this investigation was carried using recycled water with high COD (165–170 mg/l), high alkalinity (140–150 mg/l), and high heavy metal concentrations. On the contrary, stormwater has been found to have very low concentration of these chemical constituents.

Chemical clogging has not been studied in context of stormwater systems. However, based upon the above literature and the known physical (i.e., neutral pH) and chemical characteristics (low levels) of urban stormwater (refer Table 15.1), it is hypothesized that the likelihood of chemical clogging in stormwater systems is minimal.

15.7.2 Biological clogging processes

Wastewater treatment systems have reported operational issues because of biological clogging (Chang et al., 1974; De Vries 1972; Rice, 1974). Biological clogging processes are caused by the accumulation of algae and bacterial products in water and the formation of biofilms and biomass due to the growth of microorganisms. Taylor et al. (1990) state that the growth of a biofilm in a porous medium reduces the total volume and average size of the pores. Cunningham et al. (1991) also found that biofilms start to form in porous media when microbial cells that exist in suspension adsorb to solid surfaces comprising the effective pore space. Seki and Miyazaki (2001) confirm this theory, by stating that bacteria attach to solid particles first reversibly and then irreversibly with exopolysaccharide polymers.

It has been found that several mechanisms lead to reduction of the hydraulic conductivity in filters because of ongoing biological processes.

The physical presence of the microbes and their colonies is one of the several clogging mechanisms. Vandevivere and Baveye (1992) observed a distinct connection between increasing biomass accumulation and the rate of hydraulic conductivity reduction while experimenting sand filters using bacteria. The secretion of polysaccharides by adsorbed bacteria is another mechanism that contributes to the clogging phenomenon. They also measured the occurrence of polysaccharides in sand filters and concluded that this compound had a significant contribution to the reduction in hydraulic conductivity. Polysaccharide accumulation helps in development of biofilms that eventually assist in removing pathogens (Stevik et al., 2002). In certain cases, gases such as nitrogen and methane can be produced by microorganisms that then lead to blocking of pores and creating vapour barriers to infiltration (Baveye et al., 1998; Seki and Miyazaki, 2001).

Given that stormwater flows are intermittent in nature and have high sediment load but low organic matter and nutrient content in comparison with other water types, it is normally hypothesized that biological clogging does not occur in these systems. Some studies indirectly support this hypothesis. For instance, Pavelic et al. (1998) investigated the nature and extent of clogging by injecting wetland-treated stormwater to a confined aquifer in South Australia over four years. They reported that clogging occurred due to injected sediments, especially where sediment concentrations were high, and was not a result of any biological activity. Similar results can be drawn from Bouwer and Rice (1989). They investigated the effect of clogging material, and of increasing the water depth, on infiltration rate, using two columns filled with sandy loam soil samples. They conducted two similar types of experiments with inorganic and organic suspension of clogging layers by varying water depth from 20 to 85 cm. The findings suggest that the inorganic clogging layer is more prominent in decreasing hydraulic conductivity and has relevance for urban stormwater, which is relatively low in organics.

Kandra et al. (2015b) in a laboratory investigation of biological clogging of zeolite-based stormwater filters concluded that biological clogging may be present in stormwater filtration systems. Results indicated an increased rate of clogging and better sediment treatment when the filters were dosed with high-nutrient stormwater. On the other hand, the chlorinated dosed filters lasted for the longest duration. The results of loss on ignition partially confirmed these findings.

The abundance and nature of microbes in stormwater, therefore, do imply that biological clogging should not be neglected altogether especially for contaminated catchments. Furthermore, since adsorption is a dominant removal mechanism in granular filter media, it may create an environment conducive to removal of microbes and eventually growth of microbial colonies. The effect of drying and wetting on biological clogging also needs to be investigated.

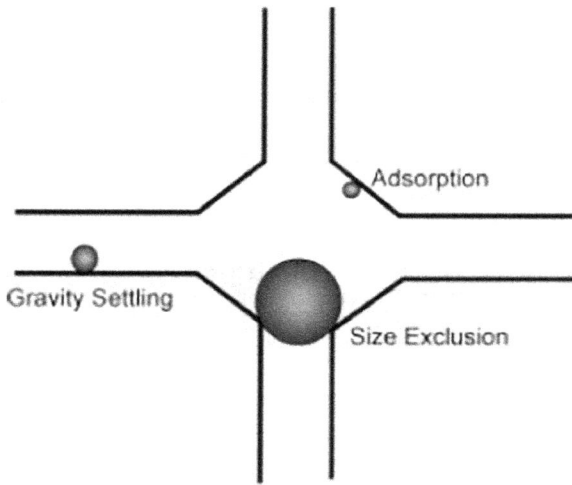

Figure 15.4 Particle capture mechanisms in porous media.
Source: Changhong (2008).

15.7.3 Physical/mechanical clogging processes

Physical/mechanical clogging may result from the migration of fine-grained materials into coarse-grained materials, consequently resulting in their entrapment and accumulation in soil pores. This entrapment of solids results mainly from three mechanisms: size exclusion, sedimentation, and adsorption, as shown in Figure 15.4 (Changhong, 2008).

Deposition of the suspended particles within the filter bed depends on many factors such as the type of filter and particle size of filter media (Perez-Paricio, 2001; Changhong 2008). Since there is a diverse range of particles in stormwater, the deposition of retained particles within the filter bed may be in more than one of the following ways:

1. *Blocking filtration* wherein the deposited particles reduce the flowing path inside the porous media, thus increasing the possibility of bridging (Figure 15.5a). This may be more pertinent for smaller-sized particles that can be removed within relatively large pores if numerous particles arrive simultaneously and block the pore by bridging (Knowles et al., 2011).

2. *Cake filtration* may occur when the particles are too large to penetrate the media or other particles are deposited on existing blocked particles. This cake layer may also be described as the clogging layer, which limits flow through the filter medium (Figure 15.5b). Depending on the age of cake, it may be compressed due to accumulation of the

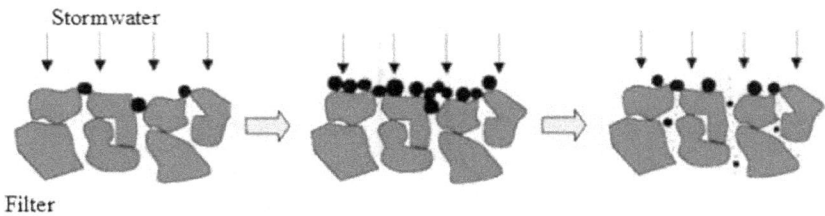

Stormwater

Filter

Figure 15.5 Clogging usually begins with blocking filtration and is followed by cake filtration and then eventually leads to deep bed filtration.

Source: Adapted from Siriwardene (2008).

particles. The hydraulic performance of such as system is a direct function of particle aggregation at the surface of the filter medium (McDowell-Boyer et al., 1986).

3. *Deep bed filtration* or internal cake formation occurs when small particles may invade the formation, bridge, and form an internal filter cake (Figure 15.5c). Straining is a common particle removal mechanism in the case of deep filters. Bradford et al. (2007) describe straining as "the retention of colloids in the smallest regions of the soil pore space formed adjacent to the points of grain-grain contact". Herzig et al. (1970) stated that straining is purely a geometric process involving mechanical removal of particles in small pore spaces. Bradford et al. (2007) found that both solution chemistry and hydrodynamic forces influence straining and, therefore, both physical and chemical mechanisms may be involved. Deep bed filtration may be preferable for stormwater applications to limit failure due to formation of an impermeable cake layer.

Siriwardene (2008), in a review of clogging processes, concluded that stormwater has fewer organic contents than the treated wastewater sources, and hence, physical nature of clogging is more prevalent for stormwater systems. In general, it is hypothesized that the importance of chemical and biological clogging in stormwater filters will be less than that of physical clogging. Furthermore, in some cases, it has been found that physical clogging initiates the process of decline in hydraulic conductivity and creates an environment conducive for biological and chemical clogging. For instance, Chang et al. (1974) concluded that although the initial reduction of porosity was caused by the sediment removal through physical processes, the growth of microbes around the trapped sediment ultimately sealed the pores in the sample. Hence, physical clogging is a very important clogging process to be considered in the context of systems treating stormwater influent that has high and variable sediment loads.

Even though physical clogging in stormwater systems has been investigated in biofilters (e.g., SEB's papers), infiltration systems (e.g., Abbott and Comino-Mateos (2001); Dechesne et al. (2005); Siriwardene (2008)), porous pavements (e.g., Pratt et al. (1995); Sansalone et al. (2011)), physical clogging processes in high-rate filtration systems need further investigation. This includes systems that use zeolites or coarse sands (like enviss™ filters that have been discussed earlier).

Given that the focus of this study is the (non-vegetated) filters that treat urban stormwater and the operational conditions are more conducive for physical clogging, the scope of this review is narrowed to physical clogging only. The next section therefore focuses on the processes and factors affecting physical clogging. However, some of the factors discussed therein may also have relevance for chemical and biological clogging as well.

15.8 FACTORS AFFECTING PHYSICAL CLOGGING

The physical process of filtering/trapping particles through filtration media is impacted by a range of factors. Firstly, it is the filter design such as the type of filtration media and its physical characteristics, in the construction of filter beds that can affect performance. Next, there are significant variations in the quantity and timing of influent applied to stormwater treatment systems, as they are only active during wet weather periods (i.e., intermittent operation) and have highly variable inflow rates (caused from variability in rainfall rates). Finally, there are significant fluctuations in the water quality in urban stormwater between wet weather events, as described in Section 15.2.2 (Table 15.1), partly attributable to the variations in rainfall rates, dry weather periods etc.

This section therefore reviews how physical clogging is influenced by filter media design and operational conditions (water quantity and quality). While these studies will provide some understanding of how these factors influence physical clogging in stormwater systems, it is noted that direct transfer of knowledge is not possible as drinking and wastewater treatment systems experience very different operational dynamics (i.e., they are continually inundated, often with a consistent water quality).

15.8.1 Filter design

Different filter medias vary from each other based on the type of material and shape of grains (such as angularity of grains and their surface texture). This may eventually affect the hydraulic and treatment performance of granular media and therefore the system. For instance, Hatt et al. (2008) compared the performance of fine sand, sandy loam, and its different combinations with hydrocell, vermiculite, perlite, compost, and mulch for stormwater treatment. All of these six media had different initial hydraulic performances (low infiltration rates ranging between 7 and 65 * 10^{-5} m/s). In their laboratory experiments, they observed a significant reduction in hydraulic performance of filter beds overtime with a different rate of decline for each media type. Similarly, Knowles et al. (2011), while studying subsurface flow wetlands for wastewater treatment, recognized that significant differences exist between sand filters and bio-retention facilities because of the differences in type of media used. However, the above-mentioned studies have been undertaken for fine filter media that have low infiltration rates. Given that various coarser filter media may have different porosity and arrangement of pores as compared to fine media filters, specific performance studies are needed before wider application.

The treatment and hydraulic performance of filter bed also depends on the shape of filter grains. For instance, improved performance of jagged and angular shaped grains over spherical grained particles has been demonstrated for drinking water filtration using fine filters (Suthaker et al., 1995; Evans et al., 2002). Knowles et al. (2011) also observed that particle shape influences media hydraulic conductivity for treatment of wastewater. This may be because the non-spherical or angular media shape of grains reduces porosity of filter bed and provides an increased specific surface area for biofilm growth. This eventually enhances pollutant removal and hence more clogging. Similarly, in a review of retention and removal of pathogenic bacteria in wastewater percolating through porous media, Stevik et al. (2002) also state that the rate of biological clogging due to accumulation of bacteria is dependent on the size and shape of the filter media and the bacterial cells. However, these experiments did not investigate the role of captured particles on the filtration and clogging performance. Further, the effect of filter grain shape needs investigation especially in the context of filters used for stormwater treatment.

The size of filter media has also been found to impact the hydraulic performance of a system as it controls the flow-through rates. Knowles et al. (2011) found that the hydraulic conductivity of porous media was very sensitive to media size for wastewater treatment in vertical and horizontal subsurface flow treatment wetlands. Indeed, the smaller the particle size, the higher specific surface area available for biofilm establishment, and surface chemistry; and the greater the likelihood of suspended solids interception due to narrower pore diameters. This therefore made the fine media prone to rapid clogging by pore occlusion from filtration and bridging of surface accumulations. Similarly, in a study of leachate collection systems, McIsaac and Rowe (2007) found that the coarse gravel bed (38 mm) performed much better than the 19-mm gravel bed and maintained a higher hydraulic conductivity even after operating for twice as long. In a study of natural sedimentation in laboratory-based water columns, Skolasinska (2006) found that trapping of influent particles in case of riverbeds was controlled by infiltration rate, suspension viscosity, and shape and size of both the porous media.

Rodgers et al. (2011) studied the effect of sand filter depth on wastewater treatment performance of four intermittently dosed fine sand filters with a grain size of 1mm. The performance of the shallower filter (0.3 m deep) appeared to diminish over time as the effluent COD and SS concentrations rose gradually as the study progressed. However, the 0.4-m-deep filter did not exhibit the same effects. Farizoglu et al. (2003) studied the effect of filter depth on wastewater pollutant removal using sand and pumice as a filtration media under rapid filtration conditions (7.64 and 15.28 m/h). The change of suspended solids removal rate with time for 1000, 750, and 470 mm of bed depths suggests that increasing bed depth increases the available surface area for the capture of particulate matter. Thus, more particulate matter can be retained in the filtration bed, so the removal rate of the particulate matter is improved by increasing the bed depth. Even though these studies are in context of water and wastewater treatment, these do highlight the significance of filter depth on pollutant removal and clogging.

Changhong (2008) observed that high flow rate (velocity) can carry particles further inside a porous medium, but no studies have been undertaken to assess the effect of infiltration rate on clogging performance. This may be consequential for nature/extent of filter bed clogging (whether it is at the surface or deep interstitial in nature) and hence could guide maintenance protocols of these systems. However, most of studies to date are limited to media that have relatively low conductivity; Suthaker et al. (1995) tested filter media for drinking water filtration at filtration rates of 12.5, 8, and 3.5 m/h; Pavelic et al. (2011) studied soil aquifer treatment at filtration rates of 0.004 m/h (loam) and 0.96 m/h (sand). This therefore necessitates the need to undertake specific studies for urban stormwater filtration using filters with comparatively higher infiltration rates.

Kandra et al. (2014) found that the angularity and smoothness of filter media particles and therefore the selection of filter media impacted the pollutant removal and clogging performance of stormwater treatment filters with infiltration rates. While comparing four different filter media (zeolite, scoria, river sand, and glass beads), the authors found that scoria-based filters had a highly variable performance, most likely due to breakdown of the particles. Further, while the selection of filter media may be guided by local availability, it is necessary to investigate the structural stability of filter media. Filter media may break within the filter bed over time, which could affect the stability of filter bed. The authors acknowledged the need to study filtration processes specifically in context of application given the diverse nature and sizes of pollutants, especially sediments, in different types of wastewaters.

In another related study, Kandra et al (2014a) investigated the effect of the design of filter bed on clogging processes in granular filters using zeolite with high infiltration rates. Design configurations were experimented: different filter media particle sizes, varied depths of filter bed, and different filter media packing configurations (single layer of a specific filter media size and multiple layers of different particle sizes). Deeper filtration systems were found to treat a significantly larger volume of stormwater possibly because they had greater ponding depth while the shallower systems clogged at the surface. Use of filter media with smaller particle sizes reduced hydraulic conductivities of the filter but improved the sediment removal performances. It was observed that filter bed with a coarser media had one-third sediment removal efficiency yet treated more than 30 times the amount of stormwater treated by the design using fine media. Layered filters were found to offer greater resistance to clogging and removed much more sediment before clogging as compared to filters using single-layered particle sizes. This may be because of the protection to the bottom layers and most of removal processes/sediment removal occurring in the upper layer. Therefore, the choice of filter media's particle size must be guided by the objective of treatment, whether it is focussed more on flow control or treatment of stormwater.

It can therefore be concluded from the above review that several design factors affect filtration and clogging processes. Most of the stormwater treatment systems are currently being designed using knowledge and experience acquired from other water treatment systems. Given the specific nature of clogging, it is important to understand clogging processes in the context of non-vegetated filter material that can be potentially used for stormwater treatment. Simple modifications to design of stormwater filtration systems using above knowledge can help improve the sediment removal performance and/or reduce maintenance intervals significantly.

15.8.2 Operational conditions – water quantity characteristics

Several studies have investigated the influence of hydraulic loading rates on hydraulic performance of filtration systems but the results are conflicting. For instance, Ruppe (2005) observed that sand filters treating wastewater with identical hydraulic loading rates but different dosing frequencies clogged at different times. On the other hand, despite that particle deposition is generally known to be less likely at higher flow rates, Reddi et al. (2000) found that difference in flow rates did not cause any noticeable change in the clogging behaviour of sandy soils (using fluids containing polystyrene or kaolinite particles). Other wastewater studies, for instance, Leverenz et al. (2009), also highlight significance of different operational variables: time of operation, hydraulic loading rate, and dosing frequency. Knowles et al. (2011) also recognized significant differences between sand filters and bio-retention facilities. It was found that sand filters usually have relatively steady inflow rates and ponding heads, while the variability of incoming runoff renders bio-retention behaviour much more dynamic. Similarly, flow rate and duration of injection were found to affect the rate of clogging when injection of wetland-treated stormwater to a confined aquifer was studied by Pavelic et al. (1998).

Hatt et al. (2008) compared six different fine filter media using stormwater. The authors found that the infiltration capacity of the filters recovered following an extended dry period, before declining again during wet periods. This may indicate that clay particles and organic matter swell during wet periods (reducing the porosity of the filter media) and develop cracks and macropores as water content decreases during dry periods (increasing porosity). Similar patterns were evident for all other filter media types, although the variation in infiltration capacity was much less for some of the filters. Li and Davis (2008) also found that bio-retention filters exhibited higher solids loading capacity under intermittent flow conditions before clogging than under continuous flow conditions. Similarly, Knowles et al. (2011) state that intermittent operations may be beneficial for reversing clogging in the wetlands. The periodicity of loading to resting determines the ability of the system to operate without clogging, and the recovery period depends on climatic conditions. Systems in cold and wet climates will require a longer recovery period than those in hot and arid climates (Knowles et al., 2011). Studies undertaken for intermittent sand filters used in wastewater treatment also suggest that variations in hydraulic loading regimes had significant effects on clogging. However, Yong et al. (2010), while investigating the effect of drying and wetting regimes on the clogging behaviour and pollutant removal efficiency of three porous pavement types, found that drying has a direct influence on the longer lifespan of these systems with higher solid loading capacity. These studies therefore suggest that intermittent loading regimes do affect the longevity of filtration systems in different ways and need further investigation before application.

Effect of operational conditions on clogging, such as hydraulic head, has been studied. Reddi et al. (2005) compared the clogging performance of sandy soils under constant flow rate and constant head conditions. Similar permeability reduction with respect to time was observed in both cases. However, permeability reduction under constant head occurred in much fewer pore volumes as compared to under constant flow rate. Siriwardene et al. (2007) in their study of gravel stormwater filters also found that dynamics of water application, and in particular the presence of a constant water level in a gravel infiltration system, affects its hydraulic performance. However, it is not possible to operate stormwater treatment systems under a constant flow regime. The methods designed to investigate the clogging processes of systems in this study will therefore focus on constant head conditions only.

Kandra et al. (2015a) investigated effect of stormwater loading rate and regime/frequency of loading on zeolite-based non-vegetated filters. Stormwater loading rate was found to affect the sediment removal and hydraulic performance of systems with high infiltration rates. However, any variations in the stormwater dosing regime had a limited effect on these filters. Since the loading rate and dosing regime are dependent on the climatic conditions and location of treatment system in the catchment, the design of stormwater systems needs to consider this background information.

Based on this review, it can be concluded that loading rates and regimes have a mixed effect on clogging of filtration systems. Specific studies therefore need to be undertaken to investigate the effect of hydraulic loading rate and regimes on clogging in context of stormwater treatment using non-vegetated filters with high infiltration rates.

15.8.3 Operational conditions – water quality characteristics

There could be a significant variation in the stormwater characteristics depending on several factors such as the catchment type and its size, nature and extent of pre-treatment, climatic conditions, and so on (refer Section 15.2.2). These may all eventually affect the way filtration systems behave.

Haselbach (2010) found that extreme storm events deposit large quantities of clay on pervious concrete pavements leading to a reduction in the infiltration capacity of the system. The rate of this reduction was found to increase with the density of the clay suspensions. Gautier et al. (1999) studied drainage in industrial areas (two low flowrate basins) and found that the particles present in the stormwater, or their concentration (based on the land use), have an influence on the clogging process. Similarly, Pavelic et al. (1998), while studying injection of wetland-treated stormwater to a confined aquifer at very low infiltration rates (sand: 23 m/day and loam: 0.1 m/day), also suggest that the rate of clogging is dependent on concentration of suspended solids and temperature.

In a review of clogging phenomenon in wetlands for wastewater treatment, Knowles et al. (2011) found that influent characteristics, such as solids content and pollutant characteristics, are vital for understanding clogging processes in the subject systems. The size of particles in the influent has also been found to affect hydraulic performance of the system. Siriwardene et al. (2007) undertook a laboratory study to understand physical clogging processes in coarse gravel stormwater filters. It was found that physical clogging is mainly caused by the migration of sediment particles less than 6 μm in diameter. Li and Davis (2008) conducted a series of laboratory column experiments and field observations for bio-retention filters with low flow rates. It was found that clay-sized component had a greater contribution to media clogging as compared with components to particles of other sizes.

Similarly, Kaminski et al. (1997) state that influent particle size distribution plays an important role in wastewater filtration (with flow rates varying between 5 and 25 m/h) and the particle removal efficiency varies for different particle size groups. It was also observed that filters with high rate of filtration were more sensitive to particle size, as compared with low-rate filters. However, Changhong (2008) in a review of different studies (using media such as sandstone, glass beads, sand and passing particles such as alumina, clay, latex, bentonite) found that bigger influent particles caused more damage as they have higher tendency to settle down and block or bridge.

Flocculation could also affect blocking processes within the filter bed. Reddi et al. (2000) compared influents made of fine and uniform-sized kaolinite particles (size: 2–12 μm) with relatively larger particle-sized influent made of polystyrene spheres (size: 1–35 μm). Permeability reductions for both influents were observed to be comparable. This was attributed to be an effect of flocculation, which may as well have a role to play in the case of stormwater systems.

Spychala and Blazejewski (2003) studied factors affecting fine sand clogging by septic tank effluent and found that sewage temperature had a significant impact on the clogging process of fine sands with lower temperature resulting in lower hydraulic conductivity. However, for sewage ponding in study columns, the impact of temperature on the filter hydraulic conductivity was more significant for biological activity than for sewage viscosity. De Vries (1972), while testing the effect of primary wastewater effluent on different experimental columns filled with sand fractions, also found that effluent applications at the same rate, but at a temperature of $4\pm3°C$, while allowing the same daily rest periods, resulted in early failure because of pore clogging. However, a limited variation in water temperature is expected in the case of stormwater treatment systems (e.g., Emerson and Traver, 2008; Roseen et al., 2009).

Kandra et al (2015a), while investigating high-flow-rate filters treating stormwater, found that sediment concentration in influent affected both hydraulic and treatment performances. There experiments estimated that the additional presence of metals and nutrients had limited or no effect on performance of these filters but acknowledged the effect of pollutants other than sediments on clogging needs to be investigated further, especially in context of intermittent operation of stormwater filters (drying and wetting regimes). Experiments conducted by these authors also concluded that coarse filters with high infiltration rates may not be effective in trapping very fine sediment because of the difference between size of their pores and incoming sediment. Therefore, the size of sediments in stormwater and their relationship with the size of filter media grains is an important parameter to be considered in design of coarse filters with high infiltration rates that are used for stormwater treatment.

Above review stresses the significance of the effect of several influent characteristics on clogging phenomenon: concentration and type of pollutants, particle size of pollutants and their comparison with pore size of filter media. It is therefore pertinent to undertake specific studies to understand the effect of stormwater influent characteristics on clogging and treatment for different catchments and rainfall conditions.

15.9 NATURE OF CLOGGING

It is important to understand the migration of pollutants and the location of clogged materials for adequate design, operation and maintenance of filtration-based systems. However, literature provides conflicting information about the extent of clogging within the filter bed. For instance, in their comparison of six different filter media using stormwater, Hatt et al. (2008) found that the accumulation of captured sediment was at or near the filter surface and was causing a reduction in hydraulic performance. Similarly, Siriwardene et al. (2007), while studying gravel stormwater filters, found that a clogging layer forms at the interface between the filter and underlying soil. Literature on porous pavements, for instance, Haselbach (2010), also states that most of the trapped material remains on the surface of the pavement and can be removed with simple maintenance procedures such as sweeping. However, Li and Davis (2008) in their study of bio-retention filter media observed that both depth filtration and cake filtration contribute to urban particle capture.

Similar contradictions related to the nature/extent of filter bed clogging exist in the case of wastewater systems. De Vries (1972), while testing the effect of primary wastewater effluent on columns filled with sand fractions, concluded that filter failure was caused by the surface sludge layer. Skolasinska (2006), however, found that while intensity of clogging decreased with depth, most of the suspended material was trapped near the surface. Some of the field investigations of artificial recharge basins that use turbid or treated wastewater report that clogging happened at the surface (Schuh, 1990). However, other studies report that although much of the suspended material is filtered at the surface, some clay particles penetrate to greater depths (Goss et al., 1973; Schuh, 1988).

In a review, Herzig et al. (1970) found that the ratio of sediment size (d_s) to filter media's grain size (d_p) was an important parameter that guides the nature of clogging. For $d_s/d_p > 0.15$, the porous medium is irreversibly blocked and a filter cake is formed. However, when the influent particle sizes are smaller $(d_s/d_p < 0.065)$, the retention remains low. Further, for the intermediate values, a partial blocking of the porous bed may occur depending on shape of filter grains and bed porosity.

Kandra et al. (2013) undertook profiling of the clogged columns to understand the nature of clogging in granular stormwater filters. Visual analysis suggested that the particle size of filter media and depth of filter bed affect the location of clogged material in the filter bed. Even though most of the sediment removal occurred within the top-most layer, some deeper clogging still occurred, indicating a combination of cake and depth filtration; adsorption and interception contribute to the clogging of these systems. Every dosing event, these filters being subjected to drying and wetting, disturbs/re-suspends the sediments trapped within the pores of filter material and eventually allows them to move through the filter bed enhancing the life of the filter bed. The size of sediment in stormwater (and its relationship with the size of filter media grains) is an important parameter to be considered in the design of coarse filters with high infiltration rates that are used for stormwater treatment. Profiling of the clogged columns with different stormwater loading rates and stormwater dosing frequency suggests that these changes in operational conditions had a limited effect on the location of clogged material in the filter bed.

Therefore, based on the existing literature, it can be concluded that the nature of clogging is different across systems and influents. It is specific to factors like the filter media characteristics, influent characteristics, and the interplay between them.

15.10 CONCLUSIONS

As understood from this review of literature, clogging is very specific to the characteristics of filtration media, influent, and nature/dynamics of stormwater application (which depends on local conditions in the catchment such as rainfall intensity and patterns). Therefore, an understanding of the processes and factors that influence these processes (filtration media, influent, and nature/dynamics of stormwater application) is important.

More work is needed to investigate biological clogging in these filters while considering effects of dry weather, prolonged drying and natural aspects such as temperature (of water and filter), and solar energy in relationship to humidity.

It is vital to test performance of locally available filter media for stormwater treatment, such as pilot testing in demonstration facilities, design rather than just replicating knowledge from other water treatments.

Models need to be developed to be able to help designers predict the hydraulic and treatment performance of filters over their lifespan using different variables such as filter bed design, operational conditions, and stormwater quality.

One of the main applications of the new knowledge and design tools would be improved design of stormwater treatment systems based on granular filters, particularly for stormwater filtration for ecosystem protection or supply of filtered stormwater for harvesting purposes.

REFERENCES

Abbott, C.L. and Comino-Mateos, L., 2001. In situ performance of an infiltration drainage system and field testing of current design procedures. *Journal of Charted Institution of Water and Environmental Management*, 15(3), pp. 198–202.

Allison, R.A., Walker, T.A., Chiew, F.H.S. and McMahon, T.A., 1998. From roads to rivers: gross pollutant removal from Urban waterways. *Co-operative Research Centre for Catchment Hydrology. Technical Report 98/6*.

Bardin, J.P., Gautier, A., Barraud, S. and Chocat, B., 2001. The purification performance of infiltration basins fitted with pre-treatment facilities: a case study. *Water Science and Technology*, 43(5), pp. 119–128.

Barraud, S., Gautier, A., Bardin, J.P. and Riou, V., 1999. The impact of intentional stormwater infiltration on soil and groundwater. *Water Science and Technology*, 39(2), pp. 185–192.

Baveye, P., Vandevivere, P., Hoyle, L.B., DeLeo, P.C. and Lozada, D.S., 1998. Environmental impact and mechanisms of the biological clogging of saturated soils and aquifer materials. *Critical Reviews in Environmental Science and Technology*, 28(2), pp. 123–191.

Bean, E.Z., Hunt, W.F. and Bidelspach, D.A., 2007. Evaluation of four permeable pavement sites in Eastern North Carolina for runoff reduction and water quality impacts, *ASCE Journal of Irrigation and Drainage Engineering*, 133 (6), pp.583–592.

Bradford, S.A., Torkzaban S. and Walker S.L., 2007. Coupling of physical and chemical mechanisms of colloid straining in saturated porous media. *Water Research*, 41, pp. 3012–3024.

Bratieres, K., Schang, C., Deletic, A. and McCarthy, D.T., 2012. Performance of enviss™ stormwater filters: results of a laboratory trial. *Water Science and Technology*. 66(4), pp 719–727.

Birch, G. F., Fazeli M. S. and Niatthai C., 2005. Efficiency of an infiltration basin in removing contaminants from urban stormwater. *Environmental Monitoring and Assessment*, 101(1–3), 23–38.

Bouwer, H. and Rice, R.C., 1989. Effect of water depth in groundwater recharge basins on infiltration. *Journal of Irrigation and Drainage Engineering*, 115(4), pp. 556–567.

Bouwer, H., 2002. Artificial recharge of groundwater: Hydrogeology and engineering. *Journal of Hydrogeology*, 10(1), pp. 121–142.

Brown, R. and Farrelly, M. (2007). Advancing urban stormwater quality management in Australia: survey results of stakeholder perceptions of institutional drivers and barriers. Report No. 07/05, National Urban Water Governance Program, Monash University.

Burton, G.A. and Pitt, R.E., 2002. *Stormwater Effects Handbook - A Toolbox for Watershed Managers, Scientists, and Engineers*, Lewis, Boca Raton, FL.

Butler, D. and Davies, J.W., 2004. *Urban Drainage*-2nd edition, Spon Press.

Cities as Water Supply Catchments: Sustainable Technologies, Centre for water sensitive cities (CWSC), Project C1.1, 2010. Available online https://watersensitivecities.org.au/content/project-c1-1/

Chang, A.C., Olmstead, W.R., Johanson, J.B. and Yamashita, G., 1974. The sealing mechanism of wastewater ponds. *Water Pollution Control Federation*, 46 (7), pp. 1715–1721.

Changhong, G., 2008. Understanding capture of non-Brownian particles in porous media with network model. *Asia Pacific Journal of Chemical Engineering*, 3, pp. 298–306.

Cunningham, A.B., Characklis, W.G., Abedeen, F. and Crawford, D., 1991. Influence of biofilm accumulation of porous media hydrodynamics. *Environmental Science and Technology*, 25, pp. 1305–1311.

De Vries, J., 1972. Soil filtration of wastewater effluent and the mechanism of pore clogging. *Water Pollution Control Federation*, 44(4), pp. 565–573.

Dierkes, C., Angelis, G., Kandasamy, J. and Kuhlmann, L., 2002. Pollution retention capability and maintenance of permeable pavements. *Proceedings of the 9th International Conference on Urban Drainage*, Portland, Oregon.

Dechesne, M., Barraud, S. and Bardin, J.P., 2004. Indicators for hydraulic and pollution retention assessment of stormwater infiltration basins. *Journal of Environmental Management*, 71, pp. 371–380.

Duchene, M., McBean, E. and Thomson, N., 1992. Modeling of infiltration from trenches for storm-water control. *Journal of Water Resources Planning and Management*, 120(3), 276–293.

Duncan, H.P., 1999. Urban stormwater quality: a statistical overview. Co-operative Research Centre for Catchment Hydrology. *Report 99/3*, Melbourne, Australia.

Duncan, H.P., 2005. Urban stormwater pollutant characteristics. In: T. H. F. Wong (eds.), *Australian Runoff Quality Guidelines*. Institution of Engineers Australia, Sydney.

Ellis, D.V., 1993. Wetlands or aquatic ape? Availability of food resources. *Nutrition and Health*, 9, pp. 205–217.

Ellis, J. and Marsalek, J., 1996. Overview of urban drainage: environmental impacts and concerns, means of mitigation and implementation policies. *Journal of Hydraulic Research*, 34(6), pp. 723–732.

Emerson, C.H. and Traver, R., 2008. Multiyear and seasonal variation of infiltration from storm-water best management practices. *Journal of Irrigation and Drainage Engineering*, 134, Special Issue: Urban Storm-Water Management, 598–605.

Emerson, C.H., Wadzuk, B.M. and Traver, R.G., 2010. Hydraulic evolution and total suspended solids capture of an infiltration trench. *Hydrological Processes*, 24(8), 1008–1014.

Evans, G., Dennis, P., Cousins, M. and Campbell, R., 2002. Use of recycled crushed glass as a filtration medium in municipal potable water treatment plants. *Water Science and Technology: Water Supply*, 2(5–6), pp 9–16.

Facility for Advanced Water Biofiltration (FAWB), Stormwater Biofiltration Systems Adoption Guidelines, 2022. Available online https://watersensitivecities.org.au/content/facility-for-advanced-water-biofiltration-stormwater-biofiltration-systems-adoption-guidelines/

Farizoglu B., Nuhoglu A., Yildiz E. and Keskinler B., 2003. The performance of pumice as a filter bed material under rapid filtration conditions. *Filtration & Separation*, 40(3), April 2003, pp. 41–47.

Fletcher, T.D., Duncan, H.P., Poelsma, P.J and Hatt, B.E., 2004. Stormwater flow and quality and the effectiveness of non-proprietary stormwater treatment measures – a review and gap analysis. Cooperative Research Centre for Catchment Hydrology, *Technical Report 04/08*, Canberra, Australia, pp.183.

Fletcher, T.D., Mitchell, V.G., Deletic, A. and Ladson, A., 2007. Is stormwater harvesting beneficial to urban waterway flow? *Water Science and Technology*, 55(4), pp. 265–272.

Francey, M., Fletcher, T.D., Deletic, A. and Duncan, H.P., 2010. New insights into water quality of urban stormwater in Southeastern Australia. *Journal of Environmental Engineering*, 136(4), pp. 381–390.

Fuchs, S., Brombach, H. and Weiss, G., 2004. New database on urban runoff pollution. In: *Proceeding of the 5th International Conference on Sustainable Techniques and Strategies in Urban Water Management NOVATECH 2004*, Lyon, France. Volume 1, pp. 145–152.

Fujita, S., 1994. Infiltration structures in Tokyo. *Water Science and Technology*, 30(1), pp. 33–41.

Gnecco, I., Berretta, C., Lanza, L.G. and Barbera, P. La., 2005. Stormwater pollution in the urban environment of Genoa, Italy. *Atmospheric Research*, 77, pp. 60–73

Goonetillekea A., Thomas E, Ginn S. and Gilbert D., 2005. Understanding the role of land use in urban stormwater quality management. *Journal of Environmental Management*, 74, pp. 31–42.

Goss, D.W., Smith, S.J., Stewart, B.A. and Jones, O.R., 1973. Fate of suspended sediment during basin recharge. *Journal of Water Resources Research*, 9(3), 668–675.

Haselbach M. L., 2010. Potential for clay clogging of pervious concrete under extreme conditions. *Journal of Hydrologic Engineering*, 15(1), pp. 67–69.

Hatt, B.E., Fletcher, T.D., Walsh, C.J. and Taylor, S.L., 2004. The influence of urban density and drainage infrastructure on the concentrations and loads of pollutants in small streams. *Environmental Management*, 34(1), pp. 112–124.

Hatt, B.E., Deletic, A. and Fletcher, T.D., 2004a. Integrated stormwater treatment and re-use systems – inventory of Australian practice, Melbourne, Australia, Cooperative Research Centre for Catchment Hydrology, *Report 04/1*.

Hatt, B.E., Deletic, A. and Fletcher, T.D., 2006. Integrated treatment and recycling of stormwater: a review of Australian practice. *Journal of Environmental Management*, 79(1), pp.102–113.

Hatt, B.E., Fletcher, T.D. and Deletic, A., 2007a. Stormwater reuse: designing biofiltration systems for reliable treatment. *Water Science and Technology*, 55(4), pp. 201–209.

Hatt, B.E., Deletic A., Fletcher T., Poelsma P., Kolotelo P., McCarthy D.T. and Chandrasena G., 2012. Monitoring the performance of the Banyan Reserve biofiltration system: preliminary results. *Presentation made to Melbourne Water*, March 2012.

Hatt, B.E., Fletcher, T.D. and Deletic, A. 2007b. Hydraulic and pollutant removal performance of stormwater filters under variable wetting and drying regimes. *Water Science and Technology*, 56(12), pp 11–19.

Hatt B.E., Deletic A. and Fletcher T.D., 2008. Hydraulic and pollutant removal performance of fine media stormwater filtration systems. *Environment Science Technology*, 42, pp. 2535–2541.

Hatt, B.E., Fletcher, T.D. and Deletic, A., 2009. Pollutant removal performance of field-scale stormwater biofiltration systems. *Water Science and Technology* 59(8), pp. 1567–1576.

Herzig, J.P., Leclerc, D.M. and Le Goff, P., 1970. Flow of suspensions through porous media, application to deep filtration. *Journal of Industrial and Engineering Chemistry*, 62(5), pp. 8–35.

Illgen, M., Harting, K., Schmitt, T.G. and Welker, A., 2007. Runoff and infiltration characteristics of pavement structures-review of an extensive monitoring program. *Water Science and Technology*, 56(10), pp. 133–40.

J. Marsalek and B. Chocat. International report: stormwater management. *Water Science Technology*, 46(6–7) (2002), pp. 1–17

Kaminski, I., Vescan, N. and Adin, A., 1997. Particle size distribution and wastewater filter performance. *Water Science Technology*, 36(4), pp. 217–224.

Kandra, H., McCarthy D. and Deletic A., Assessing nature of clogging in zeolite based stormwater filters. *8th International Water Sensitive Urban Design Conference 2013*, Gold Coast, Australia. *Engineers Australia, Barton ACT*

Kandra, H.S., McCarthy D., Fletcher T.D. and Deletic A., 2014. Assessment of clogging phenomena in granular filter media used for stormwater treatment. *Journal of Hydrology*, 512, pp. 518–527.

Kandra, H. S., Deletic A. and McCarthy D., 2014a. Assessment of impact of filter design variables on clogging in stormwater filters. *Water Resources Management*, 28, pp. 1873–1885.

Kandra, H.S., McCarthy D. and Deletic A, 2015a. Assessment of impact of stormwater characteristics on clogging in stormwater filters. *Water Resources Management*, 29:1031–1048.

Kandra, H., Callaghan J., Deletic A. and McCarthy D., 2015b. Biological clogging in storm water filters. *Journal of Environmental Engineering, ASCE*, 141(2), 10.1061/EE.1943-7870.0000853, 04014057.

Knowles, P., Dotro, G., Nivala, J. and Garcia J., 2011. Clogging in subsurface-flow treatment wetlands: occurrence and contributing factors. *Ecological Engineering*, 37, pp. 99–112.

Lancaster, A., Dugopolski, R., Forester, K., Curtis, H., 2005. Low Impact Development Operations and Maintenance Training, Washington State Department of Ecology. Available online https://ecology.wa.gov/DOE/files/1c/1ce2dc01-8e07-4ecf-a9d0-02abaa9d9f55.pdf

Landphair, H., McFalls, J. and Thompson, D., 2000. Design methods, selection and cost effectiveness of stormwater quality structures (Report 1837-1). *Texas Transportation Institute Research Report*, College Station, TX, USA.

Larroque, F. and Franceschi M., 2011. Impact of chemical clogging on de-watering well productivity: numerical assessment. *Environmental Earth Sciences*, 64, pp. 119–131.

Le, Coustumer S. and Barraud S., 2007. Long-term hydraulic and pollution retention performance of infiltration systems. *Water Science & Technology*, 55(4), pp. 235–243.

Leverenz, H.L., Tchobanoglous, G. and Darby, J.L., 2009. Clogging in intermittently dosed sand filters used for wastewater treatment. *Water Research*, 43, pp. 695–705.

Li, H. and Davis, A., 2008a. Urban particle capture in bioretention media. I: laboratory and field studies. *Journal of Environmental Engineering*, 143(6), pp. 409–418.

Li, H. and Davis, A., 2008b. Urban particle capture in bioretention media. I: theory and model development. *Journal of Environmental Engineering*, 143(6), pp. 419–432.

Liebens, J., 2002. Heavy metal contamination of sediments in stormwater management systems: the effect of land use, particle size, and age. *Journal of Environmental Geology*, 41(3–4), pp. 341–351.

Line, D. and White, N., 2007. Effect of development of runoff and pollutant export. *Water Environment Research*, 75(2), pp. 184–194.

Lindsey, G., Roberts, L. and Page, W., 1992. Inspection and maintenance of infiltration facilities. *Journal of Soil and Water Conservation*, 47(6), pp. 481–486.

Lloyd, S.D., Wong, T.H.F., Iliebig, T. and Becker, M., 1998. Sediment characteristics in stormwater pollution control ponds, *Proceedings of the R-12 HydraStorm '98, 3rd International Symposium on Stormwater Management*, Adelaide, Australia. pp. 209–214.

Mays, D.C. and Hunt, J.R., 2005. Hydrodynamic aspects of particle clogging in porous media. *Environment Science Technology*, 39, pp. 577–584.

Mays, D.C. and Hunt, J.R., 2007. Hydrodynamic and chemical factors in clogging by montmorillonite in porous media. *Environmental Science and Technology* 41(16), 5666–5671.

Mays, D.C., 2010. Contrasting clogging in granular media filters, soils, and dead-end membranes. *Journal of Environmental Engineering*, 136(5), pp. 475–480.

McDowell-Boyer, L., Hunt J. and Sitar, N., 1986. Particle transport through porous media. *Water Resources Research*, 22(13), pp. 1901–1921.

McIsaac, R. and Rowe, R.K., 2007. Clogging of gravel drainage layers permeated with landfill Leachate. *ASCE Journal of Geotechnical and Geoenvironmental Engineering*, 133(8): pp. 1026–1039.

Mikkelsen, P., Haflinger, M., Ochs, J., Jacobsen, P. and Boller, M., 1996. Experimental assessment of soil and groundwater contamination from two old infiltration systems for road run-off in Switzerland. *The Science of the Total Environment*, 189/190, pp. 341–347.

Mikkelsen, P.S., Hafliger, M., Ochs, M., Jacobsen, P., Tjell, J.C. and Boller, M., 1997. Pollution of soil and groundwater from infiltration of highly contaminated stormwater - a case study. *Water Science and Technology*, 36(8–9), pp. 325–330

Mitchell, V.G., Mein, R.G. and McMahon, T.A, 2002. Utilising stormwater and wastewater resources in urban areas. *Australian Journal of Water Resources*, 6 (1), pp. 31–43.

Mitchell, V.G., Deletic, A., Fletcher, T.D., Hatt, B.E. and McCarthy, D.T., 2007. Achieving multiple benefits from stormwater harvesting. *Water Science and Technology*, 55(4), pp. 135–144.

Nozi, T., Mase, T. and Murata, K., 1999. Maintenance and management aspect of stormwater infiltration system, *Proceedings of the 8th International Conference on Urban Storm Drainage*, Sydney, Australia, pp. 1497–1503.

Pavelic, P., Dillon, P.J., Barry, K.E., Herczeg, A.L., Rattray, K.J., Hekmeijer, P. and Gerges, N.Z., 1998. Well clogging effects determined from mass balances and hydraulic response at a stormwater ASR site. *Conference on Artificial Recharge of Ground Water*, Balkema, Rotterdam.

Pavelic, P., Dillon, P.J., Mucha, M., Nakai, T., Barry, K.E. and Bestland, E., 2011. Laboratory assessment of factors affecting soil clogging of soil aquifer treatment systems. *Water Research*, 45(10), pp. 3153–3163.

Perez-Paricio, A., 2001. Integrated modelling of clogging processes in artificial groundwater recharge, Technical University of Catalonia, Barcelona, Spain. Available online https://www.tdx.cat/bitstream/handle/10803/6214/01Introd uction.pdf?sequence=1

Poelsma, P., McCarthy D.T. and Deletic A. Changes in the filtration rate of a novel stormwater harvesting system: impacts of clogging and moisture content. *Proceedings NOVATECH Conference*, Lyon, France, 27th June–1st July 1, 2010.

Pitt, R., Clark, S., Johnson, P.D., Morquecho, R., Gill, S. and Pratap, M., 2004. High level treatment of stormwater heavy metals, Salt Lake City, UT, United States. *American Society of Mechanical Engineers*, New York, NY 10016-5990, United States, 917–926.

Pratt, C.J., Mantle, J.D.G. and Schofield, P.A., 1995. UK Research into the performance of permeable pavement, reservoir structures in controlling stormwater discharge quantity and quality. *Water Science Technology*, 32(1), 63–69.

Raimbault, G., Nadji, D. and Gauthier, C., 1999. Stormwater infiltration and porous material clogging, *Proceedings of the Eighth International Conference on Urban Storm Drainage*, Sydney, Australia. *The Institute of Australia*, pp. 1016–1024.

Rauch, W., Seggelke, K., Brown, R. and Krebs, P., 2005. Integrated approaches in urban storm drainage: where do we stand? *Environmental Management*, 35(4), pp. 396–409.

Reddi, L.N., Ming, X., Hajra, M.G. and Lee, I.M., 2000. Permeability reduction of soil filters due to physical clogging. *Journal of Geotechnical and Geoenvironmental Engineering*, 126(3), pp. 236–246.

Rice, R.C., 1974. Soil clogging during filtration of secondary effluent. *Journal of Water Pollution Control Federation*, 46, pp. 708–716.

Rinck-Pfeiffer, S., Ragusa, S., Sztajnbok, P. and Vandevelde, T., 2000. Interrelationships between biological, chemical, and physical processes as an analog to clogging in aquifer storage and recovery (ASR) wells. *Water Research*, 34 (7), pp. 2110–2118.

Rodgers, M., Walsh, G. and Healy, M.G., 2011. Different depth sand filters for laboratory treatment of synthetic wastewater with concentrations close to measured septic tank effluent. *Journal of Environmental Science and Health, Part A: Toxic/Hazardous Substances and Environmental Engineering*, 46(1), pp. 80–85.

Roesner, L., Bledsoe, B. and Brashear, R., 2001. Are best-management-practice criteria really environmentally friendly? *Journal of Water Resources Planning and Management, May/June*, pp. 150–154.

Roseen, R.M., Ballestero, P.T., Houle, J.J., Avellaneda, P., Briggs, J., Fowler, G. and Wildey, R., 2009. Seasonal performance variations for storm-water management systems in cold climate conditions. *Journal of Environmental Engineering*, 135 (3), pp. 128–137.

Ruppe, L.M., 2005. Effects of dosing frequency on the performance of intermittently loaded packed bed wastewater filters. *Dissertation*, University of California, Davis.

Sansalone, J., Kuang, X., Ying, G. and Ranieri, V., 2012. Filtration and clogging of permeable pavement loaded by urban drainage. *Water Research*, 46(20), 6763–6774.

Schang, C., McCarthy, D.T., Deletic, A. and Fletcher, T.D., 2010. Development of the enviss filtration media. *Paper presented at the Novatech 2010 Conference*, Lyon, France, 27th June–1st July, 2010.

Schuh, W.M., 1988. In-situ method for monitoring layered hydraulic impedance development during artificial recharge with turbid water, *Journal of Hydrology*, 101(1–4), 173–189.

Schueler, T.R., Kumble, A. and Heraty, M.A., 1992. A current assessment of urban best management practices; *Techniques for Reducing Non-Point Source Pollution in the Coastal Zone*. Metropolitan Washington Council of Governments, Washington, D.C.

Schuh, W.M., 1990. Seasonal variation of clogging of an artificial recharge basin in a northern climate. *Journal of Hydrology*, 121(1–4), pp. 193–215.

Seki, K., Miyazaki, T. and Nakano, M., 1996. Reduction of hydraulic conductivity due to microbial effects. *Trans. of JSIDRE, No.* 181, pp. 137–144.

Seki, K. and Miyazaki, T., 2001. A mathematical model for biological clogging of uniform porous media. *Water Resources Research*, 37 (12), pp. 2995–2999

Skolasinska, K., 2006. Clogging microstructures in the vadose zone- laboratory and field studies. *Hydrogeology Journal*, 14, pp. 1005–1017.

Smullen, J.T., Shallcross, A.L. and Cave, K.A., 1999. Updating the U.S. nationwide urban runoff quality database. *Water Science and Technology*, 39 (12), pp. 9–16.

Spychala, M. and Blazejewski, R., 2003. Sand filter clogging by septic tank effluent. *Water Science and Technology*, 48(11–12), 153–159.

Stead-Dexter, K. and Ward, N.I., 2004. Mobility of heavy metals within fresh-water sediments affected by motorway stormwater. *Science of the Total Environment*, 334–335, 271–277

Siriwardene, N., Deletic, A. and Fletcher, T.D., 2007. Clogging of stormwater gravel infiltration systems and filters: Insights from a laboratory study. *Water Research*, 41, pp. 1433–1440.

Siriwardene, N., Deletic, A. and Fletcher, T.D., 2007. Modeling of sediment transport through stormwater gravel filters over their lifespan. *Environment Science Technology*, 41, pp. 8099–8103.

Siriwardene, N.R., 2008. Development of an experimentally derived clogging prediction method for stormwater infiltration and filtration systems. *PhD thesis*, Institute for Sustainable Water Resources, Department of Civil Engineering, Monash University, Melbourne.

Stevik, T.K., Aa, K., Ausland G., Hanssen J.F., 2004. Retention and removal of pathogenic bacteria in wastewater percolating through porous media: a review. *Water Research*, 38(6), pp. 1355–1367.

Strecker, E., Quigley, M., Urbonas, B., Jones, J. and Clary, J., 2001. Determining urban storm water BMP effectiveness. *Journal of Water Resources Planning and Management*, 127(3), pp. 144–149.

Suthaker, S., Smith, D.W. and Stanley, S.J., 1995. Evaluation of filter media for upgrading existing filter performance. *Environmental Technology*, 16(7), pp. 625–643.

Tan, S.A., Fwa, T.F. and Han, C.T., 2003. Clogging evaluation of permeable bases. *Journal of Transportation Engineering*, 129(3), pp. 309–315.

Taylor, G.D., Fletcher, T.D., Wong, H.F.T., Breen, P.F. and Duncan, H.P., 2005. Nitrogen composition in urban runoff- implications for stormwater management. *Water Research*, 39 (10), pp. 1982–1989.

Taylor, S.W., Milly, P.C.D. and Jaffe, P.R., 1990. Biofilm growth and the related changes in the physical properties of a porous medium. *Water Resources Research*, 26 (9), pp. 2161–2169.

Vandevivere, P. and Baveye, P., 1992. Relationship between transport of bacteria and their clogging efficiency in sand columns. *Applied and Environmental Microbiology*, 58(8), pp. 2523–2530.

Veenhuis, J.E., Parrish, J.H. and Jennings, M.E., 1988. Monitoring and design of stormwater control basins- Design of urban runoff quality controls, *Proceedings of an Engineering Foundation Conference on Current Practice and Design Criteria for Urban Quality Control*, Potosi, Missouri, pp. 224–238.

Wadzuk, B.M., Rea, M., Woodruff, G., Flynn, K. and Traver, R.G., 2010. Water-quality performance of a constructed stormwater wetland for all flow conditions. *Journal of the American Water Resources Association*, 46(2), pp. 385–394.

Walsh, C.J., 2000. Urban impacts on the ecology of receiving waters: a framework for assessment, conservation and restoration. *Hydrobiologia*, 431(2–3), pp. 107–114.

Walsh, C.J., Fletcher, T.D. and Ladson, A.R., 2005. Stream restoration in urban catchments through redesigning stormwater systems: looking to the catchment to save the stream. *The North American Benthological Society*, 24(3), 690–705.

Warnaars, E., Larsen, A.V., Jacobsen, P. and Mikkelsen, P.S., 1999. Hydrologic behaviour of stormwater infiltration trenches in a central urban area during 23/4 years of operation. *Water Science and Technology*, 39(2), pp. 217–224.

Weiss, J.D., Hondzo, M. and Semmens, M., 2006. Storm water detention ponds: Modeling heavy metal removal by plant species and sediments. *Journal of Environmental Engineering*, 132(9), 1034–1042.

Winter, K.J. and Goetz, D. (2003) The impact of sewage composition on the soil clogging phenomena of vertical flow constructed wetlands. *Water Science and Technology*, 48(5), pp. 9–14.

Wong, T.H.F. (2000) Improving urban stormwater quality - from theory to implementation. *Journal of the Australian Water Association*, 27(6), 28–31.

Wong, T.H.F., 2006a. An overview of water sensitive urban design practices in Australia. *Water Practice & Technology* 1, 1.

Wong, T. H. F., 2006b. *Australian Runoff Quality: A guide to Water Sensitive Urban Design*. Engineers Australia, Canberra.

Yaman, C., Martin, J.P. and Korkut, E., 2006. Effects of wastewater filtration on geotextile permeability. *Geosynthetics International*, 13(3), pp. 87–97.

Yi, Q.T., Yu, J. and Kim, Y., 2010. Removal patterns of particulate and dissolved forms of pollutants in a stormwater wetland. *Water Science and Technology*, 61(8), pp. 2083–2096.

Yong, C.F., Deletic, A., Fletcher, T.D. and Grace, M.D., 2008. The clogging behaviour and treatment efficiency of a range of porous pavements. *Proceedings 11th Int. Conf. on Urban Drainage*, Edinburgh, Scotland, UK.

Chapter 16

Smart buildings and their rating tools

Danny Byrne, Ashok K. Sharma,
Shobha Muthukumaran, and Dimuth Navarata
Victoria University

CONTENTS

DOI: 10.1201/9781003368335-16

16.1 INTRODUCTION

The worldwide construction industry is focusing on the objective of developing smart and sustainable infrastructure with efficient use of resources.

30% of low-energy-efficient buildings of the global are currently responsible for one-third of direct and indirect CO_2 and particulate matter emissions [1]. Past studies have demonstrated that the construction sector has the potential to improve energy efficiency with cost-effective intervention [2]. An improvement in the performance of buildings such as of energy reduction, along with reducing carbon emissions, would yield significant benefits to building owners and occupants, such as improved durability, reduced maintenance, greater comfort, lower costs, higher property values, increased habitable space, increased productivity, and improved health and safety [3].

The commitment of all sectors to improve the energy efficiency of buildings has therefore greatly increased over the past decade, with the objective of significantly reducing the energy consumption of existing buildings, ensuring that all new buildings are characterised by high energy efficiency (very low-energy buildings), and using as much as renewable energy as possible, instead of fossil fuels, to meet the energy needs of both new and existing buildings. The final objective is to develop a smart building, namely a green building designed that can be scored with exiting Green Star Building rating.

Many countries have introduced new rating tools over the past few years in order to improve the knowledge about the level of sustainability in each country's building stock. The Green Star sustainable rating system for buildings ("Green Star") and the Green Star Performance rating tool ("Green Star – Performance") have been developed by the Green Building Council of Australia ("GBCA") [22]. Green Star – Performance evaluates the operational performance of all types of existing buildings (with the exception of single-detached dwellings). Green Star and Green Star – Performance have been developed with the assistance and participation of representatives from various stakeholders.

This chapter describes about the new and existing smart buildings that can be sustainable and energy efficient, and the review of the six-star rating accordingly to Green Building Council of Australia. It takes into consideration both design and materials aspects, with particular focus on the next generation of construction materials and the most advanced products currently entering the market for energy-efficient building components. A discussion has also been included as how buildings can achieve higher rating than six star in future.

16.2 FEATURES OF SMART BUILDINGS

In this section, the major features of smart buildings are described. The analysis is performed based on literature overview. A review of smart building assessment approaches is also performed.

The definition of a smart building will take into account the recognised need for holistic and integrated design, taking into account the current themes described in literature, the drivers for building progression, and the methods through which these can be achieved.

Frank (2007) based on review of available various definitions on intelligent buildings summarised the definition as: "An intelligent building is a computer aided (automated) building that is designed and centrally managed to ensure safety, comfort and productivity for its occupants as well as energy efficiency, through sensing and communication devices, thereby enhancing long-term sustainability at minimal running cost" [4]. Intelligent buildings and smart buildings terms are interchangeably used; however, intelligent buildings can be considered more focused on building management systems, while smart building can be considered covering both management and construction aspects of a building. In this chapter, the focus will be on smart buildings.

Smart building concepts were initiated in the 1980s [5], and development of the more effective ways of using information started from the 1970s [5]. The development of communication and information technologies with available high speed in computing to analyse ever-increasing amount of data, supports the implementation of smart building concepts [6–8]. Very recent technological advancements have led to the creation of distributed wireless sensor (and actuator) networks that are candidate technologies for improved monitoring network and controlling of critical infrastructures and operating embedded technologies, which are critical components of smart buildings.

According to Lê, Nguyen, and Barnett [9], smart buildings have the following five fundamental features:

- Automation: the ability to accommodate automatic devices or perform automatic functions.
  - Multi-functionality: the ability to allow the performance of more than one function in a building.
  - Adaptability: the ability to learn, predict, and satisfy the needs of users and the stress from the external environment.
  - Interactivity: the ability to allow the interaction among users.
  - Efficiency: the ability to provide energy efficiency and save time and costs.

However, efficiency should cover all the aspects to achieve overall efficiency of any smart building. Based on the reviewed studies, as a first attempt to identify and describe the smart building key features, the latter were categorised according to four main functions (Table 16.1); they

Table 16.1 Smart buildings functions and characteristics

Basic feature	Smartness features/technology	Important characteristics/functions	Quantified benefits of smartness features
Climate response [10, 11]	• Stochastic model predictive control (SMPC) • SMPC strategy. • Online model predictive control (MPC).	• Controller uses weather predictions to select cost-effective energy sources to keep room temperature in the required comfort levels. • Integrates building thermodynamics, occupancy data, weather forecast and heating, ventilation, and air conditioning (HVAC) component for energy reduction and stabilising temperature.	• MPC resulted in a theoretical saving of 40 % of the total energy consumption. • 18% energy saving with different temperature regulation settings.
Grid response [12, 13]	• Real-time electricity pricing and applying economic model predictive control (MPC). • Intelligent Sensor Nodes for HVAC. • Random neural network (RNN) controller.	• Economic MPC for controlling heat pumps using day-ahead electricity prices. • Load shifting to periods with low electricity prices. • Inputs for the RNN model are (1) heating set point, (2) cooling set point, (3) heating error, (4) cooling error; and (5) CO_2 concentrations.	• Optimised operating strategy saves 25–35% of the electricity cost compared to the baseline case. • The total energy saving with the RNN controller is 27.12%.
User response [14, 15]	• User-building management system (BMS) communications and fuzzy predictive model. • Wireless sensor network (WSN). • Building management system (BMS).	• HVAC system based on occupants' comfort profiles. • Sensing approach for user-BMS communications. • Learns user's comfort profiles, using a fuzzy predictive model. • Identify the optimal locations for different sensor types and gateways.	• User control modes showed a 39% reduction in daily average airflow rates of HVAC (compared to the conventional system). • Building engineering and maintenance services (BEMS) increases the overall occupant comfort by 2.2% with respect to the base case and saves 19% of the energy.
Monitoring and supervision [16, 17]	• Monitoring, measurement, and verification. • HVAC system fault detection and diagnostics. • Distribution system operators in the distribution network. • Building energy scheduling agents.	• Fault detection or inappropriate operations of the HVAC system, and reminders to the building operators to address these issues. • Smart coordination and aggregation method reduces building electricity costs and satisfies all distribution system operating constraints.	• Four pilot buildings showed an average energy saving of 15%, with a payback of less than 12 months. • Bi-level building load aggregation methodology resulted in an electricity cost reduction of 13% through a price-based MPC algorithm.

represent the macro-categories that describe the mandatory functions that a smart building must have. It is important to note that the four functions work synergistically. Table 16.1 reviews some representative studies, with quantified benefits, and categorises them considering the basic functions, elaborates the smartness, and highlights the achievable results.

16.3 ROLE OF INSTITUTIONS AND IT COMPANIES

Intelligent building concept has spread out to the city and society [18–21]. IT giants such as Microsoft, Google, or Amazon look for new niche in digitalised built environment industry as well as software developers who sit in start-ups developing new building-related software. They intend to enter the market that has been traditionally occupied by real estate developers, designers, constructors or asset, property, and facilities management companies, etc. who have not expanded their business to wider or open concepts.

Academia, industry, and education work for research and innovation in smart technology. For example, the Intelligent Building Master course at the Reading University has a holistic transdisciplinary approach, while, for example, the infrastructure and built environment research at Victoria University took the technology push approach. The focus in education has turned towards smart buildings management, for example, at Victoria University building in Melbourne CBD as well as Victoria University Sunshine Construction Futures with six-star Green Star rating.

The Green Building Council of Australia (GBCA) established in 2002 (www.gbca.org.au) has had since 2003 first released Green Star rating tool and subsequently released rating tools for office interiors, education, healthcare and industrial facilities, public buildings, multi-unit residential developments, and retail centres. Today, GBCA has developed tool covering railway station, supermarket, and restaurants [22].

The GBCA tool assesses and rates buildings, fit-outs, and communities against a range of environmental impact categories, and aims to encourage leadership in environmentally sustainable design and construction, showcase innovation in sustainable building practices, and consider occupant health, productivity and operational cost savings.

Smart/intelligent buildings are building blocks of a smart city. However, the concept of a smart city has been changing over the time from "intelligent" to "digital" to "smart" and to "sustainable smart" or "smart sustainable"; many different definitions and main dimensions of a smart city have been identified without unified consensus through a literature review, which defined the required components of a smart city and related aspects of urban life for each one of these components (Table 16.2).

Table 16.2 Smart city components

Component of a smart city	Related aspect of urban life
Smart economy	Industry, innovation, and competitiveness
Smart people	Education creativity, and social capital
Smart governance	E-democracy, participation, and empowerment
Smart mobility	Logistics and infrastructures, transportation
Smart environment	Efficiency and sustainability, resources
Smart living	Security and quality, culture

Source: Adapted from Lombardi et al. [23]

Applying modern Information and Communication Technologies (ICT) tends to ensure the fulfilment of the needs of current and future generations and responds to the challenges associated with innovations, efficiency, and competitiveness of public services and urban structure. The latest United Nations Economic Commission for Europe (UNECE) definition highlights smart and sustainability city concepts integration: "A smart sustainable city is an innovative city that uses ICT and other means to improve quality of life, efficiency of urban operation and services, and competitiveness, while ensuring that it meets the needs of present and future generations with respect to economic, social, environmental as well as cultural aspects" [24].

The rapid development of new technologies is shaping the smart cities of the future. The set of technologies for building the smart cities of the future was recently described by Fourtané [25], which includes 5G technologies, sensors, the Internet of Things (IoT), geospatial technology including geographic information systems (GIS) and global positioning systems (GPS), artificial intelligence (AI), robotics, virtual reality (VR), augmented reality (AR), and blockchain technology. Smart and sustainable city urban planning affects everyone, and it's crucial to know and understand what the technologies involved in building smart cities are and how they can help achieve the ultimate goal of urban transformation into the truly smart cities of the future. The same technologies ensure the full integration of smart buildings into the smart city platform [26]. Table 16.3 describes smart building technologies in relation to smart buildings.

16.4 SMART BUILDING CENTRIC DESIGN

According to the study of the United Nations (UN) Environment Programme (Global Status Report 2017) [27], buildings at a global level utilise about 40% of energy, 25% of water, 40% of resources, and emit about 33% of greenhouse gas (GHG) emissions of global status;

Table 16.3 Smart building application examples in relation to smart city domains (review on green building rating tools used in Australia [23])

Smart building technologies	Smart city domains				
	Smart energy	Smart mobility	Smart life	Smart environment	Smart data
Smart building materials	Possibility to generate energy (e.g., photovoltaic cells). The internet and electricity supply by the same cable (e.g., low-energy lighting).	Wireless power transmission.	Fire resistant, non-toxic, natural and close to nature (biomimicry, biophilia, green walls, roofs, etc.).	Recycled and recyclable. Adaptation to light flow (e.g., glass). Ability to change form (e.g., shading).	Information transmission through wired and wireless networks.
Smart building services	Connection to smart energy grids. Sensors. Automated control systems.	Wireless power transmission.	Video surveillance, recognition systems. Water quality, waste disposition tracking.	Water collection, filtration, and secondary use. Natural airflow control. Renewable sources.	Information transmission through wired and wireless networks.
Smart building construction	Smart lighting management, deployment of renewable energy.	Wireless power transmission, autonomous vehicles, drones.	Safety and quality trainings by using AR/VR (augmented reality/virtual reality). Progress and mobility monitoring, accidental predictive systems (GPS helmet).	Sensors for environmental monitoring and analysis (directly). Detailed analysis of the environment for design: drones. 3D scanning. Level of noise, pollution of harmful particles. Waste recycling.	Exchange of project information between participants (e.g., BIM Common Data Environment (CDE)). Digital tracking of items on site. Transport status information wirelessly.

residential and commercial buildings use about 60% of the world's electricity [18]. In Australia, according to the Department of Environment and Energy, the national total energy used for 2018 as 46% from black coal and 14% from brown coal, 19% from gas turbines, 6.7% from hydro, 6.2% from wind, 4.6% from rooftop solar PV, and 2.1 from grid-linked solar PV [27, 28].

The tendencies of building design and construction have historically changed in a way that is similar to the development path of a smart city: from energy efficient to sustainable, to green, to intelligent, and finally to smart [18]. According to the European Commission, smart buildings means buildings empowered by ICT in the context of the merging ubiquitous computing and the IoT [29]. The generalisation in instrumenting buildings with sensors, actuators, micro-chips, micro- and nano-embedded systems will allow to collect, filter, and produce more and more information locally, to be further consolidated and managed globally according to business functions and services [30]. Based on a recent forecast, the global smart buildings market is expected to demonstrate a compound annual growth rate (CAGR) of 32%, reaching $57 billion AUD by 2022 [31].

The concept of a smart building is being developed at all stages of the life cycle of a building, focusing on the design, construction, and operational phases. In many cases, a smart building combines the characteristics of a sustainable building and a green building. Smart buildings are usually confused with zero-energy buildings or passive buildings [32, 33], which are not the same things as smart buildings.

At the design stage, when the building's digital model is developed, the energy efficiency of the future building is simulated and analysed, considering the location and orientation, urban infrastructure, and other environmental conditions using the latest and most effective technological solutions of building materials and products, building services, and construction processes based on ICT [34]. It is estimated that about 10% of a building's whole CO_2 emissions originate from the building materials production (embodied energy), and 15% of total CO_2 emissions during the construction process are caused by the lack of a smart logistics approach [30, 34].

The operation stage of a smart building can be defined as the most advanced automated process, enabling different building services to interact with each other. Different sensors and controllers may regulate the performance of different systems, such as heating, air conditioning, ventilation, lighting, security, water, etc. This reduces energy consumption, to increase safety and security as well as to improve personal indoor comfort for the occupants. The smart building learns from the experience to make the most efficient real-time decisions in order to maximise comfort and productivity at the lowest energy costs. Implementing smart building solutions can reduce up to 30% of water and up to 40% of energy usage and save from 10% to 30% on costs of overall building maintenance [30].

16.5 SMART BUILDING REQUIREMENTS

The purpose of an intelligent system for building services is to create the conditions for the easy and efficient operation and management of residential and workplace areas, considering the immediate and ongoing needs of its occupants, by ensuring the long-term strategic goal of the owners in aspects such as comfort, safety, economy, energy efficiency, and representativeness [21, 35].

New technologies in construction enable developers to deliver complex projects in the most optimal manner. Technology-based ICT programmes and tools change the way companies develop the project design, its planning, and its execution. Advanced software, hardware, and analytical tools eliminate problems that have stifled the construction sector for decades, including difficulties in project design and communication [36]. These improvements have come to the market at the same time that construction projects are becoming more complex and expensive, and managers are anxious to reduce costs, meet deadlines, and improve efficiency. Real-time data capture and analytics of the condition of a construction site enable better environmental protection (site energy consumption and CO_2 emissions), safety, operational efficiency, productivity, quality, and profitability at the construction site. Thus, the deployment of smart and sustainable building technologies becomes significantly important, and building professionals have demonstrated a strong awareness about these issues [37].

The requirements for smart building materials, services, and construction based on industry practices and literature review [30] are provided in Table 16.4.

16.6 INTERNATIONAL RATING TOOLS

While it is possible to directly compare the value of an office building in New York City, Berlin, London, or Melbourne using a ten-year discounted cash flow approach, making a similar direct comparison of the sustainable features and rating of the same building is quite complex.

In the past, it appears that there has been an unwillingness to compromise on a specific rating system. It has been a barrier to develop a global rating system. Often, a rating tool can be linked back to common aspects of systems, depending largely on the particular influences on each property market. Many rating tools have been modified and adopted from earlier models that were originally developed in other countries.

While there has been fragmentation of rating systems, it can be argued that the World Green Building Council has the largest global coverage (Table 16.5). There are common links in the United States and Canada, some parts of Europe, Japan, Australia, and South Africa [38].

Table 16.4 Requirements for smart building

	Requirements for smart building material
Sustainability	Material is evaluated according to eco-friendliness, secondary use, recycling and, utilisation, purchase and installation price, transportation, produced from local resources.
Adaptability to the environment	Ability to react and adapt to the environment by changing colour, form or position, ability of self-protection and self-healing.
Information collection and transmission	Ability of the material to collect and transmit information in real time on characteristics, loads, and changes in environmental indicators.
	Requirements for smart building services
Sustainability	Positive effects of the system on energy savings, environmental protection, local economy, and social well-being.
Adaptability to the environment	Sensors, mobile applications, and wireless network are used for adaptation to occupants needs and habits: adjustable lighting, pre-controlled microclimate, controlled elevators, indoor mobility, etc.
Renewable resources	Use of renewable resources (solar, wind, water, geothermal) to generate electricity, energy for cooling or heating; use of rain and flood water.
Information collection and transmission	Information is collected, analysed, and transmitted to interested parties for use.
	Requirements for smart building construction
Sustainability	Peculiarities of logistics; duration, resources and costs of technological processes; impact on local economy, human and environment, social welfare of employees.
Real-time information communication	Real-time information tracking and its communication to all construction participants.
Information management	Building information modelling, risk management, environmental, quality, safety, and progress prediction and control.

16.7 THE DEVELOPMENT OF RATING TOOLS

The current era of rating tools commenced in 1990 with the introduction of the BREEAM rating tool, and five years later, this was followed by the French system, HQE (high environmental quality certification that assesses the environmental quality of green buildings), and by LEED in 2000 (Figure 16.1). Further analysis of this diagram confirms that the evolution of rating systems into different countries is largely based on the initially developed rating systems; for example, see BREEAM (the Netherlands), LEED (Emirates), and Green Star (South Africa) [39].

Table 16.5 Main rating tool [39]

U.K. and Europe	America	Rest of the world
BREEAM (inc eco-homes)	LEED (U.S. & Canada)	Green Star (Australia)
The green guide to	U.S. DOE (U.S. Department of	BEAM (Hong Kong)
specification	Energy) Design Guide (U.S.)	LEED (China and
Office scorer	WBDG (Whole Building	India)
ENVEST	Design Guide) (U.S.)	Greenmark
Sustainability checklists	HOK Sustainable Design Guide	(Singapore)
(e.g., SEEDA; BRE)	(U.S.)	GBTool (South Africa)
Environmental impact	BREEAM Canada (Canada)	
assessment (EIA)	Green Globes (U.S. & Canada)	

Footnote:

BREEAM, Building Research Establishment's Environmental Assessment Methods; BRE, Environmental Assessment Method; ENVEST: is a life cycle environmental impact assessment-based design tool for use through the earliest phases of commercial/mixed use building design; SEEDA, Secure End-to-End Data Aggregation protocol; EIUA, Environmental Impact Assessment (EIA); LEED, Leadership in Energy and Environmental Design; BEAM, Benchmarking Energy use Assessment of HK; GBTool, The Green Building Tool

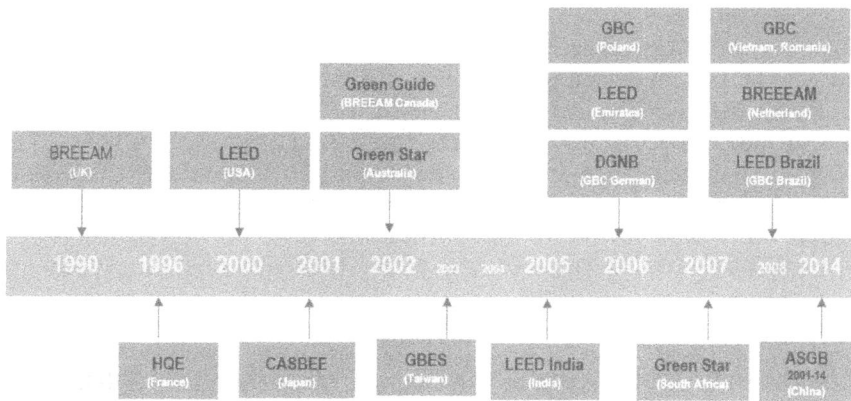

Figure 16.1 Timeline of the development of rating tools.

Source: Developed using information from Green Building Council of Australia [62] and other related literature).

16.7.1 Assessment schemes' structure comparisons and criteria between rating systems

Until now, there are a number of green building rating systems (GBRSs) implemented worldwide [40], among these are LEED (United States, since 1998), BRE Environmental Assessment Method (BREEAM, United Kingdom, since 1990), Comprehensive Assessment System for Built

Environment Efficiency (CASBEE, Japan, since 2001), GS (Australia, since 2003), Green Mark Scheme (Singapore, since 2005), ASGB (China, since 2014), DGNB (Germany, since 2007), Pearl Rating System for Estidama (Abu Dhabi Urban Planning Council, since 2008), etc. [8].

These findings from the reviewed literature indicate that the performance of green buildings has to be rated by integrating different sustainability factors. Those factors are general enough to cover all the topics but also suitable to be adapted to local situations.

The results show that energy-related requirements can be classified into three categories: energy demand reduction, renewable energy use, and environmental benefits. LEED, BREEAM, and Green Star are focused on the energy-saving effect assessment, while Green Mark and China GBRS emphasise the application of energy-saving measures [41].

Bearing this in mind, we identified potential LEED and ASGB certification levels for the certificated GS Six-Star educational building, the new Sunshine Construction Futures (SCF) at Victoria University Sunshine campus (Melbourne, Australia). The technical design phase project documentation of the SCF building was used to compare the current practice with LEED and ASGB requirements. A comparative evaluation of environmental concerns and attributes of indicators was presented with regard to the potential certification levels. Given that, this work gives useful information to design team, occupants, and decision-makers what the design of the building instructed by each of the three systems would be, and the advantages of performance-based rating on designing green. The findings contribute to understanding how and to what extent GBRSs may affect designing green. It would also make contributions to international debate on GBRSs, including standardisation [42].

16.7.2 Discussion on rating systems

Launched by the Green Building Council of Australia in 2003, GS is Australia's only national, voluntary, rating system for buildings and communities. The GS rating systems consist of four elements, namely GS Communities, GS-Design & As Built, GS-Interiors, and GS-Performance. LEED is currently considered a typical and influential GBRS worldwide [43]; the matter of environmental concern prioritisation with respect to local geographical, cultural, economic, and social parameters is integrated into it [44]. LEED V4, as the latest version, develops to accommodate the particularities of collections of buildings (communities/neighbourhoods), building types, or subsystems. ASGB is the foundation of the environmental rating systems in China, which has the largest construction market in the world. ASGB is a "one-size-fits-all civil buildings" assessment tool, consisting of ASGB-design certification and ASGB-occupancy certification.

Table 16.6 Comparison of BREEAM, LEED and GREEN STAR Rating Tools

BREEAM	LEED	GREEN STAR
Pass	Certified	One stars
Good	Silver	Two star
Very good	Gold	Three stars
Excellent	Platinum	Four stars
		Five stars
		Six stars

Source: Environmental Assessment Method BRE (2008) [63].

As criteria-based tools, these systems classify the environmental issues of buildings into several categories and assign the weight for each category. The final score is rated by a weighted scoring system and the level of certification that can be achieved by exceeding the specific point thresholds. There are four possible levels of certification for LEED: Certified (40–49 points), Silver (50–59 points), Gold (60–79 points), and Platinum (80 points and above); six rating scales for GS (from One star to Six stars). GS recognises and rewards projects and buildings that achieve a GS rating of Four (45–59 points), Five (60–74 points) and Six Stars (75 points and above). There are three levels in ASGB: One Star (50–59 points), Two Stars (60–79 points), and Three Stars (80 points and above).

In accordance with Green Star's Office Design Tool, the rating criteria for a commercial office building include management, indoor environment quality, energy, transport, water, material, land use and ecology, and emissions [44]. The LEED criteria are divided into six categories, and the point allocation in the LEED system must reach up to 34 total credits. A total of 69 points are available within the confines of the credits.

When Environmental Assessment Method assessed the schemes under normalised conditions across all the rating criteria, the following results were found as shown in Table 16.6 (BRE, 2008) [39]. LEED and green star assessments are not equivalent to BREEAM assessments. In a particular scenario, a six-star Green Star Building (the highest Green Star rating possible) is less sustainable than a Platinum LEED building (the highest LEED rating possible) and approximately equal to a "very good" BREEAM-rated building.

16.8 UPTAKE OF GREEN STAR IN AUSTRALIA

In Australia, the data are not listed on an annual basis for certified Green Star projects. According to the Green Star website, there are a total as on May 2022 the Green Building Council Australia has 3,273 Certified Projects, 1,530 Registered Projects, 748 Certified Office Projects, and more than 26 million square metre certified as green building with Green Building Council of Australia [62] (https://www.gbca.org.au/project-directory.asp?_ga=2.87889240.1338652720.1653451872-1696971073.1653451871 visited on 25 May 2022).

16.9 GREEN STAR RATING TOOLS

Green Star Buildings and Green Star – Design & As Built Guiding the sustainable design and construction of schools, offices, universities, industrial facilities, public buildings, retail centres, and hospitals.

Green Star – Communities Improving the sustainability of projects at the precinct or community scale. Communities assesses the planning, design, and construction of large-scale development projects at a precinct, neighbourhood, and/or community scale.

Green Star – Interiors Transforming the interior fit-outs in everything from offices and hotels to bank branches and shops. Interiors rate the sustainable design and construction of any building fit-out works. The interiors rating tool aims to assist the project teams to objectively rate their projects, and to achieve sustainability goals. The tool encourages a new approach to designing and constructing fit-outs by rewarding sustainability best practice and excellence. It also provides consistent and clear advice in an easy-to-use manner.

Green Star – Performance Supporting higher levels of operational efficiency within existing buildings. It is used by building owners to measure how successfully they are managing their existing assets, and helps to communicate this commitment to investors and building users. A "continuous improvement framework" has been built into the tool as the certification process is on a three-year cycle; this allows for improvements to be recognised over time [62] (https://new.gbca.org.au/green-star/evolution/ viewed on 25 May 20220).

16.9.1 Green Star rating scale

Green Star rating tools for building, fit-out, and community design and construction reward projects that achieve best practice or above sustainability outcomes (Table 16.7).

Table 16.7 Green Star rating points (Green Building Council of Australia, 2014 [62])

Green Star rating	Points
4 star	45–59 pts best practice
5 star	60–74 pts Australian excellence
6 star	75+ pts world leadership

Table 16.8 Green Star rating points and percentage reduction in GHG (Green Building Council of Australia [62])

	Average % reduction GHG
4 star	47%
5 star	57%
6 star	70%
Overall	56%

This means that Green Star – Design, As Built, Interiors and Communities projects can achieve a Green Star certification of 4–6 Green Star. Buildings assessed using the Green Star – Performance rating tool can achieve a Green Star rating from 1 to 6 Green Star [45, 62].

16.9.2 Emissions reductions Green Star

All new Green Star-rated buildings and fit-outs are required to do better than legislation by at least 10%. However, most Green Star-rated buildings do much better than that. Based on new buildings certified as of 30 June 2020, Green Star-certified buildings are designed and built to produce 56% fewer GHG emissions than standard new buildings. The savings increase with the star ratings (Table 16.8).

A net zero emission (NZEB) building refers to a building that generates at least as much energy that is emission-free as it uses emission-producing energy [46]. This definition is relatively consistent with many government policies that are promoting reduced GHG emissions such as the Kyoto Protocol. A limitation of this definition is that it advocates emission-producing energy as long as the same unit of energy is offset by emission-free energy. It is also largely dependent on the regional electricity generation techniques (i.e., coal-generated electricity use would require more emission-free energy production to offset it than nuclear-generated electricity use) [47].

16.9.3 Analysis of the Green Star rating tool

The future of Green Star is climate positive; new POSITIVE category aims to drive every building towards net-zero carbon with a focus on transformational change. This is our formula that every building should follow. These are the credits that will get you there. Figure 16.2 shows options for GHG reduction to achieve towards net-zero emissions.

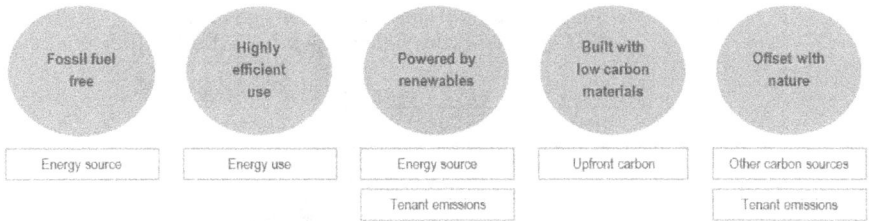

Figure 16.2 New credit structure rating tools.

Source: Based on international comparison.

Figure 16.3 Victoria University Sunshine Construction Futures (Melbourne, Australia).

Source: Danny Byrne (author).

16.10 GREEN STAR PROJECTS (CASE STUDY)

16.10.1 Victoria University Sunshine Construction Futures (6-Star Green Star rating with the final score of 76)

The design provides a loose-fit long-life shell that responds to and caters for changing course programmers by maximising flexible spaces and incorporating innovative and sustainable materials.

The building (Figure 16.3) was awarded a 6-Star Green Star rating with the final score of 76 under Education v1 as in June 2013 due to its embedded sustainability measures including optimised day lighting through highly insulated facades as well as a mix of passive and active ventilation and temperature control systems, all of which contribute to environmental benefits and provide a state-of-the-art learning environment [48].

Figure 16.4 PIXEL Building (Melbourne, Australia).

Source: Danny Byrne (author).

100% outside air ventilation is provided, while active mass in-slab hydronic tubes contribute radiant heating and cooling to teaching spaces, and a gabion rock store cooling system tempers fresh air intake to the building. The Night Sky Cooling system uses the large surface area of the roof to cool down water sprayed onto the roof, which is then used to temper the chilled water supply to the active mass cooling system, reducing load on conventional cooling systems and ongoing energy costs for the project [49].

Environmental systems are displayed on screens throughout to allow users to access and interact with building performance data, and key sustainability elements are revealed through cross-sectional "peels" of the building fabric in social and circulation spaces with the aim of promoting further conversation between the building users and visitors.

16.10.2 Pixel Building (6-Star Green Star rating with the final score of 105)

This building is located on the former Carlton Brewery site in central Melbourne (Figure 16.4). Under the Green Star rating system, 75 points is the benchmark for achieving a 6-Star Green Star rating. As well as achieving the maximum 100 points available, the Pixel Building was awarded with an extra five points for innovation.

The innovation points recognised the building's carbon neutral operations, a vacuum toilet system, an anaerobic digestion system, and reduced car parking. The water initiatives in the project also mean the building could be self-sufficient for water.

Pixel has been constructed using a new type of concrete, called PIXEL-Crete, which has half the embodied carbon in the mix compared with standard concrete. The building also features a living roof planted with native grasses, while tracking PV solar panels and wind turbines on the roof offset the carbon use of the building.

Sunshades on the façade allow natural daylight into the office space while also protecting the interior from the glare of the sun, ensuring the building stays cool in summer. Reed beds on each level filter the building's greywater and help to keep the building cool.

16.11 IMPROVE AND UPDATE TO SEVEN GREEN STARS

At current, two complete projects in Australia that have received six green stars rating with more than 100 score are PIXEL Building and Barangaroo South (www.barangaroosouth.com.au). These buildings received a final score of 105 out of 110. Both of these projects have an extra score for innovation in NZEB.

During the last decade, substantial developments have been made in the field of building energy saving through recovering waste to energy and utilising various energy conservation measures. The first generation of green buildings emerges due to increasing concern about the negative impact of building to the natural environment in the 1990s after the introduction of longstanding certification schemes such as BREEAM and LEED. Historically, one of the first attempts to conceive a Net-Zero-Energy House was done in 1977 by Esbensen and Kors-gaard "Dimensioning of the solar heating system in the zero-energy house in Denmark". Ever since then, the trend towards reducing the energy consumption of the building started with a rapid pace; in European Union, the concept of NZEB was proved after several success stories and eventually caused changes in building codes in 2010 to push builders out of their traditional practices [50]. The next generation of NZEB was emerged in the US and Japan, mainly followed by improvement of solar PV technology and reduction of cost of PV modules.

16.11.1 Net-zero-energy buildings (NZEB) and improvement to seven Green Stars rating

When considering the inconsistency of NZEB models, two questions to be considered to calibrate design goals for a NZEB. These questions involve determining the methodology by which "net-zero" is defined and determining the subsequent limitations for the energy generation options. In the context of designing a NZEB under current NZEB models, the four elements would be relatively easy to adopt a definition of NZEB that is less holistic to make a business model more feasible and more attractive to investors with consideration of a seven-star rating.

16.11.2 Net-zero site-energy buildings

A net-zero site-energy building (Site-NZEB) produces as much energy as it consumes within the building site [46]. Compared to the other definitions, Site-NZEB is easily verifiable and is subject to fewer external fluctuations, which are sometimes hard to estimate and project. It is also one of the most common and repeated definitions for net-zero energy buildings in the literature. However, from a Site-NZEB point of view, all units of energy are equal, i.e., 1 unit of electrical energy is equal to 1 unit of thermal energy, which is generally not true in terms of GHG emissions.

16.11.3 Net-zero source-energy buildings

A net-zero source-energy building (Source-NZEB) produces renewable energy in an amount such that its primary energy (source energy) equivalent is equal to the amount of primary energy it consumes. Unlike Site-NZEB, Source-NZEB accounts for the difference in energy values by converting all the energies used and produced in the building to primary energy using respective Site-to-Source Factors (StSF) [51]. In many places, reaching the Source-NZEB goal is easier compared to Site-NZEB, since in general StSF for electricity is about three times the StSF for natural gas. This means that for every three units of natural gas energy used within the building, the renewable energy system just needs to produce 1 unit of electricity. In this definition framework, all-electric Site- and Source-NZEBs are equivalent as the electricity generation efficiency and transmission losses are constant, because in that case any imported electricity that is used in place of natural gas, and exported electricity are multiplied by the same StSF [52].

The disparity in site-to-source factors of electricity and natural gas encourages using more natural gas compared to electricity in the designed Source-NZEB. In addition, if the gap between electricity and natural gas StSFs shrinks, the designed building will not meet the Source-NZEB goal anymore, except in all-electric buildings. Determining these StSFs is maybe the biggest challenge in using this definition. Significant amounts of data and projections that capture the temporal and spatial variation of these factors are needed to be able to properly design a true Source-NZEB.

16.11.4 Net-zero energy-cost buildings

A net-zero energy-cost building (Cost-NZEB) receives as much financial credit for exported energy as it is charged on the utility bills [46]. Thus, this definition is easily verifiable through utility bills. However, since utility rates can vary over the lifespan of the building, it could meet the net-zero energy cost goal during a particular year and not during the next year. Similar to Source-NZEB, in the case of cheap natural gas and expensive exported electricity, the Cost-NZEB goal encourages using natural gas in buildings [53].

16.11.5 Net-zero emission buildings

A net-zero emission building (Emission-NZEB) exports emission-free renewable energy, and displaces fossil-fuel-based grid electricity, in an amount sufficient to offset the annual carbon emission associated with its operation. Emission factors for different energy sources are needed in the design process of this goal. These factors, similar to StSFs, especially for electricity, are subject to short-term (daily and seasonal) and long-term (over the building lifespan) temporal variation, as well as spatial variation. Emission-NZEB can encourage or discourage the use of natural gas in the design process, depending on the relative emission factors for electricity and natural gas [54]. Based on Site, Source, Cost, and Emission-NZEB, additional credit can be added on the current Green Star credit as shown in Table 16.9. Thus, extra 40 points can be distributed across the four elements above the current Green Star credit system.

Table 16.9 Presents a summary of the definitions discussed and allocated points

Title	Definition	Points
Site-NZEB Produces as much energy as it consumes within the building site	— Easily verifiable — Subject to fewer external fluctuations	10
Source-NZEB Produces as much as primary energy (source energy) as it consumes within the building site	— Account for the difference in energy values — Usually easier goal to reach compared to Site-NZEB	10
Cost-NZEB Receives as much financial credit for exported energy as it is charged on the utility bills	— Easily verifiable through utility bills	10
Emission-NZEB Produces emissions-free renewable energy in an amount sufficient to offset the annual carbon emission associated with its operation	— Encourages using electricity where electricity emission factor is smaller than the same factor for natural gas	10

16.11.6 Few key ways to move to net-zero energy-source buildings

In 2014, the City of Melbourne introduced the zero net emissions by 2020 strategy. This strategy sets out targets that focus on the City of Melbourne council operations, commercial buildings, residential buildings, energy supply, transport, and waste [55]. In terms of commercial buildings, this strategy outlines plans to increase the average National Australian Built Environment Rating System (NABERS) rating to "4" by 2018 (this roughly equates to an increase in energy efficiency of 40% per building) [55].

The first strategy is to minimise energy usage by limiting the amount of heat gain and loss (i.e., appropriate insulation), considering internal energy-efficient design and building services systems such as heating, cooling, and utilities. The good examples of the National Australian Built Environment Rating System of NZEBs tend to use energy efficiency and passive design as a base for their NZEB, which can be used to support both lower energy requirements and increased comfort of the building users through maintaining desirable temperatures and supplying sufficient daylighting. Secondly, the use of renewable energy technologies should be considered to supplement the energy usage needs that cannot be minimised. Well-established renewable energy technologies such as PVs, wind turbines, solar thermal, heat pumps, and district heating and cooling can be employed [56, 57].

The other options to achieve net-zero energy buildings are as follows:

1. Minimise loads of the building
2. Use passive building design strategies
3. Implement efficient services
4. Use renewable energy generation techniques and green infrastructure inclusion

16.11.7 Strategies to achieve net-zero energy goal

To meet the project's net-zero-energy goal, a combination of strategies needs to conserve and generate enough power to meet the building's demands.

These strategies have been divided into four categories: reduction, reclamation, absorption, and generation necessary to quantify the outputs and check what it needs to balance the net-energy consumed. Then, simulate the energy consumption using various energy modelling techniques and tools to optimise the following [58, 59]:

- Building orientation
- Glazing area, exposure, and shading
- Heat island reduction
- Lighting systems and capacities
- Temperatures, humidity, and relative humidity levels

- Landscaping
- Natural resources
- The overall system efficiency

All factors should be considered together by employing passive heating or cooling strategies, such as solar chimney and direct heat gain through south-facing glazing and/or isolated gain or sunspace, considering all possible exterior wall construction that avoids thermal bridging and increasing the R-value (measure of the thermal resistance of a material of specific thickness) in all roof construction, using efficient lighting system, utilising daylighting sensors and occupancy sensors, and lastly using energy-efficient office equipment for commercial buildings and energy-efficient utilities for residential houses and buildings.

To determine how much energy needs to be generated, which requires measurement of on-site usage. Measurements are divided into four categories [60]:

1. **Envelope measurement:** includes exterior wall insulation, roof insulation, roof albedo, window overhangs, window properties, skylight glazing property, window-to-wall ratio, and skylight-to-roof area ratio.
2. **Energy load measurement:** includes lighting power density, equipment power density, daylight sensors, occupancy control, and plug receptacle control.
3. **Heating, ventilation and air conditioning (HVAC) measurement:** includes chiller efficiency, energy recovery, demand control ventilation (DCV), variable speed drives, HVAC control adjustment, evaporative cooling, cooling tower efficiency, and boiler efficiency.
4. **Renewable energy measurement:** on-site renewable energy, such as photovoltaic system, solar water heating, and wind turbine or off-site renewable energy.

16.12 CONCLUSION

Today, in a year marred by climate change, population growth, and urbanisation aspects, it is with hope and a clear vision that we share our new target for net-zero emission (New Green Star Buildings rating tool). This next step for Green Star Building is a sustainability rating system for new buildings that ensures they can respond to changes in climate, protect natural surroundings, and focus on the health of the people who use them. Industry asked for a user-friendly tool that can help our sector move the needle with climate change.

The study focused on the analysis at the category level of the net-zero energy buildings, because differences exist in the subcategory level across the various GBCA Green Star rating tools. Further research is required to focus on a specific sector (e.g., the most popular office buildings or the slow uptake in the retail, healthcare, and industrial sectors). Future research opportunities also exist for investigating the relative environmental performance of projects with design and as-built ratings.

The rating tool also goes further and highlights opportunities to improve on current Green Star – Performance and explores how to bring occupants closer to nature. The rating tool then expands on our classic definition of health to introduce issues around community resilience, social cohesion, and design for inclusion amongst other things.

Future work needs to explore how the definition of net zero can influence optimal retrofit solutions and assess the unintended consequences of these alterations in terms of grid stability, economics, emissions, and comfort. The methods used to calculate should also be the focus of future work, as these have an important impact on the calculation of emissions offsets from exported electricity. It is known that many electrical systems use a mix of different generators to meet the varying loads of buildings. As different generators come online, the associated importing electricity will also change. Finally, district energy systems were not considered in the case study. Such systems can offer several economic and technical benefits, and future work needs to explore and evaluate retrofits of such systems within the existing building stock.

Given the opportunities presented by NZEBs, it is initiative to ask why you would not design every building to be a NZEB. An immediate response might be to discount NZEBs due to the perceived cost of the building. However, evidence of NZEB progression may suggest that this is an excuse rather than a justifiable response. Therefore, the question that requires answering is "What is required to ensure that every building is a NZEB?"

In terms of the Australian context, it is explained that very little support for NZEBs has been demonstrated in Australia, particularly in mainstream community [61]. This is reinforced by a lack of specific policy support. These came to the conclusion that in Australia, separate sets of targeted policies are required for residential and non-residential buildings. The policies should advocate stronger building code and ensure much higher levels of compliance. Given that Australia has already had a higher penetration of renewables, better codes could likely get it to near zero-energy residential buildings.

REFERENCES

[1] K. Ahmed Ali, M. I. Ahmad, and Y. Yusup, "Issues, impacts, and mitigations of carbon dioxide emissions in the building sector, " *Sustainability,* vol. 12, no. 18, p. 7427, 2020.
[2] A. A. Parker, "The 2nd Submission to the Productivity Commission regarding the Draft Report on Energy Efficiency April 2005," 2005. https://www.pc.gov.au/inquiries/completed/energy-efficiency/submissions/subdr112/subdr112.pdf (viewed on 20/2/2023)

[3] C. Russell, B. Baatz, R. Cluett, and J. Amann, "Recognizing the value of energy efficiency's multiple benefits," *American Council for an Energy-Efficient Economy, Washington, DC, USA,* 2015.

[4] O. L. Frank, "Intelligent Building Concept: the challenges for building practitioners in the 21st century," Journal of the Association of Architectural Educators in Nigeria (AARCHES J), vol. 6, no. 3, pp. 107–113, 2007.

[5] J. K. Wong, H. Li, and S. Wang, "Intelligent building research: a review," *Automation in Construction,* vol. 14, no. 1, pp. 143–159, 2005.

[6] A. Haapio and P. Viitaniemi, "A critical review of building environmental assessment tools," *Environmental Impact Assessment Review,* Article vol. 28, no. 7, pp. 469–482, 2008, doi: 10.1016/j.eiar.2008.01.002.

[7] R. J. Cole and M. J. Valdebenito, "The importation of building environmental certification systems: International usages of BREEAM and LEED," *Building Research and Information,* Article vol. 41, no. 6, pp. 662–676, 2013, doi: 10.1080/09613218.2013.802115.

[8] S. Pfledderer, "Sustainability standards faceoff: we compare and contrast how the LEED and Green Globes rating systems," *Landscape Management,* vol. 54, no. 3, pp. 44–45, 2015.

[9] Q. Lê, H. B. Nguyen, and T. Barnett, "Smart homes for older people: positive aging in a digital world," *Future Internet,* vol. 4, no. 2, pp. 607–617, 2012.

[10] F. Oldewurtel et al., "Energy efficient building climate control using stochastic model predictive control and weather predictions," in *Proceedings of the 2010 American Control Conference,* 2010: IEEE, pp. 5100–5105.

[11] T. Zhang, M. P. Wan, B. F. Ng, and S. Yang, "Model predictive control for building energy reduction and temperature regulation," in *2018 IEEE Green Technologies Conference (GreenTech),* 2018: IEEE, pp. 100–106.

[12] R. Halvgaard, N. K. Poulsen, H. Madsen, and J. B. Jørgensen, "Economic model predictive control for building climate control in a smart grid," in *2012 IEEE PES innovative smart grid technologies (ISGT),* 2012: IEEE, pp. 1–6.

[13] A. Javed, H. Larijani, A. Ahmadinia, R. Emmanuel, M. Mannion, and D. Gibson, "Design and implementation of a cloud enabled random neural network-based decentralized smart controller with intelligent sensor nodes for HVAC," *IEEE Internet of Things Journal,* vol. 4, no. 2, pp. 393–403, 2016.

[14] F. Jazizadeh, A. Ghahramani, B. Becerik-Gerber, T. Kichkaylo, and M. Orosz, "User-led decentralized thermal comfort driven HVAC operations for improved efficiency in office buildings," *Energy and Buildings,* vol. 70, pp. 398–410, 2014.

[15] D. Sembroiz, D. Careglio, S. Ricciardi, and U. Fiore, "Planning and operational energy optimization solutions for smart buildings," *Information Sciences,* vol. 476, pp. 439–452, 2019.

[16] W. Shen, H. H. Xue, G. Newsham, and E. Dikel, "Smart building monitoring and ongoing commissioning: A case study with four Canadian federal government office buildings," in *2017 IEEE International Conference on Systems, Man, and Cybernetics (SMC),* 2017: IEEE, pp. 176–181.

[17] Y. Liu et al., "Coordinating the operations of smart buildings in smart grids," *Applied Energy,* vol. 228, pp. 2510–2525, 2018.

[18] A. H. Buckman, M. Mayfield, and S. B.M. Beck, "What is a Smart Building?," *Smart and Sustainable Built Environment,* vol. 3, no. 2, pp. 92–109, 2014, doi: 10.1108/sasbe-01–2014-0003.

[19] V. Angelakis, A. Bassi, A. Kapovits, H. C. Pöhls, and E. Tragos, *Designing, Developing, and Facilitating Smart Cities : Urban Design to IoT Solutions,* 1st ed. Cham: Springer International Publishing : Imprint: Springer, 2017, pp. 1 online resource (XIV, 336 pages 52 illustrations, 36 illustrations in color.

[20] K. L. Brown, C. F. Cummings, R. M. Vanacore, and B. G. Hudson, "Building collagen IV smart scaffolds on the outside of cells," (in eng), *Protein Science,* vol. 26, no. 11, pp. 2151–2161, 2017, doi: 10.1002/pro.3283.

[21] E. Janhunen, L. Pulkka, A. Säynäjoki, and S. Junnila, "Applicability of the smart readiness indicator for cold climate countries," *Buildings,* vol. 9, p. 102, 04/25 2019, doi: 10.3390/buildings9040102.

[22] I. C. S. Illankoon, V. W. Tam, K. N. Le, and C. N. Tran, "Review on green building rating tools used in Australia," in *INNOVATIVE PRODUCTION AND CONSTRUCTION: Transforming Construction Through Emerging Technologies*: World Scientific, 2019, pp. 165–184.

[23] P. Lombardi, S. Giordano, H. Farouh, and W. Yousef, "Modelling the smart city performance," *Innovation: The European Journal of Social Science Research,* vol. 25, no. 2, pp. 137–149, 2012/06/01 2012, doi: 10.1080/13511610.2012.660325.

[24] U. ECOSOC, "The UNECE–ITU smart sustainable cities indicators," United nations Economic and Social Council, ECE/ HBP/2015/4 2015. https://unece. org/DAM/hlm/projects/SMART_CITIES/ECE_HBP_2015_4.pdf (viewed on 20/2/2023)

[25] S. Fourtané, "The technologies building the smart cities of the future," *Interesting Engineering,* 2018. https://interestingengineering.com/innova-tion/the-technologies-building-the-smart-cities-of-the-future (viewed on 20/2/2023)

[26] M. Batty et al., "Smart cities of the future," *The European Physical Journal Special Topics,* vol. 214, no. 1, pp. 481–518, 2012.

[27] T. Abergel, B. Dean, and J. Dulac, "Towards a zero-emission, efficient, and resilient buildings and construction sector: Global Status Report 2017," *UN Environment and International Energy Agency: Paris, France,* vol. 22, 2017.

[28] E. M. Natsheh, "Hybrid power systems energy management based on Artificial Intelligence,"PhD Thesis, Manchester Metropolitan University, 2013. https://e-space.mmu.ac.uk/id/eprint/314015 (viewed on 20/2/2023)

[29] P. Moseley, "EU support for innovation and market uptake in smart buildings under the horizon 2020 framework programme," *Buildings,* vol. 7, no. 4, p. 105, 2017.

[30] T. Crosbie, M. Crilly, N. Dawood, J. Oliveras, and N. Niwaz, "Visualising the 'Big Picture': Key Performance Indicators and Sustainable Urban Design," in *1st Workshop organised by the EEB Data Models Community ICT for Sustainable Places,* 2014.

[31] A. Kylili and P. A. Fokaides, "European smart cities: The role of zero energy buildings," *Sustainable Cities and Society,* vol. 15, pp. 86–95, 2015/07/01/ 2015, doi: https://doi.org/10.1016/j.scs.2014.12.003.

[32] P. A. Fokaides, E. A. Christoforou, and S. A. Kalogirou, "Legislation driven scenarios based on recent construction advancements towards the achieve-ment of nearly zero energy dwellings in the southern European country of Cyprus," *Energy,* vol. 66, pp. 588–597, 2014.

[33] P. A. Fokaides, E. Christoforou, M. Ilic, and A. Papadopoulos, "Performance of a Passive House under subtropical climatic conditions," *Energy and Buildings,* vol. 133, pp. 14–31, 2016/12/01/ 2016, doi: https://doi.org/10.1016/j.enbuild.2016.09.060.

[34] R. Apanaviciene, A. Vanagas, and P. A. Fokaides, "Smart building integration into a smart city (SBISC): development of a new evaluation framework," *Energies,* vol. 13, no. 9, p. 2190, 2020.

[35] "Smart Building Market Research Report," in "Global Forecast till 2025," vol. 2021. [Online]. Available: https://www.marketresearchfuture.com/reports/smart-building-market-1860.

[36] R. Edirisinghe, "Digital skin of the construction site," *Engineering, Construction and Architectural Management,* vol. 26, no. 2, pp. 184–223, 2019, doi: 10.1108/ecam-04-2017-0066.

[37] W.-M. To, P. K. C. Lee, and K.-H. Lam, "Building professionals' intention to use smart and sustainable building technologies – An empirical study," *PLOS ONE,* vol. 13, no. 8, p. e0201625, 2018, doi: 10.1371/journal.pone.0201625.

[38] R. Reed, A. Bilos, S. Wilkinson, and K.-W. Schulte, "International comparison of sustainable rating tools," *Journal of Sustainable Real Estate,* vol. 1, no. 1, pp. 1–22, 2009.

[39] R. Reed, S. Wilkinson, A. Bilos, and K.-W. Schulte, "A comparison of international sustainable building tools–an update, " in *The 17th Annual Pacific Rim Real Estate Society Conference, Gold Coast,* 2011, pp. 16–19.

[40] R. C. Retzlaff, "Green building assessment systems: a framework and comparison for planners," *Journal of the American Planning Association,* vol. 74, no. 4, pp. 505–519, 2008/10/21 2008, doi: 10.1080/01944360802380290.

[41] Smart Building Market, " 2018 Global trends, market share, industry size, growth, opportunities and forecast to 2023," Reuters: Canary Wharf, UK, 2018. https://news.marketersmedia.com/smart-building-market-2018-global-trends-market-share-industry-size-growth-opportunities-and-forecast-to-2023/405332 (viewed on 20/2/2023)

[42] J. Reinecke, S. Manning, and O. Von Hagen, "The emergence of a standards market: multiplicity of sustainability standards in the global coffee industry," *Organization Studies,* vol. 33, no. 5–6, pp. 791–814, 2012.

[43] Y. He, T. Kvan, M. Liu, and B. Li, "How green building rating systems affect designing green," *Building and Environment,* vol. 133, pp. 19–31, 2018.

[44] F. Cappai, D. Forgues, and M. Glaus, "The integration of socio-economic indicators in the CASBEE-UD evaluation system: a case study," *Urban Science,* vol. 2, no. 1, p. 28, 2018.

[45] G. Star, A decade of green building,"Green buidling council, Australia," ed, 2012. https://www.gbca.org.au/uploads/170/34474/A_decade_of_green_building_.pdf (viewed on 20/2/2023)

[46] P. Torcellini, S. Pless, M. Deru, and D. Crawley, "Zero energy buildings: a critical look at the definition," National Renewable Energy Lab.(NREL), Golden, CO (United States), 2006.

[47] K. Caldeira, A. K. Jain, and M. I. Hoffert, "Climate sensitivity uncertainty and the need for energy without CO_2 emission," *Science,* vol. 299, no. 5615, pp. 2052–2054, 2003.

[48] SCF, "Victoria University Sunshine Construction Futures", Melbourne Australia 2012 https://archiroots.net/project/victoria-university-sunshine-construction-futures/c341916d-5137-48a9-a52c-6c9905fd97ec (viewed on 20/2/2023)

[49] Cox Architecture. "Victoria University Sunshine Construction Futures." 2012 https://archello.com/project/victoria-university-sunshine-construction-futures
viewed on 20/2/2023).

[50] A. J. Marszal et al., "Zero energy building–a review of definitions and calculation methodologies," *Energy and Buildings,* vol. 43, no. 4, pp. 971–979, 2011.

[51] B. Heard, C. J. Bradshaw, and B. W. Brook, "Beyond wind: furthering development of clean energy in South Australia," *Transactions of the Royal Society of South Australia,* vol. 139, no. 1, pp. 57–82, 2015.

[52] A. H. Wiberg et al., "A net zero emission concept analysis of a single-family house," *Energy and Buildings,* vol. 74, pp. 101–110, 2014.

[53] J.-H. Kim, H.-R. Kim, and J.-T. Kim, "Analysis of photovoltaic applications in zero energy building cases of IEA SHC/EBC task 40/annex 52," *Sustainability,* vol. 7, no. 7, pp. 8782–8800, 2015.

[54] C. S. de Silva Lokuwaduge and K. de Silva, "Emerging corporate disclosure of environmental social and governance (ESG) risks: an Australian study," *Australasian Accounting, Business and Finance Journal,* vol. 14, no. 2, pp. 35–50, 2020.

[55] City of Melbourne, "Zero-net emissions by 2020 (update 2014): A collaborative approach to the next four years of action," City of Melbourne Melbourne, 2014. https://www.melbourne.vic.gov.au/SiteCollectionDocuments/zero-net-emissions-update-2014.pdf (viewed on 20/2/2023)

[56] S. A. Kalogirou, *Solar Energy Engineering: Processes and Systems.* Academic Press, 2013.

[57] R. Best, P. J. Burke, and S. Nishitateno, "Evaluating the effectiveness of Australia's small-scale renewable energy scheme for rooftop solar," *Energy Economics,* vol. 84, p. 104475, 2019.

[58] K. Parvin, W. Hafiz, M. Abdullah, M. Hannan, and M. Salam, "Estimation of building energy management toward minimizing energy consumption and carbon emission." Journal of Advanced manufacturing Technology, 14(2.2), 2020 https://jamt.utem.edu.my/jamt/article/view/6053 (viewed on 20/2/2023)

[59] M. Raugei, M. Kamran, and A. Hutchinson, "A prospective net energy and environmental life-cycle assessment of the UK electricity grid," *Energies,* vol. 13, no. 9, p. 2207, 2020.

[60] W. Guo, X. Qiao, Y. Huang, M. Fang, and X. Han, "Study on energy saving effect of heat-reflective insulation coating on envelopes in the hot summer and cold winter zone," *Energy and Buildings,* vol. 50, pp. 196–203, 2012.

[61] L. Wells, B. Rismanchi, and L. Aye, "A review of Net-Zero Energy Buildings with reflections on the Australian context," *Energy and Buildings,* vol. 158, pp. 616–628, 2018.

[62] GBCA, Green star rating tools, Green Building Council of Australia (https://new.gbca.org.au/green-star/rating-system/) (viewed on 21/2/2023)

[63] T. Saunders (2008) A discussion document comparing international environmental assessment methods for buildings, BRE Global https://tools.breeam.com/filelibrary/International%20Comparison%20Document/Comparsion of_International_Environmental_Assessment_Methods01.pdf. (viewed on 21/2/2023)

Chapter 17

Review of water quality monitoring using Internet of Things

*T. A. Choudhury, Harpreet Singh Kandra,
Khang Dinh, and Suryani Lim*

Federation University Australia

CONTENTS

17.1 INTRODUCTION

Water is a vital natural resource. Waterways, including both ground and surface water systems, catchments, and estuarine and marine water bodies, are complex ecological systems. The quality of water resources is equally essential to sustain human and aquatic life and to sustain water security in this world struggling with increased pressures from a changing climate and more decentralised facilities. A compromise in water quality can lead to a wide range of complications, including the resilience of the ecosystem, tourism, the volume of extractions, treatment costs, etc.

Water quality monitoring and management across different sources and water bodies is primarily carried out by local water bodies and catchment

DOI: 10.1201/9781003368335-17

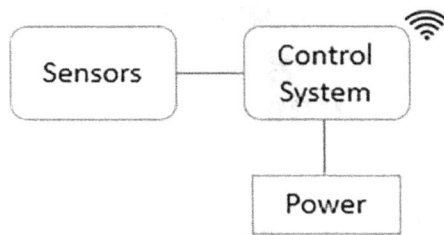

Figure 17.1 A typical setup of an IoT system with sensors that are connected to a control system capable of transmitting data.

management authorities. Traditional methods for water quality monitoring require manual collection of water samples from different locations for laboratory-based analysis of critical water quality parameters. The methodology is often inefficient as it is labour-intensive, time-consuming, and administratively challenging. At times, the locations of sampling sites can further pose challenges such as access during severe weather conditions, travel times, and costs of data collection. In addition, such methodology lacks spatiotemporal coverage and real-time water quality information to enable fast critical public health decisions. These can potentially affect the reliability and robustness of the entire process.

With the advancements of sensor and communication technologies over recent years, Internet of Things (IoT)-based platforms have become more common to replace the traditional methods to monitor and maintain key water quality parameters in real time. Furthermore, the power requirements of such sensor-based systems have steadily declined over the years, and with the availability of advancements in renewable energy and recharge technology, feasibility, reliability, robustness, and sustainable use of such IoT-based monitoring system have become very efficient. The IoT-based systems, therefore, allow for fast and real-time monitoring of key water quality parameters, allowing the public health team and key stakeholders to respond faster to any irregularities, contamination, or degradation of water quality, including cause-effect and trend analysis. Figure 17.1 shows a typical IoT setup: sensors, as required, are connected to a control system capable of transmitting data.

This review aims at providing a better understanding of the current research trends in the space of IoT-based water quality monitoring. This is achieved through a brief review of past literature to identify key water quality parameters and different aspects of IoT platforms, including sensors, cloud, and graphical user interface (GUI) platform, control system and wireless network type, unmanned surface vehicle (USV) technology, and power usage and recharge technology. Based on the reviewed literature, the study identifies key research gaps and provides recommendations for future research scopes.

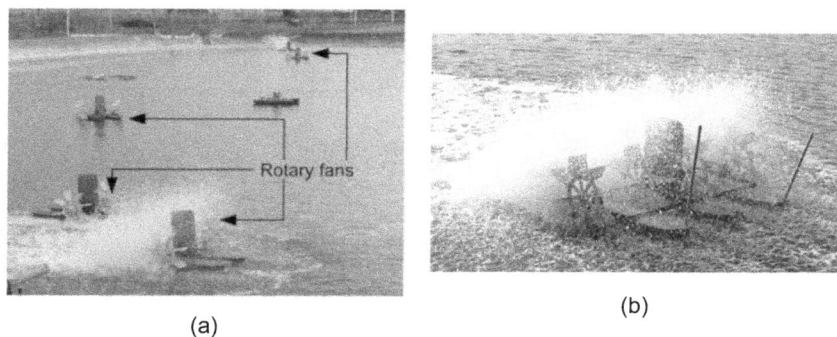

(a)

(b)

Figure 17.2 (a) Automated rotary aeration system to control dissolved oxygen concentration in water [3]. (b) The aeration system in operation is controlled based on dissolved oxygen in water using paddle wheel aerators [4].

17.2 REVIEW OF IoT APPLICATIONS IN WATER QUALITY MONITORING

Within the water treatment area, the aeration plant, SUE "Vodokanal of St. Petersburg", developed a multi-sensor array for continuous real-time monitoring of processes' water quality. This plant employs both physical and biological treatment methods of wastewater. The literature covers the need for the development of analytical methods that are capable of online implementation in physical world industrial conditions [1].

In agriculture, there have been advancements in supporting the monitoring of a diverse variety of water quality parameters. For example, in India, a "cuckoo" search algorithm has been developed to monitor various water quality parameters utilising an IoT platform. The system aims to observe the environmental and performance factors surrounding the soil and water conditions, including temperature, and use the extracted information and insights to influence the farmers' approach towards crop handling. The authors concluded that such a low-cost approach could become economically applicable and viable for farming and agricultural systems of various sizes [2].

In aquaculture, several experimental studies have been performed. Two such studies are shown in Figure 17.2. The author in one of the studies focused on the development and implementation of real-time water quality monitoring systems for a number of shrimp aquaculture centres in Indonesia. The study aimed at reducing energy consumption through optimal water condition monitoring and control for shrimp aquaculture. For example, the developed system works together with the automatic aeration system, ensuring an optimal concentration of dissolved oxygen (DO)

Figure 17.3 IoT-based monitoring of strawberry crop [5].

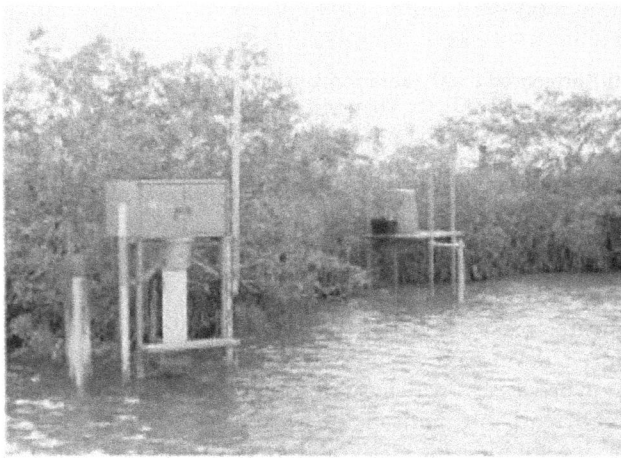

Figure 17.4 Current automated water quality-sensing devices by remote delay-tolerant water quality monitoring in the Peruvian Amazon [6].

is present [3, 4]. The implemented system overcomes several challenges associated with conventional/traditional water quality monitoring typically implemented in shrimp aquaculture. These include labour intensiveness and human errors.

A similar system was developed for a strawberry farm in Columbia that focused on disease mitigation upon the crop due to environmental changes related to soil and water (Figure 17.3). The IoT platform developed and tested by the author aimed to improve the productivity of strawberry crops using a technological solution, which attempted to control the variables (altitude, temperature, brightness, relative humidity, precipitation, and hail) that affect the development of plants [5].

In remote monitoring systems for secluded or faraway locations, such as the Napo River in the Peruvian Amazon, a system of remote delay-tolerant water quality sensors was tested (Figure 17.4). This solution addressed the

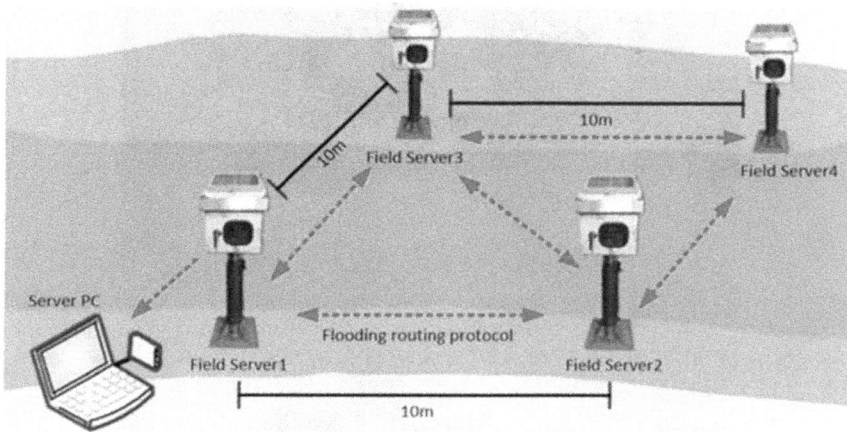

Figure 17.5 Design of the wireless sensor network (WSN) [7].

remote access nature of the area where the local community relied on the surrounding water for daily usage as well as agricultural cultivation. The author attached the sensors control system to the water treatment tower and recorded observations. A secure digital memory card (SD card) was used for storage of data in the event of a power outage, ensuring that previous data are kept intact for physical collection when needed [6].

Authors in Ref. [7] experimentally tested and monitored water quality in a wide rural area (Figure 17.5). The field servers were positioned at a range of 10 m from each other for the design of the wireless sensor networks (WSN). Each field server consists of (i) a water quality sensor module for collecting the water quality data, (ii) a sensor node for wireless data transmission, (iii) a flash memory for the data averaging process, (iv) an interface between the sensor node and the water quality sensor module, and (v) a battery charged using solar cell panels. The testing measured a variety of water quality data in relation to pollutants contributing to water-related diseases. These parameters include DO, pH, conductivity, turbidity, depth of water, and temperature. The application of the system shows potential in expansion for scattered field monitoring over a wide area, using a self-contained solar-powered system.

Related to the large inland aquatic area, a buoy float wireless sensor water quality monitoring system was developed in the Philippines (Figure 17.6) [8]. The utilisation of WSN allows greater mobility of the platform to be deployed on a wider range of water, addressing issues where the deployment can be hard to reach or hazardous to the equipment itself. A warning system in the form of a global system for mobile (GSM) communication (SMS to phone) served as the main communication method between users

Figure 17.6 The buoy float sensor water quality monitoring system in the Philippines [8].

and sensor nodes. Data were also collected both locally in an SD card and transmitted wirelessly to nearby computers [8].

The IoT-based water quality monitoring system installed in Lake Victoria, Uganda (Figure 17.7) [9], utilises a global positioning system (GPS), GSM, and wireless network. These attributes are low-cost, field-tested, and user assembly-friendly including solar panel recharge.

Analysis of IoT applications for water quality monitoring has also been undertaken. A review of published literature for different applications is presented in Figure 17.8. The highest number of applications is found in the monitoring of inland water quality, including ponds, rivers, and reservoirs [7–17]. This is expected as a large body of inland water serves as either a drinking water reserve or concentrated fauna and flora area. Monitoring of water quality for commercial food production [3, 4, 18–22] and drinking water [6, 23–29] also appears widely as these areas of application have a direct impact on the essential human supply chain.

There are a few other areas where IoT has been successfully applied to monitor water quality, such as the urban water systems [6, 30–34], water distribution networks [35–39], agriculture [2, 5, 40, 41], oceans [42–45] and coastal areas [7, 46], sewage treatment [47] and aeration plant [1], hydrologic [48] and biological oxygen demand analysis [49], and interplanetary space flights [50].

Figure 17.7 Field testing equipment for the wireless sensor network in Lake Victoria, Uganda (WSN) system [9].

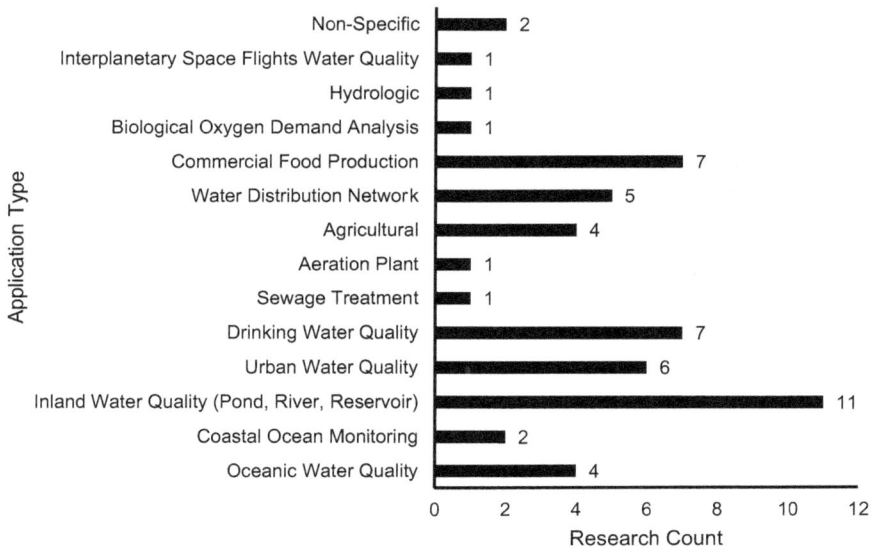

Figure 17.8 Research count of different applications of IoT for monitoring water quality.

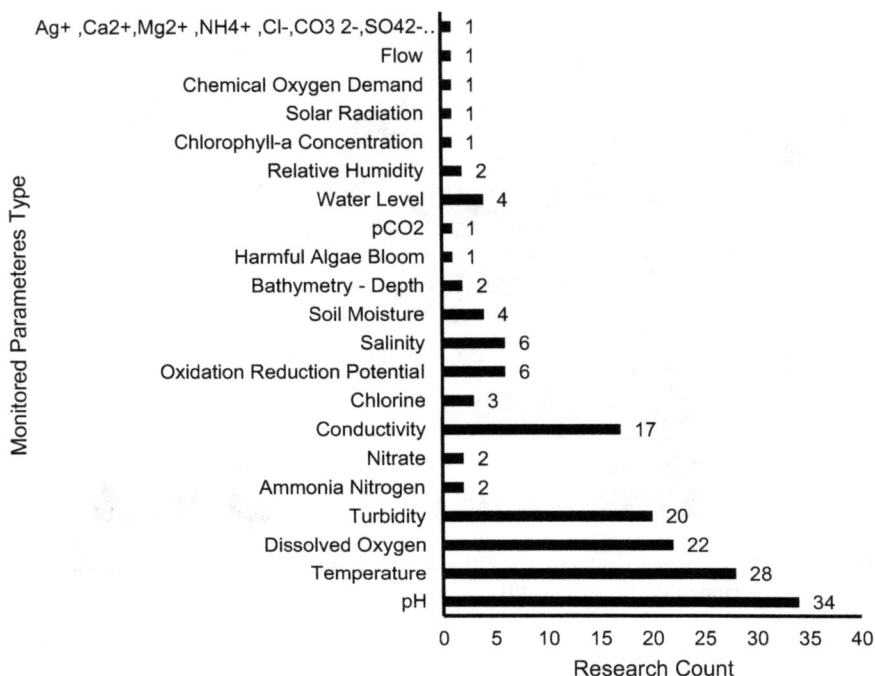

Figure 17.9 Research count for water quality parameters monitored using IoT.

17.3 KEY WATER QUALITY PARAMETERS

Water quality is indicated through the physical, chemical, and biological characteristics of the water. There are several parameters that define water quality and need to be monitored and then compared with the benchmarks related to the use of water. A distribution of the literature count, covering different parameters primarily used to monitor water quality through IoT, is presented in Figure 17.9.

17.3.1 pH

The pH level is a measurement of the hydrogen ion concentration in water. A level pH of 7 is neutral, greater than 7 is alkaline or basic, and less than 7 is acidic. The pH level of water is key to determining the health of water infrastructure within the natural environment and wildlife ecosystem. The pH level can have an indirect effect on treatment processes such as coagulant dosage, disinfection and bacteriological quality, and effect on the solubility of heavy metals, in particular lead and copper [51]. This information is critical as a high fluctuation of pH within the water can have an adverse effect on the flora and fauna living in and around the water source. The pH level of water has been monitored in IoT through the majority of the literature reviewed [2–11, 14, 17, 19, 22–25, 27–29, 31, 32, 34–39, 42, 43, 45, 47, 48, 52].

17.3.2 Turbidity

The general description of turbidity is the measure of the light-scattering property of water because of the presence of fine suspended matter such as clay or silt. This can affect disinfection processes and cause aesthetic concerns. While consuming water with a relatively high NTU doesn't necessarily mean it is unsafe to consume, the particulate itself may harbour living microorganisms [51]. The degree to which light scatter depends on the amount, size, and type of particulate matter present. When testing turbidity in a laboratory, the preferred method is the use of a nephelometric turbidity meter. The results of the test are expressed in nephelometric turbidity units (NTU) and are calibrated against a prepared formazan standard. A turbidity level of 5 NTU is recommended for the consumer.

The suggestion of potential living microorganisms in turbid water by the NHMRC indicates that it is important to know the environment that the sensor will be deployed in. There can be a huge variation in the turbidity levels from a lake to a river to a wetland. There are several references that have measured turbidity as part of the IoT-based water quality monitoring system [6, 7, 10, 11, 13, 18, 19, 23, 27–29, 31, 32, 34, 37–39, 43–45].

17.3.3 Electrical conductivity

Electrical conductivity is a measure of the ability of water to pass an electrical current, which is indicative of salinity in water which tends to increase the conductivity of water. The presence of dissolved salts and inorganic chemicals contributes to the salinity. Temperature also affects conductivity: the warmer the water, the higher the conductivity. Each water body tends to have a relatively constant range of conductivity that, once established, can be used as a baseline for comparison with regular conductivity measurements. Significant changes in conductivity could then be an indicator that a discharge or some other source of pollution has entered the aquatic resource. Several studies have included electrical conductivity in IoT-based water quality monitoring systems [5–7, 9, 12, 17, 19, 31, 34–37, 39, 44, 47, 48, 52].

17.3.4 Temperature

Water temperature is a measure of the kinetic energy of water and is expressed in degrees Fahrenheit (F) or Celsius (C). Temperature also affects the water's ability to dissolve gases, including oxygen, and may affect treatment processes in several ways. The lower the temperature, the higher the solubility. Water temperature pollution, called thermal pollution, is the artificial warming of a body of water, either through nearby industry, a city heat island, or waste. This means that temperature is a good indicator for both direct and indirect measurements of pollution in water quality measurement and has been included in many IoT-based water quality monitoring systems [2, 3, 5–11, 17, 19–25, 31, 34, 36, 39–41, 44, 45, 47, 48, 52].

17.3.5 Dissolved oxygen (DO)

DO level is another parameter most common when testing water quality. Oxygen is without a doubt, the most critical chemical for a living being. The amount of oxygen present indicates the ability of the flora and fauna to survive in the water; the importance of this parameter is closely tied to that of pH where fluctuations have an adverse effect on the ecological system [51]. A low value of DO implies the presence of excess organic activities such as harmful algae blooms [53] or high bacterial activities [54]. Given its significance, it has been widely monitored using IoT-based water quality monitoring systems [3, 4, 6–9, 17, 19, 20, 22, 24, 27, 31, 32, 36, 38, 42, 43, 45, 47, 52].

17.3.6 Other parameters

The other key parameters monitored using IoT-based water quality monitoring systems, as reported in the literature, include:

- Salinity is used for monitoring water quality in (i) the sea, natural environment, and ecosystem; (ii) drinking and consumption; and (iii) farming and agriculture [44, 45, 55]
- Oxidation-reduction potential [6, 17, 28, 31, 36, 39]
- Soil moisture [2, 40, 41, 48]
- Ammonia and nitrogen content [1, 24]
- Nitrate content [1, 15]
- Chlorine [25, 26, 37]
- Bathymetry – Depth [7, 46]
- Harmful algal bloom [30]
- pCO_2 [42]
- Water level [12, 19, 34, 48]
- Relative humidity [5, 41]
- Chlorophyll-a concentration [44]
- Solar radiation [48]

- Chemical oxygen demand [49]
- Flow [39]
- Ag^+, Ca^{2+}, Mg^{2+}, NH^{4+}, Cl^-, CO_3^{2-}, SO_4^{2-}, NO^{3-} ions [50]

17.4 INTERNET OF THINGS (IoT)

The Internet of Things (IoT) presents the idea of a connected device embedded with the internet and has been around for quite some time. Any IoT platform is a set of technology-enabled entities, including but not limited to physical smart objects (e.g., sensors, cameras, smart home assistance) and software services. An IoT platform usually collects, and processes, high-frequency data generated by either user input, sensors, or digital triggers [56].

With the advancement in high-speed internet and reliable wireless networks at long ranges, the use and demand for smart systems in domestic and industrial areas have skyrocketed. In recent years, it is evident that adopting and developing IoT base systems is essential in any type of information system management and research. Qualities like autonomous and real-time data collection are essential in any IoT platform, which is a key outcome for water quality monitoring platforms when compared with traditional methods.

Toward the end of the 20th century, the need for a real-time monitoring system capable of handling high-frequency data became a focus of research for various industries. From manufacturing, energy sector and environmental science and water treatment plant, the potential benefits that such a system could deliver are limitless. Emerging IoT technologies and research/development tend to address the following challenges:

- Size: physical footprint of equipment can often hinder its placement availability. Industrial equipment is often bulky, large and specifically build for the task. IoT attributes of interchangeable modularity and compact sizes increase the degrees of adaptability for a multitude of scenario, spaces, and geometrical challenges.
- Power Management: being independent of local power grid increases mobility and allowable operating range. Balancing a large portable power source, size, weight, and recharge ability is a hard equation that modern IoT researchers is usually confronted with.
- Communication: the most important aspect of IoT and what makes it so unique is its large arrays of communication choices. Robust short-range Wi-Fi connection enables big data packet to be used at high sampling frequency. Long-range radio wave can overcome geography variations with sacrifice to data transmission reliability.
- Sensors: as concluded by Lynggaard-Jensen [57], a suite of online in situ sensors may need to be incorporated into the ever-expanding wastewater treatment network, which may depend on local conditions.

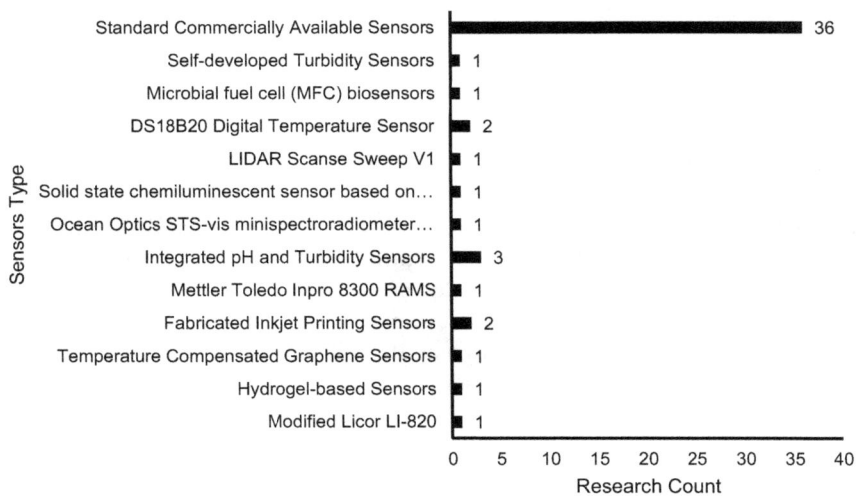

Figure 17.10 Research count of different types of sensors used for water quality monitoring through IoT.

It is important that IoT becomes an integrated part and life cycle considerations with engineering approaches, practices, and design decisions are incorporated in design [58]. To address this issue, advancement toward streamline and cost-effective method has been developed over the last couple of decades. As technology becomes smaller and smarter, new methodologies emerge to overcome limitation imposed by size, transmission range, and power consumption.

17.4.1 Evolution of sensors technology

The sensor technologies are the eyes and ears of an IoT platform, just like visual imaging or sound recording. In the case of water quality monitoring, the sensors collect water quality data for various parameters from water sources. Distribution of different types of sensors employed with the IoT technology is shown in Figure 17.10. The majority of the research uses off-the-shelf sensors in their IoT platform [1–14, 17, 19, 21–23, 27–29, 31, 32, 36–45, 47, 48]. Self-fabricated or sensor advancement in recent years has been heading toward inkjet technology for its cheap and easy-to-fabricate attributes. Other specialised sensors include (i) modified Licor LI-820 [42], (ii) hydrogel-based sensors [35], (iii) temperature-compensated graphene sensors [15], (iv) fabricated inkjet printing sensors [25, 52], (v) Mettler Toledo Inpro 8300 RAMS [18], (vi) integrated pH and turbidity sensors [16, 59, 60], (vii) ocean optics STS-vis mini spectroradiometer system [30], (viii) solid-state chemiluminescent sensor based on electropolymerised luminol [26],

(ix) LIDAR Sweep V1 [46], (x) DS18B20 digital temperature sensor [11, 45], (xi) microbial fuel cell (MFC) biosensors [49], and (xii) self-developed turbidity sensors [39].

A review done by Zhuiykov [61] expanded upon the previous research in water quality parameter monitoring, suggesting a greater focus on the novel method of measuring pH level using a solid-state sensor; the design is based on the thin or thick film of the semi-conductor sensing electrode (SE). This system aims to minimise the biofouling difficulties as well as stepping away from the traditional pH glass electrode [62]. Even though the pH electrode is currently still the most commercially successful sensor, the pH glass electrode is highly fragile and can experience significant interference from fluoride ions [63].

Biofouling is an ongoing difficulty that any waterborne sensors must overcome. This is when the biocontaminant builds up on the sensors over time, affecting the reading of the sensors. Corrosion and oxidation of equipment are also a problem when electronic equipment is in contact with water, regardless of materials [64]. In this case, biofouling is the limiting factor in creating a compact system of sensors that can operate for an extended duration; thus, the method of laboratory analysis is still a reliable method to get results despite its inefficiencies [57].

The potential solution problem created by biofouling was explored by Serge et al. [64]. The research focuses on using submicron Cu_2O-doped RuO_2 SE for potentiometric detection of DO. The aims were to validate the sensitivity and accuracy improvements over the traditional sensor's conductor material as well as its ability to resist biofouling. The field trial took place over Karkarook Park Lake with two different sensors, one RuO_2-SE and another 10 mol% Cu_2O-doped RuO_2-SE. After a three-month period, the two sensors were examined under SEM (scanning electron microscope), the sensors without the 10 mol% Cu_2O-doped showed that biofouling has started accumulating on the RuO_2 grains. The other sensors showed no traces of biofouling. This validates Cu_2O as a known effective antifouling pigment, also used in the chemically active paint systems for marine vessels [65].

With a focus on the miniaturisation and utilisation of an optical sensor to monitor turbidity, Murphy et al. [13] developed a low-cost optical sensor for monitoring the aquatic environment. The optical sensor was created from photodiodes (OPT101P) operated using Raspberry Pi (Model B) and LEDs housed within a small PVC and copper tube. This sensor measured and collected data about turbidity quality such as transmission and side-scattering of the light using a multi-wavelength light source with two photodiode detectors.

In 2016, similar research in semi-conductive solid-state sensors conducted by Xu et al. [52] utilised the new fabrication method – inkjet printing technology (IPT) to apply the membrane to the semi-conductor housing. The aim of the research was not to improve upon the solid-state research

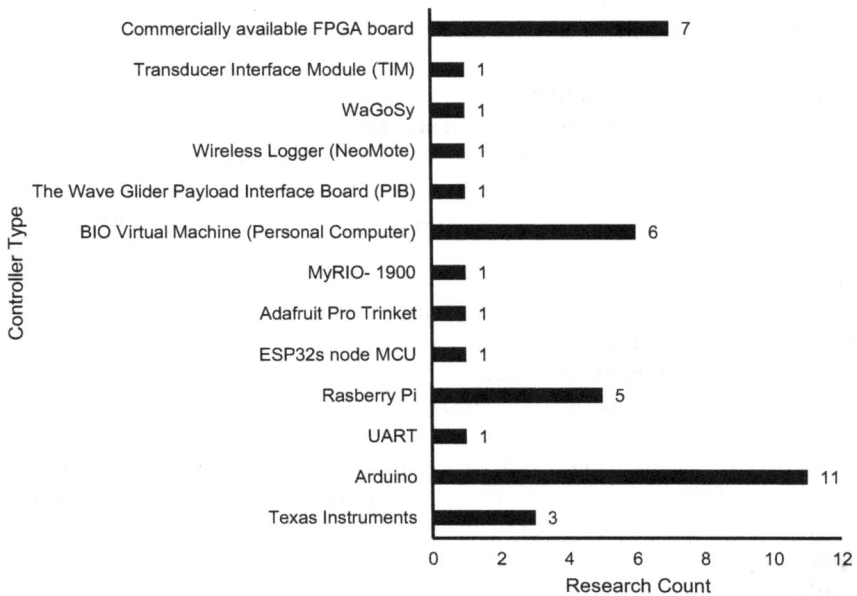

Figure 17.11 Research count of different types of control system used for water quality monitoring through IoT.

but to make it more accessible for the user to create these sensors with limited equipment and knowledge. The benefit achieved by the study is, however, offset by the high initial cost of obtaining an IPT machine capable of printing onto a semi-conductor using precious metal. The individual cost of each unit is quite low, a quality that should be taken into consideration.

17.4.2 Control system

A control system is the heart of any water quality monitoring system through IoT. The control typically exists as physical hardware, ranging from multi-input output microcontroller, standard PC, and smart devices such as mobile phones and tablets. Some unique control systems in this review are designed and manufactured specifically for a USV platform; those will be covered further down.

The most common control system used in water quality monitoring as found in the reviewed literature (Figure 17.11) is Arduino [2, 6, 8, 9, 17, 29, 36, 38, 40, 45, 47], followed by commercially available FPGA boards [7, 15, 22, 25, 27, 32, 39], BIO virtual machines [3, 4, 14, 28, 31, 33], and Raspberry Pi [5, 13, 17, 21, 23]. Other systems used include (i) UART [16], (ii) ESP32s node MCU [41], (iii) Adafruit Pro Trinket [10], (iv) NI MyRIO 1900 [46], (v) Wave Glider Payload Interface Board (PIB) [42], (vi) wireless logger [12], (vii) WaGoSy [9], (viii) Transducer Interface Module (TIM) [44], and (ix) Texas Instrument system [20, 24, 49].

Arduino is the most popular controller technology to be utilised by researchers. It is a stable hardware platform with an extensive library of codes and compatibility for a wide range of sensors and wireless communication. The hardware power consumption is relatively low, making it a good choice to centre an IoT platform around.

17.4.3 Wireless network type

Wireless communication is one of the most important components of any IoT platform; it dictates the application and degree of flexibility a system can support. Most often, short-range wireless networks, typically seen in households, offices, and indoor applications, are robust and large amounts of data to be transmitted at once. Longer-range communication in IoT applications like coastal monitoring is more challenging as they transmit data over long distances, and possibly through physical barriers. Traditionally, wireless communications have relied on radio transmission or satellites, which can increase start-up and upkeep costs.

Ideally, IoT communications are reliable even over long distances, inexpensive setup, and operational costs, as well as low-powered to ensure long periods of running time. Real-time communication of sensors depends on variables such as sampling interval, distances (between the sensor and data platform), and how power efficient the wireless communication method is if the IoT platform happens to be battery operated. As shown in Figure 17.12, we can see that Wi-Fi module and Zigbee, both relatively short-range communication protocol, are the most implemented in IoT research. It is not desirable to increase the distance between the sensor's platform and the data platform.

Wong and Kerkez [12] explored the use of a web-based service provider to create an IoT platform for a real-time environmental sensor network. The studies cited that the advancement in sensing, computation, and communications enabled a large amount of diversity in the suite of low-cost, low power-connected devices [66, 67]. Combining a variety of in situ hardware with hybrid solar and wired powered connection, the team created a real-time monitoring system utilising available software suites such as Xively Internet of Things platform, Flask webserver, and Amazon Web Services. The sensor node measured water and flow parameters such as flow rate, depth, and temperature. The system hardware housing and architecture prove handy during the field implantation phase. The free web service data logger and GUI trivialise the need for constant monitoring from users. The ability shown by the research to leverage online resources to do complicated networking means more time and attention can be spent on hardware development and implementation.

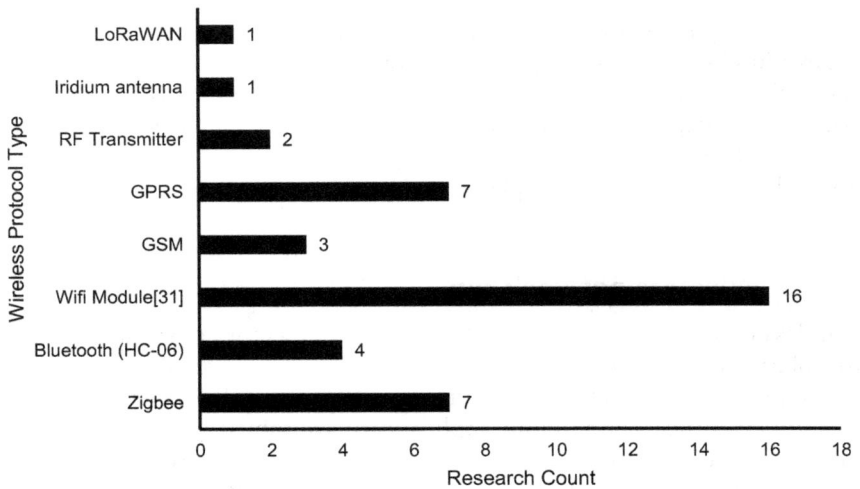

Figure 17.12 Research count of different types of wireless networks used for water quality monitoring through IoT.

A distribution of literatures of wireless networks used for water quality monitoring through IoT is presented in Figure 17.12. Off-the-shelf Wi-Fi module [28] is most widely used [3, 5, 7, 11–15, 21, 22, 24, 40, 41, 44, 47, 48] due to their widespread inclusion in every day smart portable technology. This is followed by GPRS [4, 9, 16, 33, 38, 43, 44] and Zigbee [8, 9, 20, 23, 27, 32, 39]. The other technologies used are (i) Bluetooth [10, 14, 41, 47], (ii) GSM [4, 8, 17], (iii) RF transmitter [9, 48], (iv) Iridium antenna [42], and (v) LoRaWAN [19].

Short-range (less than 2 km) communication methods are reliable as these methods can maintain real-time connection for data collection. Reliable long-range communication remains a challenge.

17.4.4 Unmanned surface vehicles (USVs) technology

Drone or autonomous mobile platform is currently on the forefront of water quality monitoring, moving from the traditional static sensor's node. USVs have had a long history of development and implementation by militaries since World War I and II, though early USVs used remote control (RC) and lacked autonomous navigation capabilities [68]. The mobility afforded by the platform means that real-time monitoring can take place at a remote location. These locations can be places where it is not practical to spend the resources to send personnel or to set up a sensors station.

An autonomous surface vehicle was developed by Chavez et al. [42] from the Monterey Bay Aquarium Research Institute (MBARI) to monitor and measure pCO2 and pH in a coastal upwelling system. The system was proven to be capable of long-range operation (>5000 km) and long-duration (>500 days) deployments. The compact size of the vehicle allows it to operate under challenging weather conditions where it would be impractical for a vessel or a monitoring buoy to be set up. Dubbed the Wave Glider ASV, the vehicle consists of three primary components: a buoyant surface float, a submerged diving plane (the glider), and a 4–7 m tether connecting the float and glider. To solve the limitation of the power source on a long journey, the Wave Glider makes use of wave energy as well as a solar-powered thruster to be operated during calm ocean conditions or strong currents. The Wave Glider was shown to be more than capable in its primary function of collecting data sets; however, with small and compact size comes a cost of durability. The research team stated that the pump and water valve suffered malfunction as well as drifting off course, especially during harsh currents.

Similar research into the autonomous surface vehicle was also done by Carlson et al. [46] aimed at creating an affordable system using obstacle avoidance light detection and ranging (LIDAR). Greenland's extensive coastlines are poorly charted, especially in shallow areas creating risk for the manned vessel. The vehicle, called The Arctic Research Centre Autonomous Boat (ARCAB), is cited by the researchers as "simple to construct, yet durable and it can operate manually, using a remote controller, and autonomously, where it follows predefined GPS waypoints". What makes the vehicle stand out from the rest is its LIDAR-integrated technology, capable of avoiding obstacles such as rock, shallow reefs, and possible seafaring vessels. However, the drawback of this technology is that the LIDAR is dependent upon the orientation of the vehicle. If the vehicle experiences rough sea conditions or sea spray, the LIDAR would give a false positive, steering the vehicle unnecessarily and wasting power.

Jo et al. [10] presented a "new, fully open-source, low-cost, and small-sized unmanned surface vehicle (USV) for measuring near-surface water quality in real-time in various environments". The research aimed at continuous water quality monitoring in the southwestern region of Peru due to a shortage of rainfall and contamination from the mining industry [69]. The proposed system called SMARTBoat 3 was equipped with sensors to monitor water quality such as turbidity, temperature, and pH. The small USV platform could be monitored and controlled using a GPS-based system. The software development of this platform was fully open source except for the housing design. Validation testing took place in pool and irrigation retention ponds. While successfully collecting data as intended, the limitation of the system was because of its small battery size and lack of operational range (92 m from the wireless node).

Aerial drone technology has seen a significant jump in technological advancement. From being a specialist tool in the research and development field in the last two decades to commercial success, accessible to the public of all ages and budgets. This technology, however, continued to be further explored to augment surveillance and monitoring techniques. Compared to the conventional manned aerial and ground system, they represent a reduced operational cost and improvement in safety and repeatability. The use of unmanned vehicles has rapidly evolved to be used in various studies such as gathering oceanographic and meteorological data [55, 70–72].

Prototype development by Becker et al. [30] used an unmanned aerial system-based spectro-radiometer for monitoring harmful algal blooms. The system's main area of monitoring is located in the Maumee River at Toledo where algae bloom represents a real nuisance to recreational users of the river. Prototypes were developed in two different locations with two distinct methods: the first prototype had a drone created in-house with inexpensive water-proof equipment, while the second prototype used a different drone made by a commercial manufacturer. The two drones cost US$2000 and US$7000, but the researcher stated, "The cost of both sUAS platforms was extremely low for the high-quality data they produce". The advantages of this system are unhindered by water surface conditions, rapid deployment, mappable path. However, the limitation of this research is flight duration and payload weight limits, a constraint that can be minimised or even eliminated as aerial drone technology improves over time.

17.4.5　Power usage and power recharge technology

Electronic equipment operating in a remote area or independently from the power grid relies on either a battery with large storage or a rechargeable power source such as a solar panel. There has been a large focus on the optimisation of power consumption for an off-the-grid wireless quality monitoring system, balancing between the data collection frequency, hardware, operating systems, and solar panel efficiency.

The longevity and usefulness of wireless sensors network depend greatly upon power sources. Yue and Ying [28] utilised SunSPOT (Sun Small Programmable Object Technology) to create a sensor network makeup of sensor nodes and a base station, powered by a 12 V solar panel. The technology was chosen for its low-powered background process. This, however, limited the number of data sets that can be obtained on any given day, especially during night-time.

Kulkarni Amruta [27] explored the concept of solar-powered water quality monitoring using WSN technology like Zigbee. They designed and implemented a prototype model using a node, powered by a solar cell. To make the system more flexible to be deployed in a remote area where wire connection is extremely impractical and unsafe, solar panels supply power to the system and excess power is used to recharge batteries for night-time usage.

Non-rechargeable power sources are still the most widely used for IoT-based water quality monitoring, as shown by the distribution of literature in Figure 17.13. This includes non-rechargeable lithium polymer batteries [4, 10, 17, 20, 25, 30, 38, 41, 46], followed by standard AC power [12, 13,

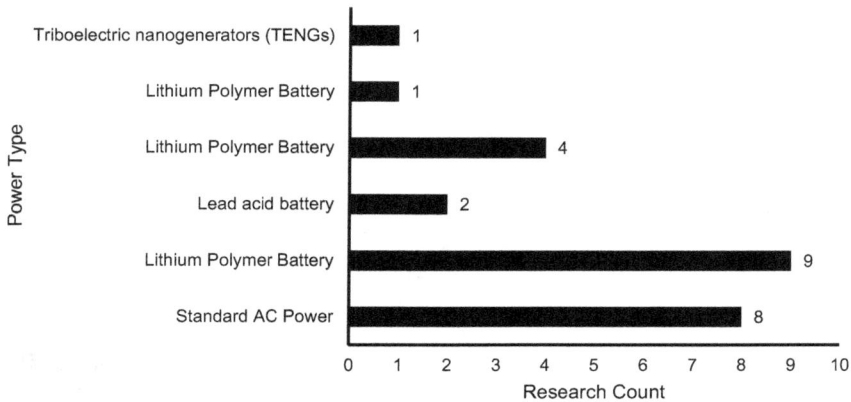

Figure 17.13 Research count of different types of power sources used for water quality monitoring through IoT.

15, 19, 22, 23, 27, 36]. When considering the rechargeable power sources, solar-powered rechargeable lithium polymer battery [7, 9, 16, 24] is the most common, followed by solar-powered rechargeable lead acid battery [8, 31], wave-powered rechargeable triboelectric nanogenerators [14], and wave and solar-powered rechargeable lead acid battery [42].

17.5 DISCUSSION

This section discusses the research gaps and therefore potential research directions. The importance of sensors cannot be understated, it is the eyes and ears of the IoT platform. One of the most challenging prospects in maintaining a reliable reading from a sensor is its prolonged exposure to dynamic water quality changes and flow rate variations. Over time, depending on the application of the sensor, its quality can deteriorate due to biofouling, chemical contaminant, or even physical wear and tear [61]. To overcome this issue, some researchers use multiple sensors to establish the trustworthiness of their image data [73]. Such a multimodal approach might be useful for IoT in water quality monitoring, and it is worth exploring but implies additional costs.

Short-range (about 2 km) wireless communication appears to be stable enough; however, data on the reliability of wireless transmission over long distances are still scant. LoRaWAN is an unlicensed protocol that is developed specifically to meet the need for IoT communications. LoRaWAN is low powered with low setup and running costs as well as long range. It is a promising technology in the area of IoT due to "networking autonomous architecture and an open standard specification" [74]. According to LoRa

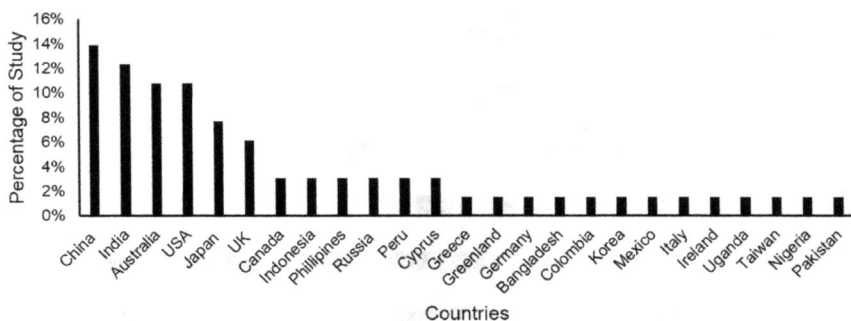

Figure 17.14 Country-wise breakdown of the number of literature reviewed in this study.

Alliance, LoRaWAN has a battery life of 105 months (nearly 10 years), very high interference immunity and mobility compared to other IoT communication systems [75]. The use of LoRaWAN in water quality monitoring is, however, limited at the moment. It is also unclear what range LoRaWAN can reliably cover for continuous real-time applications. The range of transmissions without physical barriers (line of sight) or with physical barriers is also not clear. Some applications attempt to transmit water quality data through LoRaWAN, but their testing distance was not significant enough to fully utilise its potential. It also does not address data loss and reliability with online availability. To avoid losing data transmission due to a drop in connection, some research utilised built-in memory storage for periodic collection.

From all the literatures studied, a country-wise breakdown is provided in Figure 17.14. Majority of the studies, considered in this chapter, are reported to be conducted in countries like China (14%), India (12%), Australia (11%), USA (11%), Japan (8%), and UK (6%). These countries are facing water stress and several water management problems but also technologically well placed to try and test IoT in the space of water and environmental management. There is significant scope for future research and applications of IoT-based water quality monitoring to be extended to other countries, especially for decentralised facilities.

17.6 CONCLUSIONS

The review has shown that there is a vast developing interest across the globe in the field of water quality monitoring using IoTs with applications in farming to large surface water systems. An IoT system is made up of sensors that are connected to a local system capable of sending data. The local system is powered either by batteries or by other power sources. The type of sensors used depend on the parameters required in the applications. This review has four components: the parameters monitored, the control system, power source, and communications.

Most sensors are static but advancement in UAV/USV has made IoT even more versatile as sensors can now be deployed in challenging locations. The reliability of sensors is extremely important, and in the water applications, some sensors are exposed to varying water conditions, which could reduce the reliability of the sensors. In this context, multimodal data is a promising solution worth exploring.

Regardless of which parameters the sensors monitor, they need to be connected to a control system. Arduino appears to be the most popular system. It is a stable hardware platform with an extensive library of codes and compatibility for a wide range of sensors and wireless communication. The hardware power consumption is relatively low, making it a good choice to centre an IoT platform around. Most of the research and applications currently use short but robust communication like Wi-Fi to maintain a reliable connection between the sensors and the local controller. Long-range communication or warning system utilises GSM network to send out a notification to users when a predefined parameter or threshold is reached. A large area of communication between controllers and nodes relies on existing city wireless network infrastructure. However, such an approach is impractical for some applications, such as those in geographically isolated and/or decentralised locations.

LoRaWAN has been specifically created to meet the challenges in IoT communications, but it is unclear what is the maximum range for reliable transmission over long distances for continuous real-time data and it needs to be thoroughly evaluated. While some applications have attempted to transmit water quality data through LoRaWAN, their testing distance was not significant enough to fully utilise its potential. It also does not address data loss and reliable online time. This limitation arises from remote location with poor line of sight and rechargeable power source. To avoid losing data transmission due to drop in connection, some research utilised built-in memory storage for periodic collection. The use of short-range communication method has been thoroughly explored (less than 2 km) as founded in the review. These prove much more reliable in their ability to maintain a real-time connection for data collection but there is lack of flexibility of the range a LoRaWAN device can provide (10 km plus).

Monitored parameters and application graphs show a very predictable trend across them. There were some applications with minimal appearance such as aeration plant and sewage treatment. Sensors technology in the review tends to be commercially available sensors. This could potentially be due to the ease of setup and well-known quality. Setting up and designing new sensors, such as inkjet technology, can become an even lower-cost alternative.

The importance of sensors cannot be understated, it is the eyes and ears of the IoT platform. One of the most challenging prospects in maintaining a reliable reading from a sensor is prolonging exposure to dynamic water quality. Over time, depending on the application of the sensor, its quality can deteriorate due to biofouling, chemical contaminant, or even physical wear and tear.

REFERENCES

[1] V. Belikova, V. Panchuk, E. Legin, A. Melenteva, D. Kirsanov, and A. Legin, "Continuous monitoring of water quality at aeration plant with potentiometric sensor array," *Sensors and Actuators B: Chemical,* vol. 282, pp. 854–860, 2019/03/01/ 2019.

[2] A. Pathak, M. AmazUddin, M. J. Abedin, K. Andersson, R. Mustafa, and M. S. Hossain, "IoT based smart system to support agricultural parameters: a case study," *Procedia Computer Science,* vol. 155, pp. 648–653, 2019/01/01/ 2019.

[3] A. G. Orozco-Lugo et al., "Monitoring of water quality in a shrimp farm using a FANET," *Internet of Things,* vol. 18, p. 100170, 2020/01/31/ 2020.

[4] G. Wiranto, Y. Maulana, I. D. P. Hermida, I. Syamsu, and D. Mahmudin, "Integrated Online Water Quality Monitoring an Application for Shrimp Aquaculture Data Collection and Automation." in *International Conference Smart Sensor and Application,* 2015.

[5] A. D. Juan Carlos et al., "Monitoring system of environmental variables for a strawberry crop using IoT tools," *Procedia Computer Science,* vol. 170, pp. 1083–1089, 2020/01/01/ 2020.

[6] C. Ritter, M. Cottingham, J. Leventhal, and A. Mickelson, "Remote delay tolerant water quality montoring," in *IEEE Global Humanitarian Technology Conference (GHTC 2014),* 2014, pp. 462–468.

[7] W.-Y. Chung and J.-H. Yoo, "Remote water quality monitoring in wide area," " *Sensors and Actuators B: Chemical,* vol. 217, pp. 51–57, 2015.

[8] M. V. J. Alexander T. Demetillo, and Evelyn B. Taboada, "A system for monitoring water quality in a large aquatic area using wireless sensor network technology," (in English), *Demetillo et al. Sustainable Environment Research,* Sustainable Environment Research no. 29, p. 9, 23/04/2019 2019, Art. no. 12.

[9] A. Faustine, A. Mvuma, H. Mongi, M. Gabriel, A. Tenge, and K. Samuel Baker, "Wireless sensor networks for water quality monitoring and control within Lake Victoria basin: prototype development," *Wireless Sensor Network,* vol. 06, pp. 281–290, 01/01 2014.

[10] W. Jo, Y. Hoashi, L. L. Paredes Aguilar, M. Postigo-Malaga, J. M. Garcia-Bravo, and B.-C. Min, "A low-cost and small USV platform for water quality monitoring," *HardwareX,* vol. 6, p. e00076, 2019/10/01/ 2019.

[11] M. S. U. Chowdury et al., "IoT based real-time river water quality monitoring system," *Procedia Computer Science,* vol. 155, pp. 161–168, 2019/01/01/ 2019.

[12] B. P. Wong and B. Kerkez, "Real-time environmental sensor data: an application to water quality using web services," *Environmental Modelling & Software,* vol. 84, pp. 505–517, 2016/10/01/ 2016.

[13] K. Murphy et al., "A low-cost autonomous optical sensor for water quality monitoring," *Talanta*, vol. 132, pp. 520–527, 2015/01/15/ 2015.

[14] Z. Zhou et al., "Wireless self-powered sensor networks driven by triboelectric nanogenerator for in-situ real time survey of environmental monitoring," *Nano Energy*, vol. 53, pp. 501–507, 2018/11/01/ 2018.

[15] M. E. E. Alahi, A. Nag, S. C. Mukhopadhyay, and L. Burkitt, "A temperature-compensated graphene sensor for nitrate monitoring in real-time application," *Sensors and Actuators A: Physical*, vol. 269, pp. 79–90, 2018/01/01/ 2018.

[16] T. Kageyama, M. Miura, A. Maeda, A. Mori, and S. Lee, "A wireless sensor network platform for water quality monitoring," in *2016 IEEE SENSORS*, 2016, pp. 1–3.

[17] S. S. Rizky Dharmawan, R. W. Sudibyo, W. Sarinastiti, A. Sasono, A. A. Saputra, S. Sasaki, "Design and development of a portable low-cost COTS-based water quality monitoring system," *International Seminar on Intelligent Technology and its Application, Conference Paper* vol. 1, no. 1, p. 6, 30/06/2016 2016.

[18] G. Skouteris, D. P. Webb, K. L. F. Shin, and S. Rahimifard, "Assessment of the capability of an optical sensor for in-line real-time wastewater quality analysis in food manufacturing," *Water Resources and Industry*, vol. 20, pp. 75–81, 2018/12/01/ 2018.

[19] G. Gao, K. Xiao, and M. Chen, "An intelligent IoT-based control and traceability system to forecast and maintain water quality in freshwater fish farms," *Computers and Electronics in Agriculture*, vol. 166, p. 105013, 2019/11/01/ 2019.

[20] B. Shi, V. Sreeram, D. Zhao, S. Duan, and J. Jiang, "A wireless sensor network-based monitoring system for freshwater fishpond aquaculture," *Biosystems Engineering*, vol. 172, pp. 57–66, 2018/08/01/ 2018.

[21] P. B. Bokingkito and O. E. Llantos, "Design and implementation of real-time mobile-based water temperature monitoring system," *Procedia Computer Science*, vol. 124, pp. 698–705, 2017/01/01/ 2017.

[22] X. Zhu, D. Li, D. He, J. Wang, D. Ma, and F. Li, "A remote wireless system for water quality online monitoring in intensive fish culture," *Computers and Electronics in Agriculture*, vol. 71, pp. S3–S9, 2010/04/01/ 2010.

[23] L. Raja, G. Shanthi, and P. S. Periasamy, "Internet of things based real time water monitoring system," *International Journal of Recent Technology and Engineering*, Article vol. 8, no. 2, pp. 1368–1372, 2019.

[24] J. Sheng, W. Weixing, Y. Jieping, and H. Zhongqiang, "Design a WSN system for monitoring the safety of drinking water quality," *IFAC-PapersOnLine*, vol. 51, no. 17, pp. 752–757, 2018/01/01/ 2018.

[25] Y. Qin et al., "Integrated water quality monitoring system with pH, free chlorine, and temperature sensors," *Sensors and Actuators B: Chemical*, vol. 255, pp. 781–790, 2018/02/01/ 2018.

[26] N. Kato, N. Hirano, S. Okazaki, S. Matsushita, and T. Gomei, "Development of an all-solid-state residual chlorine sensor for tap water quality monitoring," *Sensors and Actuators B: Chemical*, vol. 248, pp. 1037–1044, 2017/09/01/ 2017.

[27] M. T. S. M. Kulkarni Amruta, "Solar powered water quality monitoring system using wireless sensor network," *presented at the 2013 International Multi-Conference on Automation, Computing, Communication, Control*

and Compressed Sensing (iMac4s), Kottayam, India, 23/03/2013, 2013. Available: https://ieeexplore.ieee.org/abstract/document/6526423 https://ieeexplore.ieee.org/ielx7/6520973/6526372/06526423.pdf?tp=&arnumber=6526423&isnumber=6526372&ref=

[28] R. Yue and T. Ying, "A novel water quality monitoring system based on solar power supply & wireless sensor network," *Procedia Environmental Sciences,* vol. 12, pp. 265–272, 2012/01/01/ 2012.

[29] S. Shakhari and I. Banerjee, "A multi-class classification system for continuous water quality monitoring," *Heliyon,* vol. 5, no. 5, p. e01822, 2019/05/01/ 2019.

[30] R. H. Becker et al., "Unmanned aerial system based spectroradiometer for monitoring harmful algal blooms: a new paradigm in water quality monitoring," *Journal of Great Lakes Research,* vol. 45, no. 3, pp. 444–453, 2019/06/01/ 2019.

[31] Y. Chen and D. Han, "Water quality monitoring in smart city: a pilot project," *Automation in Construction,* vol. 89, pp. 307–316, 2018/05/01/ 2018.

[32] M. Offiong, A. A, N. Chile-Agada, R.-L. Y, and N. O, "Real time monitoring of urban water systems for developing countries," *IOSR Journal of Computer Engineering,* vol. 16, pp. 11–14, 01/01 2014.

[33] W. Dehua, L. Pan, L. U. Bo, and G. Zeng, "Water quality automatic monitoring system based on GPRS data communications," *Procedia Engineering,* vol. 28, pp. 840–843, 12/31 2012.

[34] S. G. S. Geetha, "Internet of things enabled real time water quality monitoring system," in *Geetha and Gouthami Smart Water,* S. Water, Ed., ed. India, 2017, p. 19.

[35] M. H. Banna, H. Najjaran, R. Sadiq, S. A. Imran, M. J. Rodriguez, and M. Hoorfar, "Miniaturized water quality monitoring pH and conductivity sensors," *Sensors and Actuators B: Chemical,* vol. 193, pp. 434–441, 2014/03/31/ 2014.

[36] S. M. Aravinda S. Rao, J. Gubbi, M. Palaniswami, R. Sinnott, and V. Pettigrovet, "Design of low-cost autonomous water quality monitoring system," *International Conference on Advances in Computing, Communications and Informatics (ICACCI), Conference Journal* vol. 1, no. 1, p. 6, 25/08/2013 2013. https://ieeexplore.ieee.org/xpl/conhome/6621059/proceeding.

[37] A. Aisopou, I. Stoianov, and N. J. D. Graham, "In-pipe water quality monitoring in water supply systems under steady and unsteady state flow conditions: a quantitative assessment," *Water Research,* vol. 46, no. 1, pp. 235–246, 2012/01/01/ 2012.

[38] K. Rajalashmi, N. Yugathian, S. Monisha, and N. Jeevitha, "IoT based water quality management system," *Materials Today: Proceedings* vol. 45, pp. 512–515, 2021.

[39] T. P. Lambrou, C. C. Anastasiou, C. G. Panayiotou, and M. M. Polycarpou, "A low-cost sensor network for real-time monitoring and contamination detection in drinking water distribution systems," *IEEE Sensors Journal,* vol. 14, no. 8, pp. 2765–2772, 2014.

[40] M. S. Munir, I. S. Bajwa, and S. M. Cheema, "An intelligent and secure smart watering system using fuzzy logic and blockchain," *Computers & Electrical Engineering,* vol. 77, pp. 109–119, 2019/07/01/ 2019.

[41] J. Doshi, T. Patel, and S. K. Bharti, "Smart farming using IoT, a solution for optimally monitoring farming conditions," *Procedia Computer Science,* vol. 160, pp. 746–751, 2019/01/01/ 2019.

[42] F. P. Chavez, J. Sevadjian, C. Wahl, J. Friederich, and G. E. Friederich, "Measurements of pCO_2 and pH from an autonomous surface vehicle in a coastal upwelling system," *Deep Sea Research Part II: Topical Studies in Oceanography,* vol. 151, pp. 137–146, 2018/05/01/ 2018.

[43] K. S. Tew, M.-Y. Leu, J.-T. Wang, C.-M. Chang, C.-C. Chen, and P.-J. Meng, "A continuous, real-time water quality monitoring system for the coral reef ecosystems of Nanwan Bay, Southern Taiwan," *Marine Pollution Bulletin,* vol. 85, no. 2, pp. 641–647, 2014/08/30/ 2014.

[44] F. Adamo, F. Attivissimo, C. G. C. Carducci, and A. M. L. Lanzolla, "A smart sensor network for sea water quality monitoring," *IEEE Sensors Journal,* vol. 15, no. 5, pp. 2514–2522, 2015.

[45] C. Tziortzioti, D. Amaxilatis, I. Mavrommati, and I. Chatzigiannakis, "IoT sensors in sea water environment: Ahoy! Experiences from a short summer trial," *Electronic Notes in Theoretical Computer Science,* vol. 343, pp. 117–130, 2019/05/04/ 2019.

[46] D. F. Carlson et al., "An affordable and portable autonomous surface vehicle with obstacle avoidance for coastal ocean monitoring," *HardwareX,* vol. 5, p. e00059, 2019/04/01/ 2019.

[47] S. S. Himanshu Jindal, S. S. Kasana, "Sewage water quality monitoring framework using multi-parametric sensors," (in English), *S.S. Wireless Pers Commun,* Journal vol. 97, no. 1, p. 33, 27/05/2017 2019.

[48] L. Zhang, S. Thomas, and W. J. Mitsch, "Design of real-time and long-term hydrologic and water quality wetland monitoring stations in South Florida, USA," *Ecological Engineering,* vol. 108, pp. 446–455, 2017/11/01/ 2017.

[49] G. Pasternak, J. Greenman, and I. Ieropoulos, "Self-powered, autonomous biological oxygen demand biosensor for online water quality monitoring," *Sensors and Actuators B: Chemical,* vol. 244, pp. 815–822, 2017/06/01/ 2017.

[50] G. Y. Grigoriev, A.S. Lagutin, S.S. Nabiev, B.K. Zuev, V.A. Filonenko, A.V. Legin and D.O. Kirsanov, "Water quality monitoring during interplanetary space flights," *Acta Astronautica,* vol. 163, pp. 126–132, 2019/03/26/ 2019.

[51] NHMRC, NRMMC, "Australian Drinking Water Guidelines Paper 6 National Water Quality Management Strategy", National Health and Medical Research Council, National Resource Management Ministerial Council, Commonwealth of Australia, Canberra, 2011.

[52] Z. Xu et al., "Real-time in situ sensing of multiple water quality related parameters using micro-electrode array (MEA) fabricated by inkjet-printing technology (IPT)," *Sensors and Actuators B: Chemical,* vol. 237, pp. 1108–1119, 2016/12/01/ 2016.

[53] A. W. Griffith and C. J. Gobler, "Harmful algal blooms: a climate change co-stressor in marine and freshwater ecosystems," *Harmful Algae,* vol. 91, p. 101590, 2020/01/01/ 2020.

[54] R. L. Spietz, C. M. Williams, G. Rocap, and M. C. Horner-Devine, "A dissolved oxygen threshold for shifts in bacterial community structure in a seasonally hypoxic Estuary," (in eng), *PloS one,* vol. 10, no. 8, pp. e0135731-e0135731, 2015.

[55] C. C. Eriksen et al., "Seaglider: a long-range autonomous underwater vehicle for oceanographic research," *IEEE Journal of Oceanic Engineering*, vol. 26, no. 4, pp. 424–436, 2001.

[56] M. Fahmideh and D. Zowghi, "An exploration of IoT platform development," *Information Systems*, vol. 87, p. 101409, 2020/01/01/ 2020.

[57] A. Lynggaard-Jensen, "Trends in monitoring of waste water systems," *Elsevier*, Journal Article vol. 50, no. 4, pp. 707–716, 15/11/1999 1999.

[58] I. M. Diaconescu and G. Wagner, "Towards a general framework for modeling, simulating and building sensor/actuator systems and robots for the web of things," *CEUR Workshop Proceedings*, vol. 1319, pp. 30–41, 01/01 2014.

[59] R. Komiyama, H. Miyashita, T. Kageyama, K. Ohmi, S. Lee, and H. Okura, "A microfluidic device fully integrated with three pH sensing electrodes and passive mixer for nanoparticle synthesis," in *2015 IEEE SENSORS*, 2015, pp. 1–4.

[60] R. Komiyama, T. Kageyama, M. Miura, H. Miyashita, and S.-S. Lee, *Turbidity Monitoring of Lake Water by Transmittance Measurement with a Simple Optical Setup*. 2015, pp. 1–4.

[61] S. Zhuiykov, "Morphology of Pt-doped nanofabricated RuO2 sensing electrodes and their properties in water quality monitoring sensors," *Sensors and Actuators B: Chemical*, vol. 136, no. 1, pp. 248–256, 2009/02/02/ 2009.

[62] S. Zhuiykov, "Solid-state sensors monitoring parameters of water quality for the next generation of wireless sensor networks," *Sensors and Actuators B: Chemical*, vol. 161, no. 1, p. 21, 03/01/2012 2012.

[63] X. Bin and Z. Wei-De, "Modification of vertically aligned carbon nanotubes with RuO2 for a solid-state pH sensor," *Elsevier, Academic Journal* vol. 55, no. 8, pp. 2859–2864, 01/03/2010 2010.

[64] Z. Serge, K. Eugene, and M. Donavan, "Potentiometric sensor using submicron Cu2O-doped RuO2 sensing electrode with improved antifouling resistance," (in English), *Elsevier*, vol. 82, no. 2, pp. 502–507, 15/07/2010 2010.

[65] D. Szczuko, J. Werner, S. Oswald, G. Behr, and K. Wetzig, "XPS investigations of surface segregation of doping elements in SnO_2," *Applied Surface Science*, vol. 179, no. 1, pp. 301–306, 2001/07/16/ 2001.

[66] S. Christodoulou, A. Agathokleous, A. Kounoudes, and M. Milis, "Wireless sensor networks for water loss detection," *European Water*, vol. 30, 01/01 2010.

[67] J. Ning, M. Renzhi, L. Yunfeng, L. Xizhong, and W. Qingjian, "A novel design of water environment monitoring system based on WSN," in *2010 International Conference On Computer Design and Applications*, 2010, vol. 2, pp. V2-593-V2-597.

[68] H. R. Everett and T. Michael, "Unmanned surface vehicles," in *Unmanned Systems of World Wars I and II*: MITP, 2015, pp. 76–179.

[69] M. Ertsen, T. Swiech, and C. Pererya, "Reservoir storage and irrigation in Arequipa, Peru," 05/01 2010.

[70] R. B. Wynn et al., "Autonomous underwater vehicles (AUVs): their past, present and future contributions to the advancement of marine geoscience," *Marine Geology*, vol. 352, pp. 451–468, 2014/06/01/ 2014.

[71] D. Meyer, "Glider technology for ocean observations: a review," *Ocean Science Discussions*, vol. 2016, pp. 1–26, 07/01 2016.

[72] M. Funaki and N. Hirasawa, "Outline of a small unmanned aerial vehicle (Ant-Plane) designed for Antarctic research," *Polar Science,* vol. 2, no. 2, pp. 129–142, 2008/06/25/ 2008.

[73] G. K. Mohammad Manzurul Islam, J. Kamruzzaman, and M. Murshed, "Measuring trustworthiness of IoT image sensor data using other sensors' complementary multimodal data," *presented at the 18th IEEE International Conference On Trust, Security And Privacy In Computing And Communications/13th IEEE International Conference On Big Data Science And Engineering (TrustCom/BigDataSE),* 2019.

[74] M. A. M. Almuhaya, W. A. Jabbar, N. Sulaiman, and S. Abdulmalek, "A survey on LoRaWAN technology: recent trends, opportunities, simulation tools and future directions," *Electronics,* vol. 11, no. 1, 2022.

[75] LoRa-Alliance, *LoRaWAN™ What is it? A technical overview of LoRa® and LoRaWAN™.* https://lora-alliance.org/ 2015 (accessed 19 Feb, 2023)

Index

For Product Safety Concerns and Information please contact our EU
representative GPSR@taylorandfrancis.com
Taylor & Francis Verlag GmbH, Kaufingerstraße 24, 80331 München, Germany